COMPUTER MATHEMATICS

Also in this series

1 **An Introduction to Logical Design of Digital Circuits**
C. M. Reeves 1972

2 **Information Representation and Manipulation in a Computer**
E. S. Page and L. B. Wilson, Second Edition 1978

3 **Computer Simulation of Continuous Systems**
R. J. Ord-Smith and J. Stephenson 1975

4 **Macro Processors**
A. J. Cole, Second Edition 1981

5 **An Introduction to the Use of Computers**
Murray Laver 1976

6 **Computing Systems Hardware**
M. Wells 1976

7 **An Introduction to the Study of Programming Languages**
D. W. Barron 1977

8 **ALGOL 68 – A First and Second Course**
A. D. McGettrick 1978

9 **An Introduction to Computational Combinatorics**
E. S. Page and L. B. Wilson 1979

10 **Computers and Social Change**
Murray Laver 1980

11 **The Definition of Programming Languages**
A. D. McGettrick 1980

12 **Programming via Pascal**
J. S. Rohl and H. J. Barrett 1980

13 **Program Verification using Ada**
A. D. McGettrick 1982

14 **Simulation Techniques for Discrete Event Systems**
I. Mitrani 1982

15 **Information Representation and Manipulation using Pascal**
E. S. Page and L. B. Wilson 1982

16 **Writing Pascal Programs**
J. S. Rohl 1983

17 **An Introduction to APL**
S. Pommier 1983

Cambridge Computer Science Texts · 18

Computer mathematics

D. J. COOKE and H. E. BEZ

Department of Computer Studies, Loughborough University of Technology

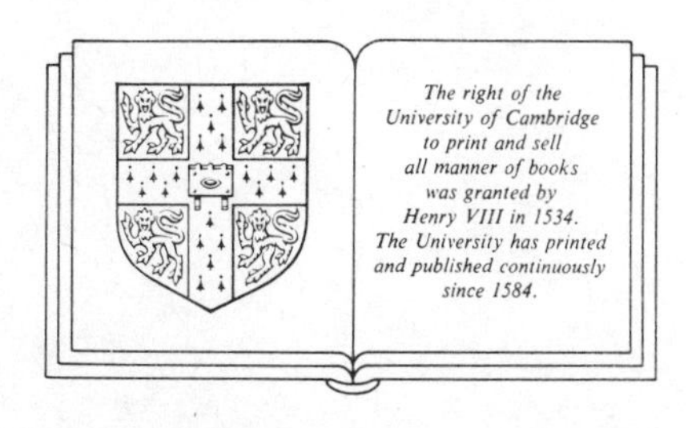

CAMBRIDGE UNIVERSITY PRESS

Cambridge

London New York New Rochelle

Melbourne Sydney

Published by the Press Syndicate of the University of Cambridge
The Pitt Building, Trumpington Street, Cambridge CB2 1RP
32 East 57th Street, New York, NY 10022, USA
296 Beaconsfield Parade, Middle Park, Melbourne 3206, Australia

First published 1984

Printed in Great Britain at the University Press, Cambridge

Library of Congress catalogue card number: 83-7588

British Library cataloguing in publication data

Cooke, D. J.
Computer mathematics – (Cambridge computer
science text; 18)

1. Mathematics – 1961-
I. Title II. Bez, H. E.
510 QA36

ISBN 0 521 25341 1 hard covers
ISBN 0 521 27324 2 paperback

AS

CONTENTS

PREFACE

Computing is an exact science and the systematic study of any aspect, including such diverse topics as database design, systems verification and computer animation, necessarily involves the use of mathematical models. In view of this many computing curricula in universities and polytechnics now include specialized courses to acquaint students with the appropriate mathematical structures and techniques. The content of this book is derived from such a course and is taught, in approximately one hundred lecture hours, to honours students in the Department of Computer Studies at Loughborough University of Technology. The work is covered in the first two years of the course and all students attend the first year lectures. The material in the later chapters is covered in the optional second year lectures and provides a basis for some courses in the final year. The contents of the book have evolved over a number of years. During that time we have enlisted the help of many around us, including numerous colleagues (at Loughborough and elsewhere) and students, who sometimes unknowingly have assisted with the preparation and revision of the material. We particularly wish to thank our wives, Chris and Carys, for constant encouragement and understanding throughout the project. Special thanks are due to Chris Cooke for the many hours she spent assisting us with the production of a draft typescript, Ornella Lardner for expertly typing the final version and Alan Benson for reading the entire work and making many constructive suggestions. Any remaining errors and inaccuracies are solely our responsibility.

D. J. Cooke
H. E. Bez
Loughborough 1982

0 INTRODUCTION

The book is principally intended as a text for computer science courses at universities and polytechnics; however, it may also be of use to computer professionals seeking a deeper understanding of their subject and whose formal education did not include a mathematics course of this type. The aim is to furnish the reader with a working knowledge of the areas of modern mathematics that relate to computing and hence provide him with a tool for the concise and precise description of many problems in computer science.

Despite adherence to the overriding aim of supplying material directly related to computer science, an attempt has been made to give a reasoned and rigorous presentation that is mathematically respectable. The approach adopted in the book is a constructive one. Wherever possible each new topic is defined in terms of earlier ones and supported by exercises, discussion and worked examples. It must be emphasized that working through the examples and attempting the exercises is an integral part of studying the material presented. In pursuing this goal we need somewhere to start. Our initial undefined concept is that of a set and we describe informally the properties we wish to ascribe to sets. From this we can define all our subsequent concepts in a constructive and mathematically acceptable way. Such an approach is necessary so that any errors can easily be traced back to a false assumption somewhere in the chain of ideas. It also means that part or all of the underlying theory may be programmed.

Sets are discussed in Chapter 1 and used throughout the remainder of the book. Chapters 2, 3, 5, 6 and 7 form the foundation for a formal discussion of many topics in computing and the material covered was chosen for its broad applicability. Chapter 4 describes a mathematical model of the arithmetic system used in digital computers for computation with integers. Chapters 8, 9 and 10 relate the mathematics of earlier work directly to computing topics. The particular areas chosen are language theory, the theory of computing machines and computer geometry. These topics reflect both the authors' interests and a desire to present important areas of general and current concern, but are by no means the only possible application areas. Other topics for which the earlier material is relevant include databases, networks, program verification and numerical analysis. Chapters 8 to 10 themselves constitute a starting point for further study in other areas including compiling, systems modelling, the theory of computation, computer graphics, computational geometry and computer aided design. The logical interrelationship between the chapters and other computing topics is shown in the diagram below.

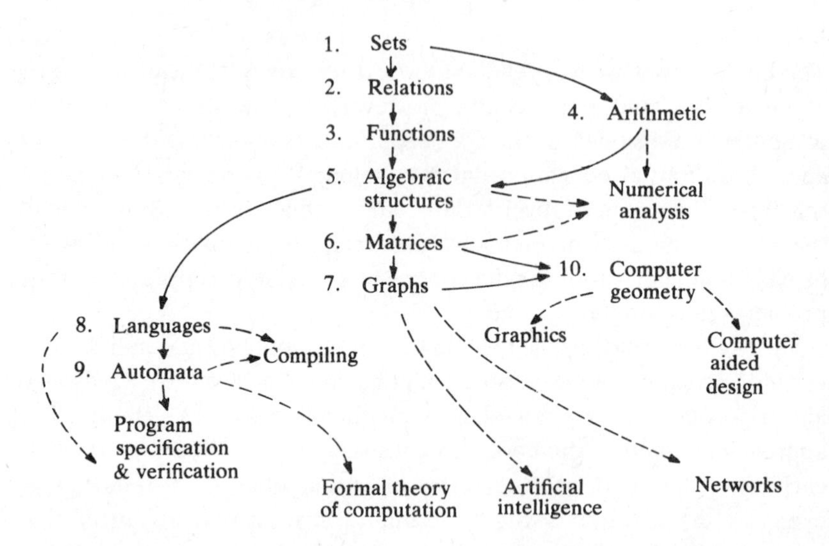

Terminology and notation will usually be formally introduced at the appropriate point in the text. However, in examples we sometimes use terms informally before they are defined. In such cases intuition will usually guide the reader in the right direction and no formal arguments will be based on these examples. One case in point is the use of the term 'finite machine' in Chapter 1 before its definition in Chapter 9. In Section 3.5 some properties of real numbers are used without proof. For example, we use the triangle inequality $|x+y| \leq |x|+|y|$, where $| \ |$ denotes the absolute value, in some proofs involving limits. The reader may wish to omit these proofs until Section 5.3.4 has been read. Throughout the book the symbol // will be used to denote the end of a definition, example, proof etc. and the symbol ⋇ will indicate the recognition of a logical error, i.e. a contradiction. Also for reasons which will become clear in Section 1.5 we usually indicate multiplication explicitly by the symbol $*$ (and not by $\times$) although when we are working with normal arithmetic the $*$ will sometimes be omitted so as not to over-complicate large expressions.

1 SETS

1.1 Sets and their specification

As already noted, we would like to build our mathematics in a sound workmanlike fashion; however, we have some difficulty with the foundations – it is impossible to use the same approach. Instead we shall begin by giving a description of our initial concept, that of a set. Simple though this may seem, the reader is expressly encouraged to study the examples at the end of this section before concluding that the subject matter is trivial.

We state that a *set* is a collection of definite, distinguishable objects such that, given a set and an object, we can ascertain whether or not the specified object is included in the set.

Subject only to this restriction a set can contain almost any kind of object, for example

the set of (all) London underground stations,
the set of left shoes,
the set of natural numbers; 1, 2, 3, 4 etc.,
the set of characters available on a specific typewriter,
the set of operation codes on a particular computer,
the set of Pascal reserved words.

We shall ultimately be interested in sets such as

the set of identifiers occurring in a specific program,
the set of operations in the same program or
the set of operations that may be carried out immediately after a given instruction in the same program.

However, these sets are too complex to develop the skills required so for most examples we shall use small abstract sets, such as sets of numbers.

A set is usually denoted by an upper case letter, for example A, and is specified in one of two ways. If the set is small (contains only a few objects) then we simply list all the objects. For instance, if we define A to be the set of all integers strictly between 6 and 10 then this can be written

$$A = \{7, 8, 9\}$$

and is read as 'A is the set containing 7, 8 and 9'. Here '=' is being used in the definitive sense; A *is* equal to the set.... Later we shall have reason to ask whether A equals.... In this case we will need to devise a procedure to ascertain the validity of the statement. Alternatively, a set can be characterized by its defining property and hence the set A can also be specified as

$$A = \{x : x \text{ is an integer and } 6 < x < 10\}$$

which is read as

A is the set of (*all*) x such that....

The only objects which are integers and are greater than 6 and less than 10 are 7, 8 and 9 and hence we have $\{7, 8, 9\}$ as before. Sets are often referred to as 'unordered collections of objects' and although it is sometimes useful to stress that, for example

$$\{7, 8, 9\} = \{8, 9, 7\} = \{9, 8, 7\} = \ldots,$$

we have not made any stipulation about the order in which the objects are considered and hence it is wrong to presume any specific order.

Now, given any integer, we can decide whether the integer belongs to the set A. If it does then we can denote the fact by the statement

'7 is an *element* of (the set) A'

which can be written

'$7 \in A$'

whereas

'6 is not an element of A',

and this is denoted by

'$6 \notin A$'.

The symbol $\in$ is derived from the Greek letter epsilon 'ε' and is negated by overwriting with a stroke /. This negation of an operation (or operation symbol) is common in mathematics and will often be used in what follows.

It must be emphasized that great care should be taken with the specification of sets. A set does not exist simply because you write down a 'specification' for it. The process of forming a set may go on for ever and yet still give a valid set. Also, a particular specification may result in a situation where we cannot decide whether an object is included or excluded.

So far we have encountered the symbols $\{:\}$, $=$, $\{,,,\}$, $\in$ and $\notin$. These seem fairly straightforward but great care is needed in their manipulation as is illustrated in the following examples.

Example 1.1.1

Which of the following are valid set definitions?

$A=\{1,2,3\}$,

$B=\{5,6,6,7\}$,

$C=\{x\colon x\notin A\}$,

$D=\{A, C\}$,

$E=\{x\colon x=1 \text{ or } x=\{y\} \text{ and } y\in E\}$,

$F=\{\text{sets that are not members of themselves}\}$

$\quad=\{x\colon x \text{ is a set and } x\notin x\}$.

Membership of the set A is easily checked and there are no duplicates; A is therefore valid; B looks to be equally valid except for the occurrence of two 6s. We can certainly check for inclusion within the set and this is surely the most important requirement. Hence we might regard this as valid and equal to $\{5,6,7\}$. However, there are problems that arise out of such a situation. If we consider the original definition of B and remove one of the 6s then we apparently have $6\notin B$ *and* $6\in B$. This conclusion is not allowed since it is inconsistent; hence we shall regard repetition within a set as referring to the same element and its duplication as being an oversight; the removal of duplicates forms the basis of several mathematical arguments later on. The set A contained numbers and numbers are very strange – they do not exist! To be precise, we use numerals; numerals are names of numbers. So B was a set of names and we usually use names to represent the objects to which they refer. In computing, names have particular significance, especially in the study of programming language semantics (the meanings of programs). Here is not the place to go into detailed discussion of these matters; it is sufficient to point out the pitfalls and the need for adequate specification of the objects under examination. Take for example the set

$X=\{$'An Introduction to Pascal',
'Fundamentals of Data Structures',
'An Introduction to Pascal'$\}$.

Is this a set of *two* book titles with one element inadvertently written down twice, or is it a set of *three* books two of which have the same title? If it is the latter then the two Pascal books need to be distinguished in some way. From the information given nobody knows the answer, so beware!

C is as valid as A since if $x\in A$ then $x\notin C$ and if $x\notin A$ then $x\in C$. The set C is very big, it contains *everything* except the numerals 1, 2 and 3. The notion of 'everything' is imprecise and mathematically dangerous as we shall soon see.

Assuming that, in the definition of D, A and C represent the sets previously defined then D is also valid. Notice that it is a set of sets (there is nothing wrong with that!), that it has only two elements, and in particular that $1 \notin D$ even though $1 \in A$ and $A \in D$. This is easy to check since $1 \neq A$, $1 \neq C$ and A and C are the only elements of D.

E is our first example of a recursively defined set; it is defined (partially) in terms of itself. The construction process goes on for ever so we must have a rule for determining the elements. We cannot write them down explicitly. Notice that E is not defined *completely* in terms of E. We must know *something* about the set which is independent of the rest of the definition; in this case $1 \in E$. The construction then proceeds:

$$\begin{aligned} 1 \in E &\quad \text{so} \quad \{1\} \in E, \\ \{1\} \in E &\quad \text{so} \quad \{\{1\}\} \in E, \\ \{\{1\}\} \in E &\quad \text{so} \quad \{\{\{1\}\}\} \in E, \text{ etc.} \end{aligned}$$

Even though the construction process is unending, given any object and enough time, we can determine whether the object is contained in E.

Now for F; this is difficult. To see why F cannot exist we first presume its existence and then demonstrate that there is a particular object, let us denote it by y, such that we cannot decide whether $y \in F$ or $y \notin F$. In general the search for an 'awkward' example by which we can show a logical flaw is not easy; however, in this case we can use F itself. To clarify matters let us call this set G. If, as we assume, F is a proper set then either $G \in F$ *or* $G \notin F$. We consider the two possible cases; (a) $G \in F$ implies that G satisfies the condition for containment, i.e. $G \notin G$, and therefore $G \notin F$; (b) $G \notin F$ says that G does satisfy the condition for containment in F and thus $G \in F$; hence starting from either situation leads us to the opposite one, so F cannot exist. Where did we go wrong? Sets of sets are certainly allowed and infinitely big sets (for example E above) are also allowed; however 'the set of all sets' cannot be dealt with by normal set theory – it requires a different sort of mathematics. This anomaly of set theory is known as Russell's paradox. If we already had the set H then we could define J by:

$$J = \{x: x \in H \text{ and } x \notin x\}. \quad //$$

Thus we shall use only sets that can be explicitly written down or constructed by well defined processes. So sets are not as trivial as they might first appear but, providing that we always follow the rules, neither are they particularly difficult. Try the following exercises yourself.

Exercises 1.1

1. Consider at least two possible interpretations of
   {Smith, Smith, Brown}.

   Define each of them as unambiguously as you can.

2. Consider the following four sets. How may they be simplified and which are equivalent?

Devise as many interpretations as possible but be precise about your assumptions.

(*a*) {1, 2, 3, 4, 5},
(*b*) {I, II, III, IV, V},
(*c*) {one, I, un, uno, ein},
(*d*) {5, V, cinq, five}.

3. Argue the validity or invalidity of the statements

'This statement is false.' and 'I am a liar.'

Which English words in the statements require proper (i.e. mathematically precise) definitions in order to make your answers water-tight?

4. Let X be the set $\{1, 2\}$ and Y be the set $\{x: x = y + z: y, z \in X\}$. Define the set Y explicitly.

What are the sets
$\{y: y = x + z: x, z \in X\}$
and
$\{y: x = y + z: x, z \in X\}$?

5. Suppose x is a specific object and is *not* a set. Consequently, $y \notin x$ for any y and hence it follows that $x \notin x$. Can the set $\{x, \{x\}, \{\{x\}\}\}$ be simplified? What about $\{x, y, \{x, y\}\}$?

6. Let A be the set of all whole numbers. Describe, in words, the set

$X = \{x: x \in A \text{ and, } x = 1 \text{ or } (x - 2) \in X\}$.

1.2 Simple set operations

As we have seen from consideration of the fallacious 'set' F in Example 1.1.1 we must take care when defining sets; however, by constructing new sets from old ones in simple ways we *are* able to obtain more interesting sets safely. Later we shall write down formal rules that govern the manipulation of sets but just to get matters going we introduce some notation. We begin with the simplest operations imaginable.

Definitions. Given two sets A and B, the *intersection* of A and B is the set of all elements which are elements of both A and B and is written $A \cap B$, thus

$$A \cap B = \{x: x \in A \text{ and } x \in B\}.$$

Similarly, the *union* of A and B is written $A \cup B$ and is defined by

$$A \cup B = \{x: x \in A \text{ or } x \in B\}.$$

The meaning of this notation is not difficult to understand but can sometimes be confusing. One way of remembering which symbol denotes each operation is to incorporate the symbols into the words and write 'i∩tersection' and '∪nion'. //

These definitions are derived from the words 'and' and 'or', and as a consequence we have

$$A \cup B = B \cup A, \quad A \cap B = B \cap A$$

and, probably less obviously,

$$A \cup B = B \cup A, \quad A \cap B = B \cap A$$

These latter identities are important for two reasons; firstly, in some of the mathematical discussion to follow, we shall need to reduce $A \cup A$ (respectively, $A \cap A$) to A or to expand A into $A \cup A$ (respectively, $A \cap A$); and secondly, because when expressed in words they may sound silly, even though they are logically correct, and hence are in danger of being overlooked or ignored.

Notice also that the definition of union uses the inclusive 'or', so called because it includes 'and' so that

$$\{1, 2\} \cup \{2, 3\} = \{1, 2, 3\}$$

and

$$\{1, 2\} \cap \{2, 3\} = \{2\}.$$

The elements in the intersection of sets, in this case the single number 2, are included in the union. This is the usual mathematical convention and is an instance where the mathematical meaning is more precise than the common usage.

Example 1.2.1

Under the presumption that every day is either wet or dry; the *mathematical* (or logical) answer to the question

'Is it wet or dry today?'

is

'Yes!' //

Definition. The *difference* of sets A and B (also called the complement of B relative to A) is written $A \backslash B$ (A not B, or A down B) and is defined by

$$A \backslash B = \{x\colon x \in A \text{ and } x \notin B\}. \quad //$$

So if

$$A = \{1, 2, 3\} \quad \text{and } B = \{2, 3, 4\},$$

then

$$A \backslash B = \{1\} \quad \text{and } B \backslash A = \{4\}.$$

The following definition is included here for the sake of completeness. Although we shall make little use of it directly, as will be seen later a similar operator is of great significance when performing computer arithmetic.

Definition. The *symmetric difference* of sets A and B, $A \triangle B$, is defined as

$$A \triangle B = (A \cup B) \backslash (A \cap B). \quad //$$

Should anyone be perplexed by the meanings of the symbols $\cup$, $\cap$, $\backslash$, $\triangle$ or believe that they are so elementary as not to be of any practical use the analogy in the following example may help.

Example 1.2.2

Suppose we have two programs, call them P and Q, and that A is the set of all data values acceptable to P and B is the set of all data values acceptable to Q. Then

$A \cap B$ is the set of all data acceptable to both P and Q,
$A \cup B$ is the set of all data acceptable to at least one of P and Q,
$A \backslash B$ is the set of all data acceptable to P but not to Q,
$B \backslash A$ is the set of all data acceptable to Q but not to P, and
$A \triangle B$ is the set of all data acceptable to exactly one of the programs P and Q.

To ensure that A and B are well defined we need to know something about the computations associated with P and Q. It suffices to say that they are to be run on a given finite machine. //

Before going further it is convenient to define two special sets. The first is the empty set.

Definition. The *empty set*, written $\varnothing$, is a set having the property that

$$x \notin \varnothing \text{ for any } x. \quad //$$

The second set, the definition of which is problem dependent, is called the universe of discourse.

Definition. The *universe of discourse*, written $\mathscr{E}$, is the set of all objects under consideration in a given problem. //

Restricting $\mathscr{E}$ in this way helps to avoid difficulties such as those associated with the 'set' F in Example 1.1.1, and in any case most objects are insignificant in any given problem, for example the dimensions of the third edition of 'The Penguin English Dictionary' are of doubtful interest when considering the behaviour of a particular Fortran program!

Definition. Two sets, A and B, are said to be *disjoint* if

$$A \cap B = \varnothing. \quad /\!/$$

Definition. In any context where $\mathscr{E}$ is understood (i.e. properly defined) we define the *complement* of any set A, written A', by

$$A' = \mathscr{E} \backslash A = \{x: x \notin A\}.$$

Equivalently, and of more practical importance, from the definitions of $\varnothing$ and $\mathscr{E}$, A' is a set that satisfies the two identities

$$A \cup A' = \mathscr{E},$$

$$A \cap A' = \varnothing. \quad /\!/$$

In Section 1.4 we demonstrate that, for a given $\mathscr{E}$, these identities are sufficient to uniquely specify A'.

Example 1.2.3

Let

$\mathscr{E} = \{1, 2, 3, 4\}$,

$A = \{1, 3, 4\}$,

$B = \{2, 3\}$,

$C = \{1, 4\}$.

From these definitions it is a simple matter to deduce, for example, A', $B \cap C$, $C \backslash A$, etc. However, we may wish to investigate more complicated expressions involving two or more operations. In such cases, so as to indicate the order in which we must perform the elementary set operations, we use brackets. Any expression enclosed in brackets must be evaluated before its result can be used in another calculation. Thus in $(A \cap B)'$, the intersection, $\{3\}$, is computed before the complement, $\{1, 2, 4\}$, can be found. This convention is obviously sufficient. However, so as to avoid too many brackets, we shall not require brackets when we wish to perform complementation before any operator in the set $\{\cap, \cup, \backslash, \Delta\}$. So $A \cap B'$ means $A \cap (B)'$, etc.

Therefore, we have

$$\begin{aligned} A \cap B' &= A \cap \{1,4\} \\ &= \{1,4\}, \\ (A \cap B)' &= \{3\}' \\ &= \{1,2,4\}, \\ (B \backslash A) \cup C\{2\} \cup C & \\ &= \{1,2,4\}. \quad // \end{aligned}$$

The reader may wonder why our presentation is built on the notion of set rather than that of number. Indeed so far we have used only numbers as elements of sets, and this was merely because we required the reader to have some familiarity with the objects he was manipulating. The fact of the matter is that sets are more fundamental than numbers; we can obtain numbers from sets but not vice versa. However, for many applications of the theory that follows, we shall need to make precise statements about some special sets of numbers. To provide a basis from which to construct such sets we postulate the set, $\mathbb{N}$, of strictly positive whole numbers, the *natural numbers*,

$$\mathbb{N} = \{1, 2, 3, \ldots\}.$$

A proper definition of the set $\mathbb{N}$, together with the arithmetic operations $+$ and $*$, and its ordering will be given later; however, in the present chapter, we shall assume the reader is familiar with the behaviour of $\mathbb{N}$. Similarly, $\mathbb{Z}$ will denote the set of all integers:

$$\mathbb{Z} = \{\ldots, -2, -1, 0, 1, 2, \ldots\}.$$

Of course the sets $\mathbb{N}$ and $\mathbb{Z}$ cannot be written out explicitly (they are too big), but for the time being you should understand '...' to mean 'and so on'. Now consider the set

$$A = \{1, 2, \ldots, n\} = \{x: x \in \mathbb{N}, 1 \leqslant x \leqslant n\}.$$

We say that the *cardinality* (or *size*, or *norm*, or *length*) of this set is n. It has n elements and this is denoted by

$$|A| = \text{card}\,(A) = n.$$

Further, *any* set B that has the same number of elements as A has equal cardinality, and of course these elements need not themselves be numbers. With small sets it is quite easy to count the elements but for other sets, for example $\mathbb{N}$, it may be impossible. Later we shall give a formalism for counting but in the meantime an informal counting procedure and the following definition will suffice.

Definition. A set X is said to be *finite* if $X = \varnothing$ or if there is a set of the form

$$\{1, 2, \ldots, n\} \quad \text{for some } n \in \mathbb{N},$$

having the same number of elements as X. If $X \neq \varnothing$ and no such n can be found then X is said to be *infinite*. //

Now that we have some notation defined we are in a position to give some exercises. To make the sets easy to write down we shall again use numbers and letters for elements but remember the same operations can be applied to any sets.

Exercises 1.2

1. Let

   $\mathscr{E} = \{1, 2, 3, 4, 5, 6\}$,

   $X = \{1, 5\}$,

   $Y = \{1, 2, 4\}$,

   $Z = \{2, 5\}$.

   Compute the sets

   (a) $X \cap Y'$,
   (b) $(X \cap Z) \cup Y'$,
   (c) $X \cup (Y \cap Z)$,
   (d) $(X \cup Y) \cap (X \cup Z)$,
   (e) $(X \cup Y)'$
   (f) $X' \cap Y'$,
   (g) $(X \cap Y)'$,
   (h) $X' \cup Y'$,
   (i) $(X \cup Y) \cup Z$,
   (j) $X \cup (Y \cup Z)$,
   (k) $X \backslash Z$,
   (l) $(X \backslash Z) \cup (Y \backslash Z)$.

2. Let

   $\mathscr{E} = \{a, b, c, d, e, f\}$,

   $A = \{a, b, c\}$,

   $B = \{f, e, c, a\}$,

   $C = \{d, e, f\}$.

   Compute the sets

   (a) $A \backslash C$,
   (b) $B \backslash C$,

(*c*) $C\backslash B$,
(*d*) $A\backslash B$,
(*e*) $A' \cup B$,
(*f*) $B \cap A'$,
(*g*) $A \cap C$,
(*h*) $C \cap A$,
(*i*) $C \Delta A$.

3. Given any two sets A and B such that $A \cap B = \varnothing$ what are $A\backslash B$ and $B\backslash A$?
4. Given any two sets C and D such that $C \cap D' = \varnothing$ what are $C \cap D$ and $C \cup D$

5. Given any set X what are
   (i) $X \cap X'$,
   (ii) $X \cup X'$, and
   (iii) $X\backslash X'$?

6. Which of the following are true?
   (*a*) $0 \in \varnothing$,
   (*b*) $\varnothing = \{0\}$,
   (*c*) $|\{\varnothing\}| = 1$,
   (*d*) $\{\{\varnothing\}\} \in \{\{\{\varnothing\}\}\}$,
   (*e*) $|\{\{\varnothing\}\}| = 2$.
   This question is deceptive. Although it may seem simple or contrived the empty set and its associated properties are very important. If you are not absolutely sure about your answers rework the question using the analogy of a bag instead of a set; so $\{\{\,\},\{\,\}\}$ is a bag containing two empty bags and hence $|\{\{\,\},\{\,\}\}| = 2$ etc.

7. Let M and N be two finite computers (see Example 1.2.2) with fixed programs. Further, let A be the set of data values acceptable to M such that if $x \in A$ and machine M is run with input x then M will halt and give a result. Similarly let B be the set of data values that cause N to halt and give a result.
   If we are told any element of A is acceptable to both M and N what does this infer for elements of B'? Explain this situation symbolically and discuss the usefulness of this information.

8. Explain, in terms of suitable sets, why Example 1.2.1 is valid.

9. On giving the definition of *union* it was emphasized that we were using the inclusive 'or'. How, in terms of sets, would you express the exclusive 'or'?

10. Frequently in numerical computing we use arithmetic operations to form new sets, thus if A and B are sets of numbers

$A+B=\{x: x=a+b, a\in A, b\in B\}$.

Similarly we can define the induced operations $*$, $-$, and $/$ between sets of numbers.

Compute the following sets:

(*a*) $\{1,2\}+\{1,3\}$,
(*b*) $\{1,2\}\cup\{1,3\}$,
(*c*) $\{1,2\}*\{1,3\}$,
(*d*) $\{1,2\}\cap\{1,3\}$,
(*e*) $\{1,2\}-\{1,3\}$,
(*f*) $\{1,2\}\backslash\{1,3\}$,
(*g*) $\{2,4\}/\{2\}$,
(*h*) $\{2,4\}\backslash\{2\}$,
(*i*) $\{2,4\}-\{2\}$.

11. In order to be able to apply set manipulation techniques to a problem we must inevitably at some stage take a 'non-mathematical' statement and translate it into mathematical notation. Usually, but not always, this will make the description more compact. The mathematical expression will, however, always be exact whereas the original might not be so. (Where this happens there will be a need to ascertain exactly how the original formulation is lacking.)

(*a*) Try to formulate the following in set notation:
  (i) Given sets A, B and C, define the set of elements included in exactly two of these sets.
  (ii) As (i) but with the knowledge that A, B and C are mutually disjoint. (Think!)
  (iii) Given sets V, W, X, Y and Z, define the set of elements in at least two of V, W, X and Y, and not in Z.

(*b*) Similarly, describe in words the sets given below:
  (i) $(J\cap(K\cup L))'\cup(H\backslash L)$,
  (ii) $(P\cup R\cup Q)\backslash(P\cap(Q\backslash R))$,
  (iii) $((E\backslash F)\cup(F\backslash E))'\cup G$.

1.3 Venn diagrams

You may have already noticed some special properties of the set operations, especially different ways of denoting the same set. Later in this chapter we shall discuss ways of proving such properties in a formal way, but it is often useful to have a diagrammatic representation of sets. Such representations cannot replace a proof but may be useful in convincing yourself either that a particular statement is true, and hence that a proof is possible, or that it is false, in which case it may indicate how to construct an example to prove that it is false. The diagrams we use are called Venn diagrams (after the English mathematician John Venn) and are constructed as described below.

First, draw a large rectangle to represent $\mathscr{E}$ (Figure 1.1). Second, draw

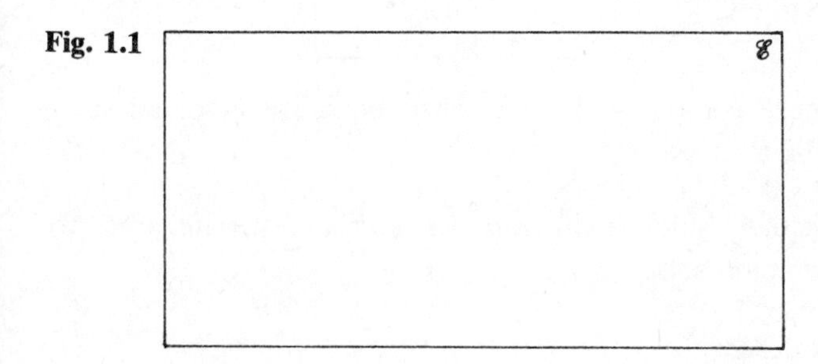

Fig. 1.1

circles (or any other appropriate closed curve) within the rectangle to represent the sets. These must intersect in the most general way required by the problem and should be suitably labelled as in Figure 1.2. Points

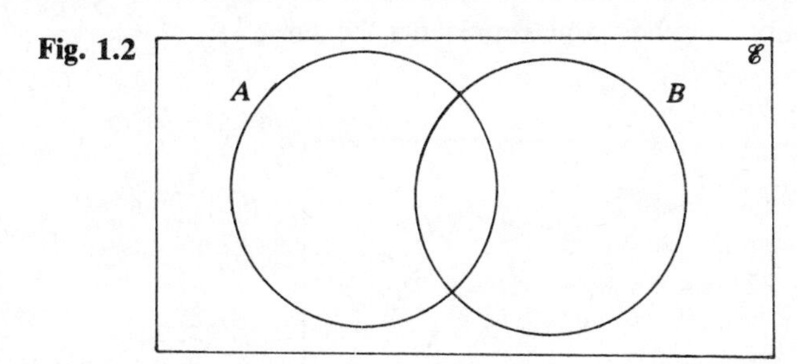

Fig. 1.2

which lie within the various regions of the diagram can now be considered to represent the elements of the respective sets. If the number of elements in the sets is small then the individual elements may be written within the appropriate regions as shown in Example 1.3.1.

Example 1.3.1

Let

$\mathscr{E} = \{a, b, c, d, e\}$,

$A = \{b, c, d\}$,

$B = \{c, e\}$.

The corresponding diagram is shown in Figure 1.3. Figure 1.3 depicts Example 1.3.1 fully, providing that we know the elements of $\mathscr{E}$. If, for instance, $A \in \mathscr{E}$ then it is not clear what the Venn diagram is supposed

Fig. 1.3

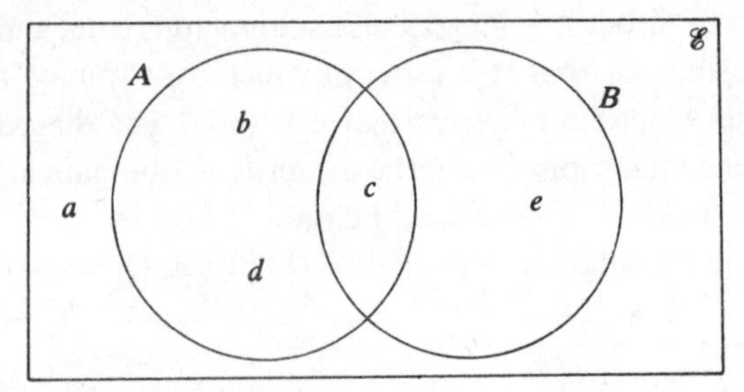

to represent. Where complicated set constructions are involved such diagrams should be avoided. //

Having constructed a suitable diagram we can shade specific areas to indicate new compound sets.

Example 1.3.2
To represent the set

$A \cup (B' \cap C)$

start with the general diagram as in Figure 1.4. Shade B' with diagonal lines in one direction and C with lines in the other direction (see Figure 1.5). The area double hatched now represents the set

$B' \cap C.$

Fig. 1.4

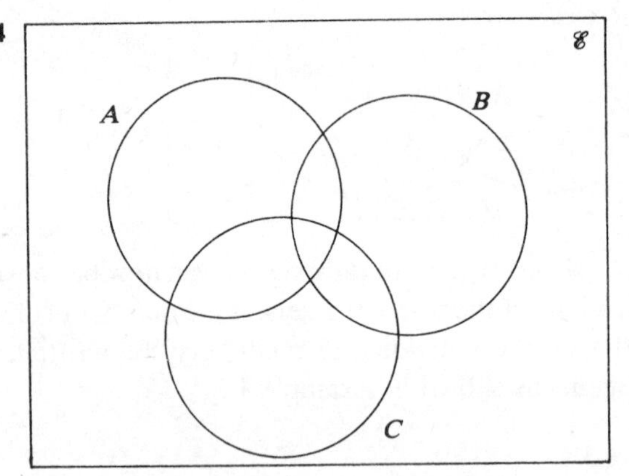

On a new copy of the diagram shade this area with horizontal lines and A with vertical lines. All the area shaded in Figure 1.6 now represents the set

$A \cup (B' \cap C).$ //

If, in a particular case, we have extra knowledge about the sets under consideration then we may be able to use this to simplify the Venn diagram.

Fig. 1.5

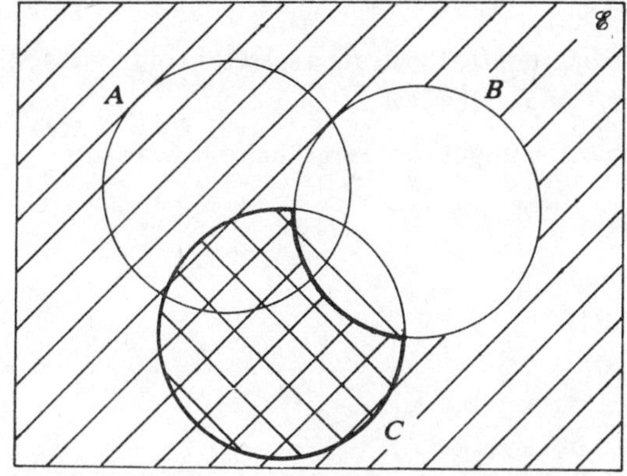

Fig. 1.6

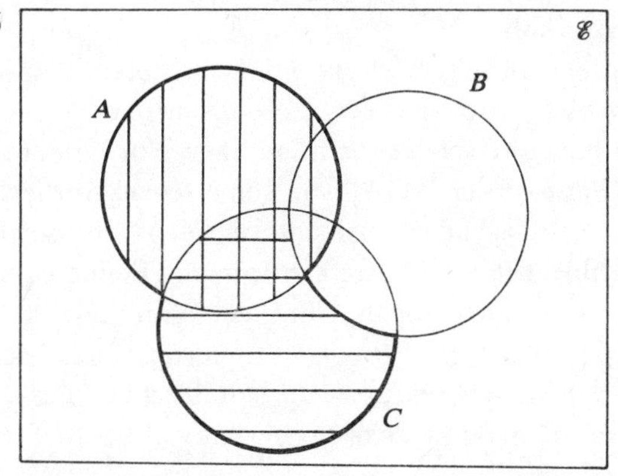

Example 1.3.3

If $A \cap B = \varnothing$ then the diagram in Figure 1.7 is adequate. //

Fig. 1.7

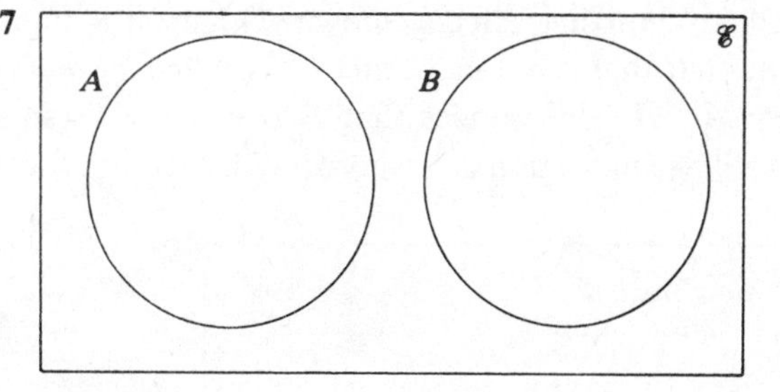

Notice also that in most cases the sets involved have very many elements and hence these elements cannot be represented individually. It is, therefore, more convenient to talk about each set as a complete unit and not to refer to specific elements.

Exercises 1.3

1. Draw Venn diagrams to illustrate a selection of the sets computed in Question 1 of Exercises 1.2.
2. Consider how you might represent the following sets using Venn diagrams:
   (i) $\{A, \{A\}\}$,
   (ii) $\{\{a\}, \{b\}\}$,
   (iii) $\{X, Y, Z\}$,
   where
   $X = \{x \colon x = 1 \text{ or } (x-2) \in X\}$,
   $Y = \{x \colon x = 3 \text{ or } (x-3) \in Y\}$,
   $Z = \{x \colon x = 2 \text{ or } (x-2) \in Z\}$.

1.4 Subsets and proofs

The operations of intersection, union, difference and complementation allow us to form new sets from given ones; however, as yet, we cannot say how one set relates to another. For instance, given sets X and Y, the intersection $X \cap Y$ is in some sense 'smaller' (or at least no larger) than X. In fact all the elements of $X \cap Y$ are also included in the set X. From this observation we can formally define equality of sets and of different expressions for the same set. With these definitions we are able to write proper logical proofs of important facts relating to sets. These results, though 'obvious', provide suitable situations in which to introduce some of the main kinds of proof that will be used later on.

Definitions. Given sets A and B which have the property that $x \in A$ implies that $x \in B$ then we say that A is a *subset* of B. This is denoted by

$$A \subseteq B$$

and the corresponding Venn diagram is as shown in Figure 1.8. If, further, there is an element of B that is not in A, then A is called a *proper subset* of B and we write $A \subset B$. This implies that B is in some sense bigger than A, but as you will see in Section 3.3, where infinite sets are concerned

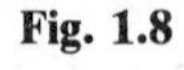
Fig. 1.8

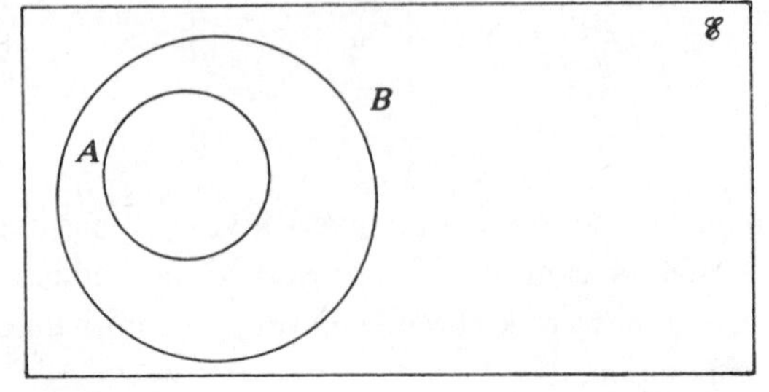

such terms can be misleading; hence our caution. These relations can be drawn in reverse as

$$B \supset A \quad \text{and} \quad B \supseteq A$$

and B may be called a (proper) *superset* of A.

Of course, for any set A we have the following three relationships:

$$\varnothing \subseteq A,$$

$$A \subseteq A,$$

and

$$A \subseteq \mathscr{E}.$$

The second of these is especially important; sets A and B are said to be *equal* (written $A = B$) if

$$A \subseteq B \quad \text{and} \quad B \subseteq A.$$

That is all the elements of A are in B and all the elements of B are in A. //

From the definition of set equality we now prove some set identities by way of a series of five detailed examples. These are our first examples of proper proofs and therefore we perhaps ought to consider briefly the question of why we need to bother with proofs within computing science. Strictly speaking, of course, just as in mathematics, we only really need to prove something once and for all; we can then proceed in the knowledge (or belief) that this particular piece of information has been validated by ourselves, or somebody else, and hence can be regarded as a fact. However, as in most aspects of computing science, the method by which we achieve a result is arguably at least as important as the result itself. Consideration of a proof highlights the assumptions made, the derivation of their consequences and provides knowledge of deductive processes which may be of use in solving structurally similar problems. (This is analogous to the gaining of experience of suitable data structures for use in programming.) Also of note is the observation that proofs of theorems form the basis of all automatic problem solving systems; but this is very advanced work. Back to our examples.

In Example 1.4.1 we use *direct proofs* to show that

(*a*) $A \cap (B \cup C) \subseteq (A \cap B) \cup (A \cap C)$ and that

(*b*) $(A \cap B) \cup (A \cap C) \subseteq A \cap (B \cup C)$.

Each of these proofs consists of a sequence of statements of the form

'if P then Q'

('if P is true then Q is true').

For convenience this is written '$P \Rightarrow Q$' and read as 'P implies Q'.

Therefore if we have a sequence $P_0, P_1, P_2, \ldots, P_n$ such that $P_0 \Rightarrow P_1$, $P_1 \Rightarrow P_2, \ldots, P_{n-1} \Rightarrow P_n$ (so P_0 implies P_1 and P_1 implies P_2 and ... and P_{n-1} implies P_n) then we have a direct proof that $P_0 \Rightarrow P_n$.

Example 1.4.1
For sets A, B and C prove that
$A \cap (B \cup C) = (A \cap B) \cup (A \cap C)$.

Proof.

$x \in A \cap (B \cup C)$

$\Rightarrow x \in A$ and $x \in B \cup C$ (This grouping is necessary because the brackets indicate that the union is computed before the intersection.)

$\Rightarrow x \in A$ and $(x \in B$ or $x \in C)$

$\Rightarrow (x \in A$ and $x \in B)$ or $(x \in A$ and $x \in C)$

$\Rightarrow (x \in A \cap B)$ or $(x \in A \cap C)$

$\Rightarrow x \in (A \cap B) \cup (A \cap C)$.

Thus

$$A \cap (B \cup C) \subseteq (A \cap B) \cup (A \cap C).$$

Now we must show the set inclusion ($\subseteq$) in the reverse direction.

$x \in (A \cap B) \cup (A \cap C)$

$\Rightarrow (x \in A \cap B)$ or $(x \in A \cap C)$

$\Rightarrow (x \in A$ and $x \in B)$ or $(x \in A$ and $x \in C)$

$\Rightarrow x \in A$ and $(x \in B$ or $x \in C)$

$\Rightarrow x \in A$ and $x \in B \cup C$

$\Rightarrow x \in A \cap (B \cup C)$.

Therefore

$$(A \cap B) \cup (A \cap C) \subseteq A \cap (B \cup C).$$

So

$$A \cap (B \cup C) = (A \cap B) \cup (A \cap C). \quad //$$

In this particular case the second part of the proof is exactly the reverse of the first and hence we could have written

$x \in A \cap (B \cup C)$

$\Leftrightarrow x \in A$ and $x \in B \cup C$ and so on.

Here the symbol '$\Leftrightarrow$' further compacts the notation, '$P \Leftrightarrow Q$' being merely shorthand for '$P \Rightarrow Q$ and $Q \Rightarrow P$'. It can be read as 'implies and is implied by' (which is often read as 'if and only if' and written 'iff') and denotes a kind of equivalence between the two statements P and Q. However, it is not always possible to simply reverse the argument and hence, in general, we need to do both parts separately. Notice also that this identity can easily be inferred (though not proved) from a suitable Venn diagram but it is not always possible to draw a diagram (which we can be sure is exactly as it should be) and so a proper proof is necessary. This proof depends on the interrelationship between the meanings of the words 'and' and 'or' and hence may be challenged for its mathematical validity. Later, when we formally define some algebraic structures, we shall demonstrate that this causes no problem.

Examples 1.4.2 to 1.4.5 also use direct proofs but they are written in slightly different ways.

Example 1.4.2

Relative to a given $\mathscr{E}$ the complement of any set A ($A \subseteq \mathscr{E}$) is unique.

Proof. Suppose there are two sets B and C each of which satisfy the requirements of a complement of A.

Thus

$$B \cap A = C \cap A = \varnothing \quad \text{and} \quad B \cup A = C \cup A = \mathscr{E}.$$

Then

$$\begin{aligned} B &= B \cap \mathscr{E} && \text{(property of } \mathscr{E}\text{)},\\ &= B \cap (C \cup A) && \text{(assumption)},\\ &= (B \cap C) \cup (B \cap A) && \text{(Example 1.4.1)},\\ &= (B \cap C) \cup \varnothing && \text{(assumption)},\\ &= B \cap C && \text{(property of } \varnothing\text{)}, \end{aligned}$$

so

$$\begin{aligned} & x \in B \\ \Rightarrow\ & x \in B \text{ and } x \in C \\ \Rightarrow\ & B \subseteq B \cap C \\ \Rightarrow\ & B \subseteq B \text{ and } B \subseteq C. \end{aligned}$$

But we know $B \subseteq B$ so we have deduced the extra fact that

$$B \subseteq C.$$

By a similar argument (swap B and C) we get

$$C \subseteq B$$

so

$$B = C \quad \text{i.e.} \quad B = C = A' \quad \text{and } A' \text{ is unique.} \quad //$$

The example above embodies the general mathematical approach employed to show uniqueness of a particular object; assume that there are two such objects and show that they must be equal.

In the next example we again resort to assumptions about the word 'and' to allow us to write the set expression $A \cup B \cup C$ as either $(A \cup B) \cup C$ or $A \cup (B \cup C)$ whichever is more convenient in the manipulation required.

Example 1.4.3

Given sets A, B and C such that

$$A \cup B \cup C = \mathscr{E}$$

and A, B and C are mutually disjoint then

$$A' = B \cup C,$$

$$B' = A \cup C$$

and

$$C' = A \cup B.$$

Proof. $A \cup B \cup C = A \cup (B \cup C) = \mathscr{E}$ and

$$\begin{aligned} A \cap (B \cup C) &= (A \cap B) \cup (A \cap C) && \text{(Example 1.4.1)} \\ &= \varnothing \cup \varnothing && (A, B, C \text{ disjoint}) \\ &= \varnothing. \end{aligned}$$

Hence $B \cup C$ satisfies the conditions for A' which is unique so

$A' = B \cup C$; similarly for B' and C'. //

Example 1.4.4

Given sets X and Y then

$$(X \cap Y)' = (X \cap Y') \cup (X' \cap Y) \cup (X' \cap Y').$$

Proof. Suppose we have sets A, B and C as in Example 1.4.3 and further that

$$C = D \cup E \quad \text{and} \quad D \cap E = \varnothing.$$

(The Venn diagram in Figure 1.9 may help you follow the dissection of $\mathscr{E}$ that is required here.)

The A, B, D and E are mutually disjoint and

$$A \cup B \cup D \cup E = \mathscr{E}.$$

Moreover

$$A' = B \cup C,$$

Fig. 1.9

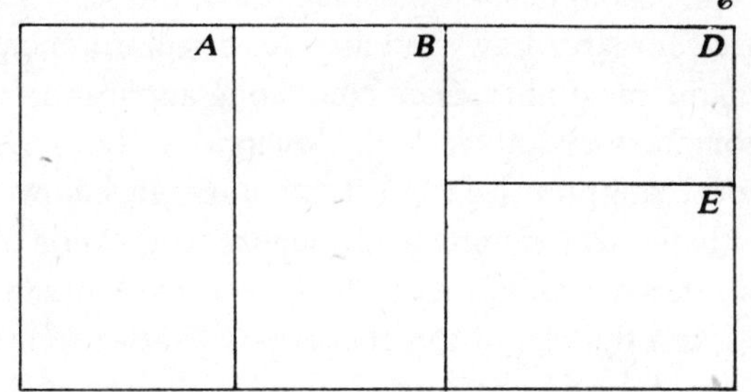

so

$$A' = B \cup D \cup E. \qquad (*)$$

Now, it is easy to show that if we set

$$A = X \cap Y,$$

$$B = X \cap Y',$$

$$D = X' \cap Y$$

and

$$E = X' \cap Y',$$

then the required conditions hold and so the result follows immediately from (*). //

Example 1.4.5

For sets X and Y

$(X \cap Y)' = X' \cup Y'$

Proof.

$x \in (X \cap Y)'$

$\Leftrightarrow x \in (X \cap Y') \cup (X' \cap Y) \cup (X' \cap Y')$ (by Example 1.4.4)

$\Leftrightarrow x \in (X \cap Y') \cup (X' \cap Y') \cup (X' \cap Y) \cup (X' \cap Y')$ (one term duplicated, allowed since $A = A \cup A$ for any A)

$\Leftrightarrow x \in ((X \cap Y') \cup (X' \cap Y')) \cup ((X' \cap Y) \cup (X' \cap Y'))$ (regroup)

$\Leftrightarrow x \in ((X \cup X') \cap Y') \cup (X' \cap (Y \cup Y'))$ (by Example 1.4.1)

$\Leftrightarrow x \in (\mathscr{E} \cap Y') \cup (X' \cap \mathscr{E})$ (definition of $\mathscr{E}$)

$\Leftrightarrow x \in Y' \cup X'$ (property of $\mathscr{E}$)

$\Leftrightarrow x \in X' \cup Y'.$

Hence

$$(X \cap Y)' = X' \cup Y'. \quad //$$

The result proved in Example 1.4.5 and a related identity (see Exercises 1.4.5) are called De Morgan's Laws and are of particular importance in mathematical logic, the most immediate computing application being in the design of combinatory circuits in logic design.

The sequence of Examples 1.4.1 to 1.4.5 illustrates how we can develop proper mathematical arguments through a progression of simple theorems and hence derive significant results such as De Morgan's Laws. Before progressing with the rest of this chapter try to rewrite the proofs of Examples 1.4.2 to 1.4.4 in the formal manner as used in Example 1.4.1. In particular, each step should be justified by quoting a relevant result from previous work or by a technical term (such as 'disjoint'). Later we shall introduce sufficient terminology to allow more concise arguments to be used. Before proceeding to consideration of further proof techniques we give two supplementary definitions.

Definition. Two sets, A and B, are said to be *unequal* if they are not equal. This property can be demonstrated by showing that one of the sets

$$A \setminus B$$

or

$$B \setminus A$$

is non-empty. //

Definition. For a given set X, the set of all subsets of X constitutes the *power set* of X, written $\mathscr{P}(X)$. (Some authors use the notation 2^X; the reason for this will become clear when the next set of exercises is studied.) Formally,

$$\mathscr{P}(X) = \{Y : Y \subseteq X\}.$$

In particular, notice that since $\varnothing \subseteq X$ and $X \subseteq X$ it follows that

$$\varnothing \in \mathscr{P}(X)$$

and

$$X \in \mathscr{P}(X). \quad //$$

Example 1.4.6

If

$$A = \{1, 2, 3\}$$

then

$$\mathscr{P}(A) = \{\varnothing, \{1\}, \{2\}, \{3\}, \{1, 2\}, \{1, 3\}, \{2, 3\}, A\}. \quad //$$

We conclude this section by mentioning two indirect proof methods.

The first is proof by *contradiction.* Recall Russell's paradox (Example 1.1.1):

$$F \in F \Rightarrow F \notin F$$

and

$$F \notin F \Rightarrow F \in F.$$

If we denote the statement $F \in F$ by P then we have

$$P \text{ true} \Rightarrow P \text{ false}$$

and

$$P \text{ false} \Rightarrow P \text{ true}.$$

Fundamental to mathematics is the assumption that no assertion can be both true and false (i.e. the logical system must be *consistent.* We rejected Russell's set because its definition was inconsistent.), and we use this as the basis for a proof by contradiction. Suppose that we have a collection of assumptions $P_1, P_2, \ldots, P_n$ and we wish to prove

$$(P_1 \text{ true and } P_2 \text{ true and} \ldots \text{and } P_n \text{ true}) \Rightarrow Q \text{ true},$$

or more simply

$$(P_1 \text{ and } P_2 \text{ and} \ldots \text{and } P_n) \Rightarrow Q.$$

If we assume (P_1 and P_2 and ... and P_n and not Q) – i.e. P_1 to P_n are true and Q is false – and from this we can show that some statement, P say, is both true and false then the logical system based on

$$(P_1 \text{ and } P_2 \text{ and} \ldots \text{and } P_n \text{ and not } Q)$$

is inadmissible. Thus provided that

$$(P_1 \text{ and } P_2 \text{ and} \ldots \text{and } P_n)$$

is sound then we can deduce that

$$(P_1 \text{ and } P_2 \text{ and} \ldots \text{and } P_n) \Rightarrow Q,$$

since assuming not Q leads to a *contradiction.* To illustrate we consider the following example.

Example 1.4.7

Prove for arbitrary sets A and B, that

$$A \subseteq B \Leftrightarrow B' \subseteq A'.$$

Proof. Assume the properties (i.e. the definitions) of $\mathscr{E}$ and $\varnothing$, etc., and that $A \subseteq B$ and $B' \nsubseteq A'$. (In terms of the general situation described above Q is the statement $B' \subseteq A'$.)

$$A \subseteq B \Rightarrow \text{if } x \in A \text{ then } x \in B, \qquad (*)$$

$B' \not\subseteq A' \Rightarrow$ there is some element, y say, such that $y \in B'$ and $y \notin A'$,

$y \notin A' \Rightarrow y \in A$,

$y \in A \Rightarrow y \in B$ (by (*)),

$\Rightarrow y \in B'$ and $y \in B$

$\Rightarrow y \in B' \cap B = \varnothing$ ※ (contradiction).

Consequently we deduce $B' \not\subseteq A'$ is false and hence $B' \subseteq A'$. Similarly we can show $B' \subseteq A' \Rightarrow A \subseteq B$ and therefore

$$A \subseteq B \Leftrightarrow B' \subseteq A'. \quad //$$

This example also provides motivation for a second kind of indirect proof.

We have just proved that

$$(x \in A \Rightarrow x \in B) \Leftrightarrow (x \notin B \Rightarrow x \notin A).$$

This bi-implication is of the form

$$(P \Rightarrow Q) \text{ iff } (\text{not } Q \Rightarrow \text{not } P)$$

and, although probably not obvious to the reader, expressions of this form are always true and hence if we can ascertain that (not Q) $\Rightarrow$ (not P) then we can deduce $P \Rightarrow Q$. This is called a *contrapositive* proof. Before moving on to the next group of exercises the reader should consider the following 'non-mathematical' example of a contrapositive deduction.

Example 1.4.8

Let P denote

'today is Tuesday'

and Q denote

'today is a weekday'.

Then

($P \Rightarrow Q$) denotes 'if today is Tuesday then it is a weekday'.

and

(not $Q \Rightarrow$ not P) denotes 'if today is not a weekday then it cannot be Tuesday'.

The reader should convince himself, by examining all possibilities, that the two implications are equivalent (i.e. that both are true or both are false). //

From this point on, although Venn diagrams may be used in an attempt to clarify a particular situation, all results should be properly derived from the assumptions given in the question. Remember, if something is true you should be able to prove it. If it really is obvious then you should be able to prove it easily, if you cannot then it probably is not as obvious as you think and may not even be true.

Exercises 1.4

1. Prove that

   $A \cap (B \cap C) = (A \cap B) \cap C.$

2. Given sets A, B and C: $C \subseteq B$ prove that

   (i) $A \cap C \subseteq A \cap B$,
   (ii) $A \cup C \subseteq A \cup B$,
   (iii) $A \backslash B \subseteq A \backslash C$,
   (iv) $C \backslash A \subseteq B \backslash A$,
   (v) $B' \backslash A \subseteq C' \backslash A$.

   Is it true that

   $A \triangle C \subseteq A \triangle B$

   or that

   $A \triangle B \subseteq A \triangle C$?

3. Prove that if

   $A \subseteq B$

   then

   $\mathcal{P}(A) \subseteq \mathcal{P}(B)$.

4. Prove that

   $A \cup (B \cap C) = (A \cup B) \cap (A \cup C)$.

5. Prove that

   $(A \cup B)' = A' \cap B'$.

   (Hint: show that $(A \cup B) \cup (A' \cap B') = \mathscr{E}$ and $(A \cup B) \cap (A' \cap B') = \varnothing$.)

6. Prove that the following statements are equivalent, i.e. each implies the others

   (i) $A \cup B = \mathscr{E}$,
   (ii) $A' \subseteq B$,
   (iii) $A' \cap B' = \varnothing$.

7. Which of the following are true?

   (*a*) $0 \in \varnothing$,
   (*b*) $\{\varnothing\} \subseteq \varnothing$,
   (*c*) $\varnothing \subseteq \{\varnothing\}$,
   (*d*) $\varnothing \subseteq \mathscr{E}$,
   (*e*) $\{\varnothing\} \subseteq \{\{\varnothing\}\}$.

Compare your answers to this question with those of Exercise 1.2.6. There is a connection between $\in$ and $\subseteq$ but they are not the same. How does the 'bag' analogy relate to $\subseteq$?

8. Show that for a finite set A

   $|2^A| = 2^{|A|}$.

   (Hint: write $A = \{a_1, \ldots, a_n\}$ and consider its subsets.)

1.5 Products of sets

So far we have been concerned mostly with creating smaller sets from existing ones; we now look at one of the common ways in which larger sets are constructed. Consider, by way of illustration, the continental system of labelling the squares of a chess board (see Figure 1.10).

Fig. 1.10

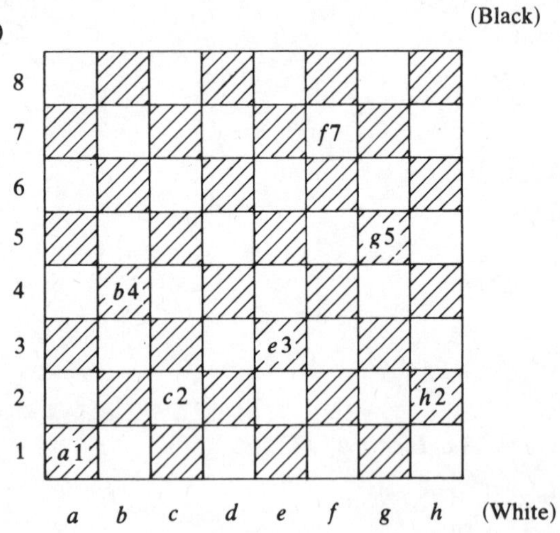

Regarding the board from White's side the columns, properly called files, are labelled $a, b, c, \ldots, g, h$ (from left to right) and the rows, or ranks, from 1 to 8 starting from White's side. Consequently, any square can be specified uniquely by giving two symbols, one from the set $F = \{a, b, c, d, e, f, g, h\}$ and one from the set $R = \{1, 2, 3, 4, 5, 6, 7, 8\}$, for example $a1$, $f7$, $e3$, etc. Thus, from the set of files, F and the set of ranks, R, we can derive the set of all squares on the board.

This example embodies all the new concepts involved in constructing products of sets, however, in order to be able to generalize the situation, we must be a little more precise.

Definitions. We denote a sequence of n objects $x_1, x_2, \ldots, x_n$ by

$$(x_1, x_2, \ldots, x_n).$$

Here round brackets are used to indicate that the order in which the objects (elements) are written is important; so if, for example, $x_1 \neq x_2$ then the sequence $(x_2, x_1, \ldots, x_n)$ is not the same as the first sequence given. Such a sequence is called an *n-tuple* and a 2-tuple is commonly called a *pair.*

Given n sets $A_1, \ldots, A_n$, the set of all n-tuples $(x_1, \ldots, x_n)$ such that $x_1 \in A_1, \ldots, x_n \in A_n$ is called the *Cartesian product* of A_1 to A_n, denoted by $A_1 \times A_2 \times \ldots \times A_n$ and read 'A_1 cross A_2 cross ... cross A_n'. Using indexed notation we can write this product even more concisely as

$$\bigtimes_{i=1}^{n} A_i. \quad //$$

Example 1.5.1

Let

$$X = \{0, 1\},$$
$$Y = \{x, y\},$$
$$Z = \{0, 1, 2\}.$$

Then

$$X \times Y = \{(0, x), (0, y), (1, x), (1, y)\},$$
$$Y \times X = \{(x, 0), (x, 1), (y, 0), (y, 1)\}.$$

So $X \times Y \neq Y \times X$. Referring back to the chess board example, it would be obvious what was meant if someone wrote $3e$ because the sets F and R were disjoint, although $(e, 3) \in F \times R$ and $(3, e) \notin F \times R$. Now

$$X \times Z = \{(0, 0), (0, 1), (0, 2), (1, 0), (1, 1), (1, 2)\}$$

so $(0, 1)$ and $(1, 0)$ are distinct elements of $X \times Z$ and therefore, mathematically, we would reject $3e$ as a valid chess board reference. //

We will often have cause to construct Cartesian products from like sets, in which case it is convenient to write

$$A \times A \times \ldots \times A$$

as

$$A^n.$$

Exercises 1.5

1. If $X = \{a, b, c\}$ and $Y = \{a, b, e, f\}$, compute $X \times Y$ and Y^2.
2. Prove that if $A \subseteq X$ and $B \subseteq Y$ then

   $A \times B \subseteq X \times Y.$

3. Prove that

$$A \times (B \cap C) = (A \times B) \cap (A \times C).$$

4. Prove that for all non-empty finite sets A and B,
   (i) $\varnothing \times A = \varnothing$,
   (ii) $\mathscr{E} \times A \neq A$,
   (iii) $A \not\subseteq A \times A$,
   (iv) $|A \times \{x\}| = |A|$,
   (v) $A \times B = B \times A$ iff $A = B$.

2 RELATIONS

Often in computing we need to pick out elements of sets which we regard as being 'related' in some way. This notion is very general and hence widely applicable; indeed items can be related simply because we choose to relate them. They *need not* be connected by any simple or obvious formula, though in situations where we are required to perform calculations some manageable description must be found.

Before tackling the topic mathematically we can get some idea of the concepts from the following simple analogy (which also gives rise to proper relations!). Suppose we have a set of programs P for a specific finite machine, a finite set of data values D and a set of results R. Now, if we select a specific value from D it will be valid input for several programs in P, and for each program in P several values in D will be valid. Here we have a relationship between data values and programs and hence there are certain elements of $D \times P$ that are of interest. Similarly, if we limit consideration to $p \in P$ then p associates suitable input values from D with results in R. One can envisage some data causing p to fail and some results not being derivable from p hence we arrive at a subset of $D \times R$. (The association suggested by the movement from Data (D) to Results (R) may prove useful in remembering some terminology to be given shortly.) With this illustration in our minds we now proceed to the formal presentation.

2.1 Fundamental concepts

An *n-place relation* R on the sets A_1 to A_n is a subset of the Cartesian product

$$A_1 \times A_2 \times A_3 \times \ldots \times A_n.$$

Put another way, the objects $x_1, \ldots, x_n$ (where $x_1 \in A_1 \ldots$) are related by R iff the ordered n-tuple $(x_1, x_2, \ldots, x_n) \in R$.

The most common relations that we shall encounter are those defined when $n = 2$ and are called *binary relations.* A (binary) relation between sets A and B is therefore simply a subset of $A \times B$. If these sets are equal ($= A$ say) then we say that a subset of A^2 defines a *relation on A*. //

Relations are not new; we can easily construct relations that will be familiar to the reader. Consider the following examples.

Example 2.1.1

If

$$A = \{1, 2, 3, 4, 5, 6, 7, 8, 9, 10\}$$

then

$$R = \{(x, y): x, y \in A \text{ and } x \text{ is a factor of } y \text{ and } x \leq 5\}$$

can be written explicitly as

$$\begin{aligned} R = \{&(1,1), (1,2), (1,3), (1,4), (1,5), \\ &(1,6), (1,7), (1,8), (1,9), (1,10), \\ &(2,2), (2,4), (2,6), (2,8), (2,10), \\ &(3,3), (3,6), (3,9), \\ &(4,4), (4,8), \\ &(5,5), (5,10)\}. \quad // \end{aligned}$$

Example 2.1.2 (Chess again.)

As before let

$$F = \{a, b, c, d, e, f, g, h\}$$

and

$$R = \{1, 2, 3, 4, 5, 6, 7, 8\}$$

and now

$$S = F \times R.$$

S is thus the set of all squares denoted by pairs (x, y) where $x \in F$ and $y \in R$. Let us now define the binary relation C (for castle!) on S such that $(s, t) \in C$ iff s and t are elements of S and a castle can go from s to t in a single move on an otherwise empty board. (For the uninitiated, a castle can change its rank or file – not both – in any move.) So

$$C \subseteq S \times S$$

and

$$\begin{aligned} C = \{((f_s, r_s), (f_t, r_t)): &(f_s = f_t \text{ and } r_s \neq r_t) \text{ or} \\ &(f_s \neq f_t \text{ and } r_s = r_t)\}. \quad // \end{aligned}$$

At first sight the definition of C may look complicated but careful investigation, taking one layer of complexity at a time, will reveal not only that the meaning is straightforward but also that all the information present is needed.

In general the number of different relations on a set A depends on $|A|$. Most of these relations will not be of particular interest. However, for a given set we can always derive three special relations.

Definition. For any set A we can define the *identity relation* on A, denoted by I_A, and the *universal relation* on A, denoted by U_A, by

$$I_A = \{(a, a): a \in A\}$$

and

$$U_A = \{(a, b): a \in A, b \in A\}.$$

So $U_A = A^2$. Also, since $\varnothing \subseteq A^2$, $\varnothing$ is a relation on A and is called the *empty relation.* //

Once a relation R has been defined between A and B (as depicted in Figure 2.1) it is quite usual to concentrate our attention on what happens at 'the ends' of R; i.e. to consider the elements of A or B that actually participate in R. These are included in subsets of A and B respectively and, as might be expected, have special names.

Fig. 2.1

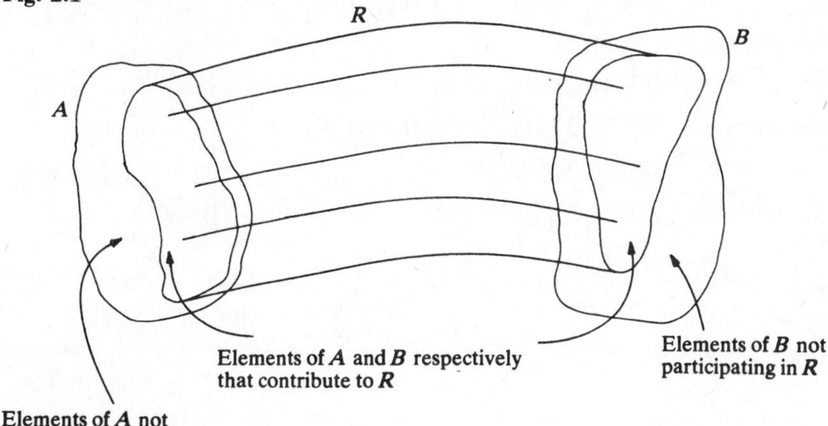

Definition. Associated with any binary relation R between A and B are two sets known as the *domain* of R, $\mathscr{D}(R)$, and the *range* of R, $\mathscr{R}(R)$. These are defined by

$$\mathscr{D}(R) = \{x: (x, y) \in R\}$$

and

$$\mathscr{R}(R) = \{y: (x, y) \in R\}. \quad //$$

Example 2.1.3

With reference to the relation R defined in Example 2.1.1 we have

$$\mathscr{D}(R) = \{1, 2, 3, 4, 5\}$$

and

$$\mathscr{R}(R) = A. \quad //$$

Example 2.1.4

Suppose we have a rather strange program. It reads in two numbers from the set $A = \{1, 2, 3, 4, 5\}$ represented by (x, y) and if $x < y$ it prints a number, z (also in A), such that $x \leq z < y$. In any case the program halts within five seconds of reading its input.

The problem defines a relation, call it P, so $P \subseteq A^2 \times A$ such that:

$$P = \{((x, y), z): x < y \text{ and } x \leq z < y\}.$$

Not all inputs give rise to an output and hence the domain of P is not the whole of A^2; explicitly we have

$$\begin{aligned} P = \{&((1, 2), 1), \\ &((1, 3), 1), ((1, 3), 2), \\ &((1, 4), 1), ((1, 4), 2), ((1, 4), 3), \\ &((1, 5), 1), ((1, 5), 2), ((1, 5), 3), ((1, 5), 4), \\ &((2, 3), 2), \\ &((2, 4), 2), ((2, 4), 3), \\ &((2, 5), 2), ((2, 5), 3), ((2, 5), 4), \\ &((3, 4), 3), \\ &((3, 5), 3), ((3, 5), 4), \\ &((4, 5), 4)\}; \end{aligned}$$

$$\begin{aligned} \mathscr{D}(P) = \{&(1, 2), (1, 3), (1, 4), (1, 5), \\ &(2, 3), (2, 4), (2, 5), \\ &(3, 4), (3, 5), \\ &(4, 5)\}; \end{aligned}$$

$$\mathscr{R}(P) = \{1, 2, 3, 4\}. \quad //$$

Although any relation is a set and hence can be denoted by a capital letter, it is also common practice to use lower case Greek letters such as ρ, σ and τ (see the glossary of symbols at the end of the text). The following alternative notations are often used when referring to relations

(1) $(a, b) \in \rho$, i.e. (a, b) is in ρ,

(2) $a\rho b$, a is ρ-related to b,

and

(3) $b \in \rho(a)$.

The first of these is the natural notation following from the set-theoretic definitions. The second makes more sense if we consider the order relation (which will be properly defined later) where

$$R \subseteq \mathbb{N}^2$$

and

$$R = \{(x, y): x < y\}.$$

Here, instead of writing $6R7$ we can write $6<7$, hence we allow any relation that is associated with a particular symbol to be represented by that symbol whenever the meaning of the resulting sequence of characters is unambiguous. The third form is new and will be developed into the more usual notation for functions in the next chapter. From a given binary relation R it is possible to derive many other relations, the most common being the inverse relation.

Definition. Given a binary relation R, we define the *inverse relation* R^{-1} by

$$R^{-1} = \{(x, y): (y, x) \in R\}.$$

Thus R^{-1} associates the same pairs of elements as R but 'goes the other way'. //

It therefore follows that if $R \subseteq A \times B$ then

$$R^{-1} \subseteq B \times A,$$

$$\mathscr{D}(R^{-1}) = \mathscr{R}(R)$$

and

$$\mathscr{R}(R^{-1}) = \mathscr{D}(R).$$

In order to avoid the use of too many brackets we shall also allow the notation $\mathscr{D}_R$ and $\mathscr{R}_R$ instead of $\mathscr{D}(R)$ and $\mathscr{R}(R)$.

Exercises 2.1

1. If $X = \{2, 4, 6, 8\}$ and $\rho = \{(x, y): x, y \in X \text{ and } x < y\}$ list all elements of ρ and ρ^{-1}.
2. If $\mathscr{E} = \mathscr{P}(\{a, b, c\})$ what are the elements of the relations $\subset$ and $\subseteq$ on $\mathscr{E}$?
3. If $\mathscr{E} = \mathbb{Z}^2$ and $\rho = \{(x, y): x < y\}$ describe the complement ρ of ρ', *without* using a negated relation $(\not<)$.

4. If $\sigma = \{(x, y): x \subset y\}$ can σ' be described in a way similar to ρ' in Question 3? Justify your answer.
5. A street contains thirty houses numbered in the usual way with odd numbers down one side and even numbers down the other side. If h_n denotes the household living in the house which is numbered n, describe symbolically the relation N on the set of households such that h_i is N-related to h_j if the households are next-door neighbours.

   How is N affected if the street is a cul-de-sac?
6. Referring back to the set S of squares on a chess board, the relation K relates squares which define a knight's move (i.e. xKy if a knight can move from x to y in a single step). Define K symbolically.
7. Similarly, let G be a relation on S such that xGy iff x is the initial position of a (white) piece and y is the square where the first move of the game ends. Describe G, $\mathscr{D}(G)$ and $\mathscr{R}(G)$.

2.2 Graphic representations

As the first stage in solving a mathematical problem it is often useful to draw a 'picture' in order to see more clearly how the components of the problem fit together. This is particularly true of relations since, written down as a set of ordered pairs, relations are not easily deciphered.

Relations are only sets which have a particular structure (their elements have several components) and so in principle we could use a Venn diagram to depict them. Although it is possible to utilize this technique, especially for relations on some large sets of numbers, better methods exist for more commonly occurring situations involving binary relations on small sets. In this section we shall briefly consider four such methods. To exemplify we shall use the set

$$X = \{a, b, c, d, e\}$$

and the relations I_X, U_X and R where

$$R = \{(a, b), (a, c), (b, d), (c, e), (e, b)\}.$$

Firstly consider a method closely allied to traditional coordinate geometry. Draw a pair of perpendicular axes (the x-axis horizontal and the y-axis vertical) and on each mark points representing the elements of the set X as in Figure 2.2(i). Now, for each element (x, y) in the relation, place a mark on the diagram above the x on the x-axis representing $\mathscr{D}_{I_X}$ and to the right of the y on the y-axis representing $\mathscr{R}_{I_X}$. The resulting diagram is shown in Figure 2.2(ii). Corresponding diagrams for the relations U_X and R are given in Figure 2.2(iii) and 2.2(iv).

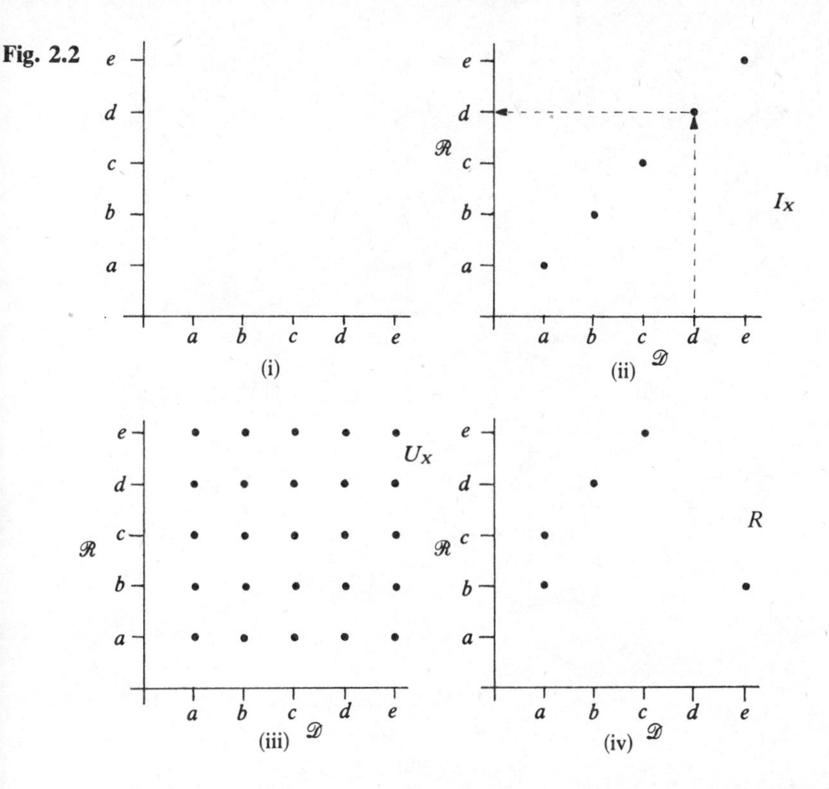

The main drawback with this method is that if $|X|$ is increased it is not easy to see which elements in the domain and range correspond to the dots that denote the relation. To overcome this we can omit the dots and join by an arrow the points $x \in \mathscr{D}$ and $y \in \mathscr{R}$ when (x, y) is in the relation. This gives the diagrams in Figure 2.3. The diagram representing U_X is clearly over-cluttered, but U_X has a lot of elements so it is to be expected that its diagram should be more complicated than the others. On the positive side the relations I_X and R are clearly represented and their domains and ranges are easily seen. The diagram of U_X is most confused near to where the two axes cross. Now that we are not using the coordinates of the domain and range to place the elements of the relation (as in the first method) the axes do not need to cross, we can draw them parallel. So, using vertical parallel lines and moving from left to right – the line on the left representing the set from which the domain is taken – we obtain the diagrams in Figure 2.4.

Here the arrowheads are not really required since we know that the relations go from their domains to their ranges. This leads to two possibilities; either we can replace the arrows by straight lines or we can replace the two lines of points by a single collection of points. (For instance, the point representing 'c' in the domain is the same point that

Fig. 2.3

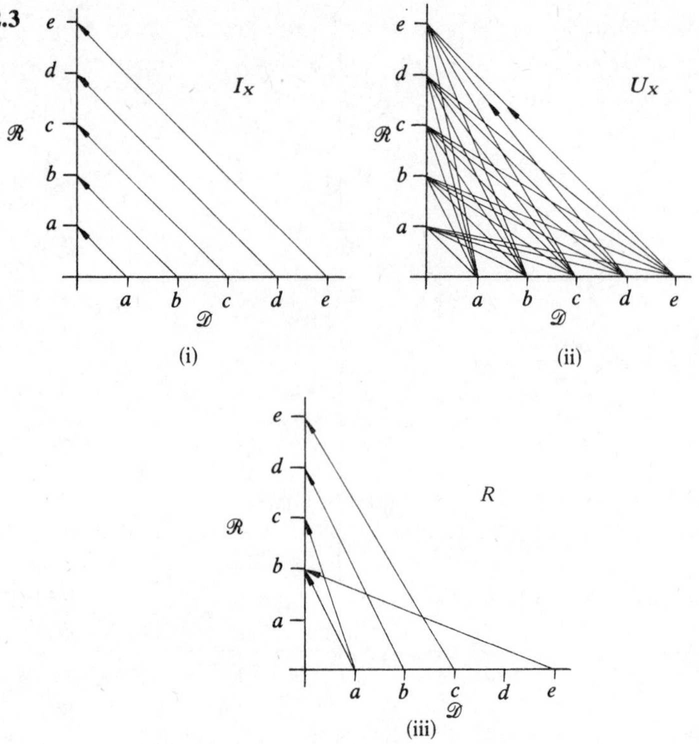

(i)

(ii)

(iii)

Fig. 2.4

(i)

(ii)

(iii)

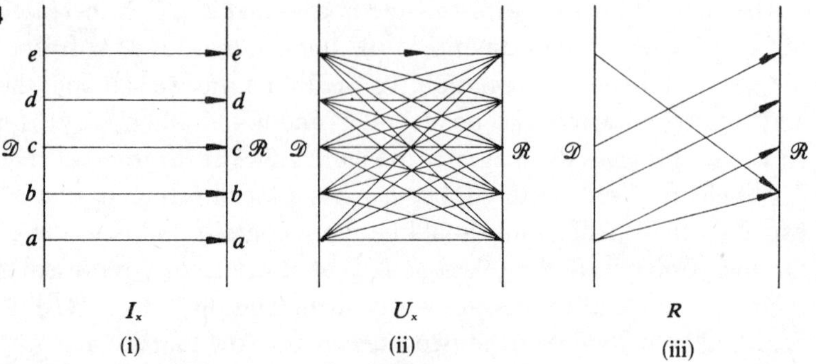

represents 'c' in the range.) This gives the diagrams in Figure 2.5. These are indicative of the most common ways in which binary relations are displayed graphically and will be used exclusively throughout the rest of the book. We shall continue discussion of graphical methods associated with relations in Chapter 7.

Exercises 2.2

1. Draw a diagram to represent the relation ρ of Question 1 in Exercises 2.1.

2. Draw a diagram to represent the relation N (see Exercises 2.1 Question 5) on a street having ten houses. How is the diagram changed if the street is a cul-de-sac? //

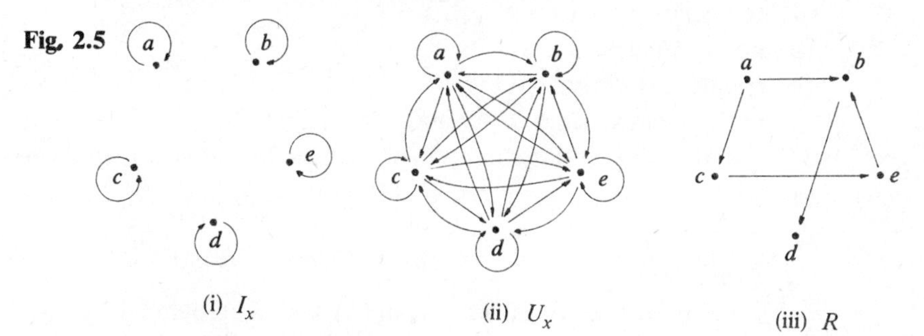

Fig. 2.5 (i) I_X (ii) U_X (iii) R

2.3 Properties of relations

Obviously general relations, being just subsets of a product of sets, are not particularly interesting as very little can be said about them. However, when the relations satisfy further conditions more significant statements can be made. In this section we consider some of the main properties with which relations may be endowed. Each property is said to be present whenever the corresponding condition is satisfied.

Definition. Let ρ be a relation on a set A, then

(a) ρ is *reflexive* if $x\rho x$ for all $x \in A$,

(b) ρ is *symmetric* if $x\rho y$ implies $y\rho x$,

(c) ρ is *transitive* if $x\rho y$ and $y\rho z$ imply $x\rho z$ and

(d) ρ is *antisymmetric* if $x\rho y$ and $y\rho x$ imply that $x = y$. //

The terminology defined here will most probably be new to the reader but he should already be familiar with the underlying concepts which are illustrated in the following examples.

Example 2.3.1

Let

$\rho = \{(x, y): x, y \in \mathbb{N}$ and x divides $y\}$,

$\sigma = \{(x, y): x, y \in \mathbb{N}$ and $x \leq y\}$,

$\tau = \{(x, y): x, y \in \mathbb{N}\backslash\{1\}$ and x and y have a common factor$\}$,

then ρ is

(a) reflexive since $x/x = 1$ for all $x \in \mathbb{N}$, is

(b) not symmetric since 2 divides 4 but 4 does not divide 2, is

(c) transitive since if $y/x \in \mathbb{N}$ and $z/y \in \mathbb{N}$, then

$z/x = (y/x)*(z/y) \in \mathbb{N}$

and is

(d) antisymmetric since if $x/y \in \mathbb{N}$ and $y/x \in \mathbb{N}$ then $x = y$.

Similarly, σ is

(a) reflexive since $x \leqslant x$ for all $x \in \mathbb{N}$, is

(b) not symmetric since $2 \leqslant 3$ but $3 \nleqslant 2$, is

(c) trivially transitive, and is

(d) antisymmetric since $x \leqslant y$ and $y \leqslant x$ implies $x = y$.

Lastly, τ is reflexive and symmetric but is not transitive or antisymmetric. //

Example 2.3.2

Let P be the set of all people and A and S be defined by

$A = \{(x, y): x, y \in P$ and x is an ancestor of $y\}$,

$S = \{(x, y): x, y \in P$ and, x and y have the same parents$\}$.

Clearly A is transitive and S is reflexive, symmetric and transitive. //

Notice that the properties of symmetry and antisymmetry are not mutually exclusive. For any set X, I_X is both symmetric and antisymmetric! Check this against the definitions. Also we may have a relation that is neither symmetric nor antisymmetric.

Example 2.3.3

Again let P be the set of all people and define the relation B such that xBy iff x is the brother of y. In a family consisting of two brothers p and q and a sister r, we have the situation depicted in Figure 2.6. The relation B is not symmetric since pBr but $r\not{B}p$. It also fails to be antisymmetric since pBq and qBp even though p and q are distinct. //

Fig. 2.6

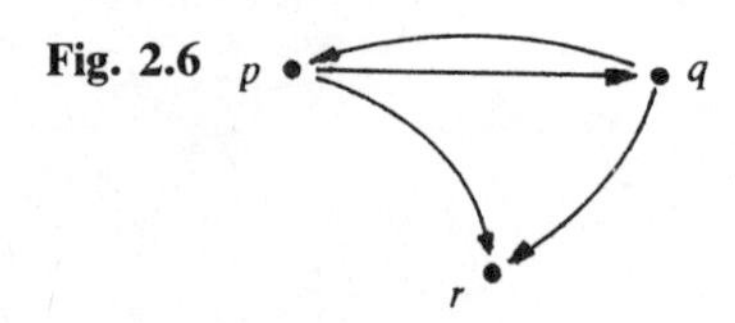

In a more general situation we may interpret the four special relational characteristics diagrammatically as

(i) a relation is reflexive iff for each node (or point) on the diagram there is an arrow which starts and finishes at that node;

(ii) a relation is symmetric iff for every arrow joining two nodes there is also an arrow joining the same nodes but directed the other way;

(iii) a relation is transitive iff each pair of nodes, x and y, connected by a sequence of arrows, x to a_1, a_1 to $a_2, \ldots, a_{n-1}$ to a_n, a_n to y, are also connected by a direct arrow from x to y; and

(iv) a relation is antisymmetric iff there are no two distinct nodes linked by a pair of arrows (one each way).

There are many other properties of relations that could be considered. However, the ones given above are the most important and will be put to use shortly.

Exercises 2.3

1. For the following relations say whether each is reflexive, symmetric, transitive or antisymmetric.
   (i) The relation on $\{1, 2, 3, 4, 5\}$ defined by $\{(a, b): a-b$ is even$\}$.
   (ii) The relation on $\{1, 2, 3, 4, 5\}$ defined by $\{(a, b): a+b$ is even$\}$.
   (iii) The relation on P, the set of all people, defined by

   $$\{(a, b): a \text{ and } b \text{ have an ancestor in common}\}.$$

2. The following statement is fallacious;

   A relation, on a set S, which is symmetric and transitive is also reflexive since aRb and bRa implies aRa.

   By careful reference to the definitions, find the fault. Construct a relation on $\{1, 2, 3\}$ that is symmetric and transitive but not reflexive.

3. If ρ is a relation between A and B, and $a \in A$ then $\rho(a)$ is defined as the set $\{b: a\rho b\}$ and is a subset of B. Let the following relations be defined on $\{-4, -3, -2, -1, 0, 1, 2, 3, 4\}$:

   $\rho = \{(a, b): a < b\}$,

   $\sigma = \{(a, b): b-1 < a < b+2\}$,

   $\tau = \{(a, b): a^2 \leq b\}$.

   What are the sets

   (*a*) $\rho(0)$,
   (*b*) $\sigma(0)$,
   (*c*) $\tau(0)$,
   (*d*) $\rho(1)$,
   (*e*) $\sigma(-1)$,
   (*f*) $\tau(-1)$?

2.4 Partitions and equivalence relations

In many aspects of computing we take large sets and subdivide them in suitable ways so as to investigate the overall situation by considering a few well chosen examples. For instance, one way of quickly gaining an appreciation of the characteristics of a programming language is to look at programs written in that language. However, any interesting language, including all conventional high-level languages such as Pascal and Fortran, gives rise to infinitely many programs and hence we need to be careful about the choice of programs selected to illustrate the language. To be more specific let us further suppose that the language has three basic control structures, four data accessing modes and, for current purposes, no other salient features. We could then take, as our sample, seven programs each including a different one of these features (although each program may include more than one feature). Investigation of these programs would then cover all the major features of the language. Mathematically this concept is embodied in the following definition.

Definition. Given a (non-empty) set A, a collection $(A_i(1 \leq i \leq n)$ for some $n \in \mathbb{N})$ of subsets of A such that

$$\bigcup_{i=1}^{n} A_i = A$$

is a *covering* of A. //

Example 2.4.1

$\{A, B\}$ is a covering of $A \cup B$,

$\{A, A \cup B, B, C\}$ is a covering of $A \cup B \cup C$. //

Use of coverings should ensure that nothing is missed since any element in a set is included in at least one of the subsets in the covering; however there may be much duplication. If we further specify that the elements of the covering must be disjoint then duplication of effort is avoided. Such a situation gives rise to the concept of a partition.

Definition. A *partition* of a (non-empty) set A is a (non-empty) subset of $\mathcal{P}(A)$ such that the union of all elements of this subset is A and all subsets are mutually disjoint; i.e. A is divided up so that any element of A is contained in exactly one set in the partition. //

Example 2.4.2

$\{A, A'\}$ is a partition of $\mathscr{E}$,

$\{A \cap B, A \cap B', A' \cap B, A' \cap B'\}$ is a partition of $\mathscr{E}$

and

$$\{A \backslash B, A \cap B, B \backslash A\} \text{ is a partition of } A \cup B. \quad //$$

Partitions uniquely determine, and are themselves induced by, a particular sort of relation called an equivalence relation. These are relations that behave in a way similar to '=' between numbers or between sets, and by extracting the essential properties of equality we obtain the following definition.

Definition. A (binary) relation on a set is called an *equivalence relation* if it is reflexive, symmetric and transitive. //

Example 2.4.3

On the set of all triangles the relation defined by $\{(x, y)$: x and y have the same area$\}$ is trivially an equivalence relation. Of more relevance to computing the relation defined on the set of all programs by $\{(a, b)$: a and b compute the same function on a specific finite machine$\}$ is also an equivalence relation. //

Recall that we are looking for simpler ways of treating large sets by breaking them into a smaller number of parts than would be obtained by taking the elements one at a time. We now have all the mathematical machinery but lack suitable simple notation. This is now given and immediately followed by some familiar examples.

Definition. Given an equivalence relation ρ on a set A we define the *equivalence class*, $[x]$, of $x \in A$ as

$$[x] = \{y: x\rho y\}.$$

$[x]$ is the set of all elements of A that are *ρ-equivalent* to x. In cases where only one equivalence relation is under consideration we can also use the notation '$\equiv$' ('is equivalent to') so

$$[x] = \{y: x \equiv y\}.$$

In certain specific instances it is also customary to use the symbol '$\sim$' to denote equivalence. Now, instead of examining the whole of a set A, we can choose *representatives* (one from each equivalence class) in any way that simplifies the calculations in a given situation. //

The next example illustrates this point.

Example 2.4.4

Let s be a fixed element of $\mathbb{N}$ and define the relation ρ_s on $\mathbb{Z}$ by

$$\rho_s = \{(x, y): x - y = ns \text{ where } n \in \mathbb{Z}\}.$$

Consider the case when $s=10$, then

$$[1]=\{11, 21, -9, 10976631, \ldots\},$$

$$[1066]=\{66, 226, -24, \ldots\}$$

and so on.

In fact there are only 10 distinct equivalence classes; the integers $0, 1, 2, \ldots, 9$ are each in a different class and hence we can use them as representatives. //

In computing, equivalence relations which are of particular interest include those associated with different algorithms for computing the same result or for carrying out the same manipulation on a data set, or representing certain information in different (equivalent?) data structures. An explanation of these apparently straightforward situations is fraught with difficulties arising from questions of decidability. Such factors, however, do not cause problems when the sets are created by well-defined processes from a finite base. Typically, even these well-behaved cases require a considerable amount of detailed description and hence would be out of place here. Notwithstanding these disclosures, we shall return to related matters in Section 2.6 and in Chapters 8 and 9.

Probably the most familiar equivalence relation known to the reader, although he may not have realized that it was in fact an equivalence relation, is that associated with fractions. Consider the set $\mathbb{Z}\times\mathbb{N}$. Given $(a, b)\in\mathbb{Z}\times\mathbb{N}$ we can regard this as the fraction $\frac{a}{b}$. (Technically these two ways of denoting elements of $\mathbb{Z}\times\mathbb{N}$ are different but 'isomorphic'. We shall consider this further in 5.1; the important fact is that we can change from one form to the other in an *exact* and *reversible* fashion.)

So far so good but there are distinct elements in $\mathbb{Z}\times\mathbb{N}$ which we should like to consider as being the same – but are written differently. To overcome this difficulty we define the (equivalence) relation on $\mathbb{Z}\times\mathbb{N}$ by saying

$$(a, b)\equiv(c, d) \text{ iff } a*d=b*c.$$

The set of all equivalence classes given by this relation on $\mathbb{Z}\times\mathbb{N}$ is called the *rational numbers*, which we denote by $\mathbb{Q}$. Typically, we shall choose the representatives which are in lowest terms.

It must also be mentioned that we assume the existence of *real* numbers (the set of all reals being denoted by $\mathbb{R}$). These can be regarded as being written in the form

$$\ldots 0d_n \ldots d_2 d_1 d_0 \cdot \delta_1 \delta_2 \ldots \delta_m \ldots$$

where each d_i and δ_j is drawn from the set $\{0, 1, \ldots, 9\}=\mathbb{Z}_{10}$ and $d_n\neq 0$, except in the cases when $n=0$ and then any $d_n \in \mathbb{Z}_{10}$ is allowed. In particular, we admit non-terminating, non-repeating representations although the 0s before d_n will usually be omitted. (Negative numbers are represented by preceding a non-zero positive number with a minus sign.)

Finally, we note that the *index* of an equivalence relation ρ on a set A is the cardinality of the partition of A induced by ρ (the number of ρ-equivalence classes).

Exercises 2.4

1. Prove that any equivalence relation induces a partition by showing that for any $x, y \in A$

   either $[x]=[y]$

   or $[x] \cap [y] = \varnothing$.
2. Let A be a finite set. What are the equivalence relations on A that give the largest and the smallest number of equivalence classes and what are their respective indices?
3. If $\{A_1, A_2, \ldots, A_n\}$ is a partition of A and A is finite show that

$$|A| = \sum_{i=1}^{n} |A_i|.$$

2.5 Order relations

Just as the notion of equality (between, say, numbers) gives rise to the mathematical concept of equivalence, certain inequalities can also be used as the model for a more widely applicable class of relations.

A (partial) *ordering* on a set A is a relation on A that is reflexive, antisymmetric and transitive. An ordering, also called an *order relation*, is a generalization of the relation $\leqslant$ on $\mathbb{N}$ and hence the three required properties are easily justified. Notice that we could have chosen to base the definition on the relation $<$ and that transitivity is the only property possessed by both of these relations and is therefore of major importance.

Having defined $\leqslant$ in a specific situation we may then define $<$ such that

$$a < b \Leftrightarrow a \leqslant b \quad \text{and} \quad a \neq b.$$

Equivalently, given $<$,

$$a \leqslant b \Leftrightarrow a = b \quad \text{or} \quad a < b.$$

Example 2.5.1

Given any set A, then the relation $\subseteq$ on $\mathscr{P}(A)$ is trivially an order relation. ($X \subseteq X$ for all X, $X \subseteq Y$ and $Y \subseteq X \Rightarrow X = Y$, and $X \subseteq Y$ and $Y \subseteq Z \Rightarrow X \subseteq Z$). //

An ordering ρ on A is said to be *total* if for any $x, y \in A$ either $x\rho y$ or $y\rho x$ (or both). Hence if $x \not\rho y$ then it follows that we must have $y\rho x$.

Example 2.5.2

The ordering on subsets of a given set is clearly not total. The natural ordering of numbers in the 'real line' $\mathbb{R}$ *is* total. //

As your study of (computer-related) mathematics progresses you should begin to see that mathematics is not a collection of disconnected ideas but of relatively few concepts that occur in many different situations; hence when a basic principle has been identified and studied it actually *solves* problems in all these situations in one fell swoop. The next example provides a simple illustration of this idea.

Example 2.5.3

From the order defined on $\mathbb{N}$, we can formally obtain the usual orderings associated with the related sets of numbers $\mathbb{Z}$, $\mathbb{Q}$ and $\mathbb{R}$. (As already noted, a proper treatment of $\mathbb{N}$ will be given later, in Section 3.3.)

First we examine $\mathbb{Z}$. To aid discussion let us partition $\mathbb{Z}$ by

$$\mathbb{Z} = \mathbb{N} \cup \{0\} \cup A.$$

So $A = \{-x: x \in \mathbb{N}\}$. Now we can define a relation (which will be a total order relation) on $\mathbb{Z}$ by considering all possible choices of elements x and y from the partition $\{\mathbb{N}, \{0\}, A\}$.

To guarantee reflexivity, we state that if $x = y$ then $x \leq y$ and $y \leq x$. Then, assuming $x \neq y$, we define all other possibilities as follows:

(i) if $x, y \in \mathbb{N}$ then the order in $\mathbb{Z}$ is the same as that in $\mathbb{N}$;

(ii) if $x, y \in A$ then

$$x \leq y \text{ iff } -y \leq -x \text{ in } \mathbb{N} \quad (\text{e.g. } -5 \leq -4 \text{ since } 4 \leq 5);$$

(iii) if $x = 0$ and $y \in \mathbb{N}$ then $x \leq y$;

(iv) if $x \in A$ and $y = 0$ then $x \leq y$;

(v) if $x \in A$ and $y \in \mathbb{N}$ then $x \leq y$, otherwise $y \leq x$.

From the ordering on $\mathbb{Z}$ and the usual arithmetic on integers we can now define the ordering on $\mathbb{Q}$ by

$$(a/b) \leq (c/d) \text{ iff } a * d \leq b * c.$$

Verification that this defines an order relation is left as an exercise. Finally we can define an order relation on the set of real numbers $\mathbb{R}$ by considering the decimal representation of the two positive real numbers

$$D = \ldots 0d_n \ldots d_2 d_1 d_0 \cdot \delta_1 \delta_2 \ldots$$

and

$$C = \ldots 0c_m \ldots c_2 c_1 c_0 \cdot \gamma_1 \gamma_2 \ldots .$$

If $d_i = c_i$ and $\delta_i = \gamma_i$ for all i then $D = C$ and hence $D \leqslant C$ and $C \leqslant D$. Otherwise

(i) $d_n \neq 0$, $c_m \neq 0$ and $n \neq m$ in which case if $n < m$ then $D \leqslant C$ and if $m < n$ then $C \leqslant D$, or

(ii) $n = m$ and $d_i \neq c_i$ but $d_j = c_j$ for all j such that $i < j \leqslant n$ then if $d_i < c_i$ it follows that $D \leqslant C$ and, conversely, if $c_i < d_i$ then $C \leqslant D$, or

(iii) $n = m$ and $d_i = c_i$ for all i but $\delta_k \neq \gamma_k$ for some k and $\delta_j = \gamma_j$ for all j such that $0 < j < k$. Then $C \leqslant D$ if $\gamma_k < \delta_k$ and $D \leqslant C$ if $\delta_k < \gamma_k$.

The reader is invited to test this for himself. Negative numbers can be treated as in $\mathbb{Z}$. //

A set, X, together with an ordering, $\leqslant$, is sometimes referred to as a *partially ordered set* or *po-set* and written $(X, \leqslant)$. If now we take two elements x, y of $(X, \leqslant)$ then any element $u \in (X, \leqslant)$ such that $x \leqslant u$ and $y \leqslant u$ is called an *upper bound* of x and y. Similarly, if $l \in (X, \leqslant)$, $l \leqslant x$ and $l \leqslant y$ then l is a *lower bound* of x and y. The set of all upper bounds of x and y is a subset of X and is ordered by $\leqslant$. If there is a unique lowest element of this set, i.e. if there is $\mu \in (X, \leqslant)$ such that $x \leqslant \mu$ and $y \leqslant \mu$ and $\mu \leqslant u$ for any upper bound u, of x and y then μ is called the *supremum* (*sup*) of x and y. Similarly, if a unique greatest lower bound of x and y exists it is called the *infimum* (*inf*). We shall put infs and sups to work in 5.5 when we study lattices.

Finally we note the use of the natural ordering on $\mathbb{R}$ to define new sets. These are called *intervals.*

$$[a, b] = \{x\colon x \in \mathbb{R}, a \leqslant x \leqslant b\}$$

– this set is the *closed interval* from a to b.

$$]a, b[= \{x\colon x \in \mathbb{R}, a < x < b\}$$

– the *open interval* from a to b.

In either case the numbers a and b are called *end points.*

A closed interval includes its end points, an open one does not. It is also convenient to define *half-open* intervals.

$$[a, b[= \{x\colon x \in \mathbb{R}, a \leqslant x < b\},$$

$$]a, b[= \{x\colon x \in \mathbb{R}, a < x \leqslant b\}.$$

As a notational convention the following will also be allowed:

$$]-\infty, a] = \{x\colon x \leqslant a\},$$

$$]-\infty, a[= \{x\colon x < a\},$$

$$[a, \infty[= \{x\colon a \leqslant x\},$$

$$]a, \infty[= \{x\colon a < x\},$$

and

$$]-\infty, \infty[= \mathbb{R}.$$

Although intervals, and sets of numbers in general, are not central to our considerations we shall find it convenient to use them from time to time.

Exercises 2.5

1. Let A be any set and ρ be a relation on $\mathcal{P}(A) \times \mathcal{P}(A)$, defined by

   $(P, Q)\rho(X, Y)$ iff $(P \triangle Q) \subseteq (X \triangle Y)$.

   Is ρ an order relation?
2. Let A be any set and σ be a relation on $\mathcal{P}(A) \times \mathcal{P}(A)$, defined by

   $(P, Q)\sigma(X, Y)$ iff $P \subseteq X$ and $Q \subseteq Y$.

   Is σ an order relation? If so, is it total?
3. Let τ and π be relations on $\mathbb{N}^2$ defined by

   $(a, b)\tau(c, d)$ iff $a \leqslant c$ and $b \leqslant d$,

   $(a, b)\pi(c, d)$ iff $a \leqslant c$ and $b \geqslant d$.

   Are τ and π order relations?
4. Let π be defined on positive elements of $\mathbb{Q}$ by

   $(a/b)\pi(c/d)$ iff $a * d \leqslant b * c$.

   Show that π is a total order relation.

2.6 Relations in databases and data structures

As we have already stated, relations are all around us; all that is required is a decidable statement, s, of n variables x_1 to x_n so that we can construct the set

$$\{(x_1, \ldots, x_n): s(x_1, \ldots, x_n) \text{ is true}\}.$$

Given any n-tuple we know whether it is included in the relation because s, evaluated using the components of that n-tuple, is decidable and hence is either true or false. Of course s need not be in the form of a 'nice' formula. Indeed, it could well be argued that instead of s defining a set of n-tuples, any set of n-tuples also defines a relation (and an associated containment property, s). These two approaches are equivalent.

Definition. In a practical data-processing situation the n-tuples are *records*, the elements of the n-tuples are *fields*, and the records that define a relation can be held in a *file*. When several files are required to hold a collection of 'relevant' relations then we have a (relational) *database.* //

Note. This is the *data-processing* usage of the word 'field' and is not to be confused with the mathematical term defined in Section 5.3.

This gives us our first tangible example of relations which are more obviously associated with computing, albeit applied computing and therefore more directly relevant to solving the problems of an end-user rather than being generated within computing itself. Nevertheless, a brief discussion of some of the simpler features of relational databases not only provides the basis for further mathematical development of relations but also highlights various operational factors, the understanding of which is necessary for the efficient administration of database systems.

Contemporary database theory involves the study of many so-called normal forms, but the justification for some of these is only apparent in fairly complex situations. We shall only give the three original forms and consider the need to

(i) insert new n-tuples,
(ii) delete n-tuples no longer required, and
(iii) update n-tuples.

Terminology is introduced as necessary. We begin with the simplest 'normal' form.

Definition. Files in *first normal form* (1NF) – or more simply, normalized files – have fixed-length records of elements drawn from sets whose elements cannot be further decomposed and, at any instant in time can be represented by an $M \times N$ array of values. Each record, being an n-tuple, can be written as a row in the array. //

Example 2.6.1

Consider the relation FAM1 in which we collect together parents and children. Each record holds, in order, the surname, and the names of the father, the mother and the children. Hence we may have

$$(\text{Smith, Joe, Joyce, (Sally, Ben)}) \in \text{FAM1}$$

and

$$(\text{Brown, Fred, Lisa, (Lucy)}) \in \text{FAM1}.$$

If we now denote by F and M the sets of fathers and mothers it follows that:

$$\text{Joe(Smith)} \in F \quad \text{Joyce(Smith)} \in M$$

$$\text{Fred(Brown)} \in F \quad \text{Lisa(Brown)} \in M.$$

Also, Lucy is the child of the Brown family, but (Sally, Ben) is not *the child* of the Smith family. Since there is more than one child in this family the corresponding record is larger and hence we violate the conditions of first normal form.

From FAM1 we can obtain FAM2, built from S, F, M and C where S is the set of surnames and C is the set of children, by constructing records:

(Smith, Joe, Joyce, Sally)

(Smith, Joe, Joyce, Ben)

(Brown, Fred, Lisa, Lucy).

FAM2 is in 1NF and can be represented diagrammatically as in Figure 2.7.

Fig. 2.7 FAM2

SURNAME	FATHER	MOTHER	CHILD
Smith	Joe	Joyce	Sally
Smith	Joe	Joyce	Ben
Brown	Fred	Lisa	Lucy

//

But what about Mr and Mrs Jones who have no children? If we need to know about these people then we must re-think the file structure; but this is leaping ahead. Firstly, let us introduce some terminology.

Definition. When using a tabular form to depict a relation/file with the n-tuples/records written as rows the (names of the) columns are known as *attributes.* //

Hence, SURNAME, FATHER, MOTHER and CHILD are the attributes of the various fields in FAM2. In order to access records in a file we use what is commonly called a key. This can be precisely defined in terms of attributes.

Definition. An attribute or (ordered) set of attributes whose values uniquely identify each record in a file is a *key* for that file. (Note that a file may have many different keys.) //

Any key for the relation/file FAM2 must involve the attribute CHILD. For other examples we now turn away from family relations.

Example 2.6.2

Any corporate computer user needs to buy spares for his machinery and hence may have a file SUP1 (suppliers 1) of the form as shown in Figure 2.8.

Fig. 2.8 SUP1

COMPANY	REGISTERED OFFICE	SALES MANAGER
ACE	LONDON	SMITH
IBL	LONDON	JONES
DATAMETZ	BIRMINGHAM	JONES
PRINTACO	MANCHESTER	BROWN
WOOLIES	BIRMINGHAM	BROWN
RTX	LONDON	SMITH
OXONDATA	OXFORD	WILSON

COMPANY is the key of SUP1 and other information in the file is accessed via the key. Thus we may access REGISTERED OFFICE *of* WOOLIES or SALES MANAGER *of* RTX. //

Definition. In this context '*of*' is a *projection* which extracts a field from a record; the record having been located by the key. In the reverse direction, we say that these attributes are *dependent* on the key. //

Diagrammatically the dependencies of SUP1 can be represented as in Figure 2.9.

Fig. 2.9

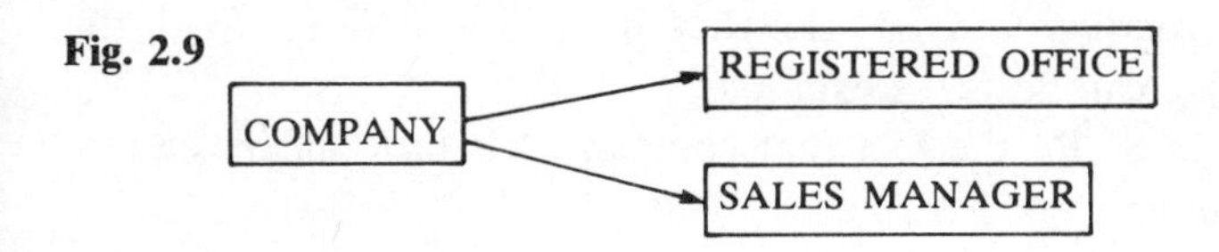

Example 2.6.3

We can now modify SUP1 to include information about the various spare parts stocked and in what quantities a particular supplier is willing to sell in a single transaction. Also included is a delivery code from which we can ascertain the speed and frequency of deliveries. To avoid some extraneous details we have introduced company numbers and part numbers. (See SUP2 in Figure 2.10). //

What can we do with SUP2 and what can we not do?

(i) *Insertion.* We cannot add a record indicating that company no. 4, PRINTACO, is in Manchester without any details of parts which it can supply.

(ii) *Deletion.* If RTX, company 6, stop supplying part no. 1 then we are obliged to remove all details of that company.

Fig. 2.10 SUP2

COMPANY	LOCATION	DELIVERY	PART	QUANTITY
1	LONDON	2	1	10
1	LONDON	2	2	1
1	LONDON	2	3	10
2	LONDON	2	2	2
2	LONDON	2	4	5
3	B'HAM	4	2	2
3	B'HAM	4	3	10
3	B'HAM	4	4	4
5	B'HAM	4	1	10
5	B'HAM	4	3	10
6	LONDON	2	1	10

(iii) *Updating.* If the delivery code for London is changed (because, for example, of transport disruption) then the appropriate field must be changed in every record with LONDON as its LOCATION.

How can we reduce or remove these problems? From a practical standpoint we need to separate out the information in SUP2 in such a way that needless repetitions are avoided, thus allowing what would be insertions/deletions of partial records in SUP2 to be carried out. One possible, and sensible, dissection gives SUP3 (Figure 2.11). The remainder of the information in SUP2 can then be held in PARTS. With this configuration we can now, for example,

(i) enter in SUP3 the fact that company 4 is in Manchester (and the delivery code is 3),

(ii) delete the reference to company 6 as a supplier of part no. 1 but leave the corresponding entry in SUP3, and

(iii) change the delivery code for LONDON to 7 by altering only 3 entries corresponding to companies rather than 6 entries denoting parts.

These alterations result in the tables given as Figure 2.12. This is better but could still be improved upon. To see what is going on we need to isolate the keys and dependencies (Figure 2.13). (Notice that PARTS requires a concatenated key.)

All non-keys depend directly on the key. This gives the next normal form property. //

Definition. A file is in *second normal form* (2NF) if it is in 1NF and all non-key attributes are fully dependent on the key (a proper subset of the key will not do). //

Fig. 2.11 SUP3

COMPANY	LOCATION	DELIVERY
1	LONDON	2
2	LONDON	2
3	B'HAM	4
5	B'HAM	4
6	LONDON	2

PARTS

COMPANY	PART	QUANTITY
1	1	10
1	2	1
1	3	10
2	2	2
2	4	5
3	2	2
3	3	10
3	4	4
5	1	10
5	3	10
6	1	10

Example 2.6.3 (cont.)

SUP3 is still over-complicated in that for a given record, DELIVERY can be ascertained by investigating either the COMPANY or LOCATION fields. This is the reason for (a) the requirement for a delivery code for Manchester before the record for company no. 4 can be entered in SUP3 and (b) the possible need to change more than one record when modifying the *single* item of data concerning a location delivery code. In practice we can remove this problem by projecting SUP3 into SUP4 and DEL (Figure 2.14). (Note that updating delivery codes in this way prevents the possibility that at any instant different records in a file hold conflicting information. In SUP3 it was possible to have 'DELIVERY *of* COMPANY 6 = 2' and 'DELIVERY *of* COMPANY 1 = 7' at some stage in updating SUP2 despite the fact that both companies are in LONDON !)

The dependency relations of SUP4 and DEL are now as in Figure 2.15. //

The non-transitivity of the dependency relation, a proper relation, gives rise to the third normal form property.

Fig. 2.12 SUP3

COMPANY	LOCATION	DELIVERY
1	LONDON	7
2	LONDON	7
3	B'HAM	4
4	M'CHESTER	3
5	B'HAM	4
6	LONDON	7

PARTS

COMPANY	PART	QUANTITY
1	1	10
1	2	1
1	3	10
2	2	2
2	4	5
3	2	2
3	3	10
3	4	4
5	1	10
5	3	10

Fig. 2.13

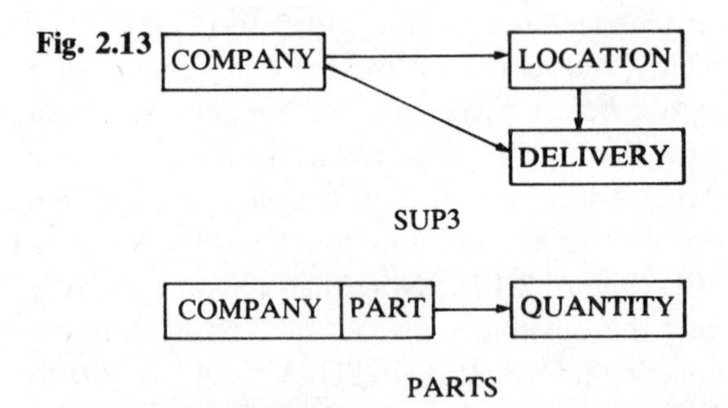

Definition. A file is in *third normal form* (3NF) if it is in 2NF and every non-key attribute is non-transitively dependent on the key. //

Put another way:

> Every non-key attribute depends on the key, the whole key and nothing but the key.

Fig. 2.14

SUP4

COMPANY	LOCATION
1	LONDON
2	LONDON
3	B'HAM
4	M'CHESTER
5	B'HAM
6	LONDON

DEL

LOCATION	DELIVERY
LONDON	7
B'HAM	4
M'CHESTER	3

Fig. 2.15

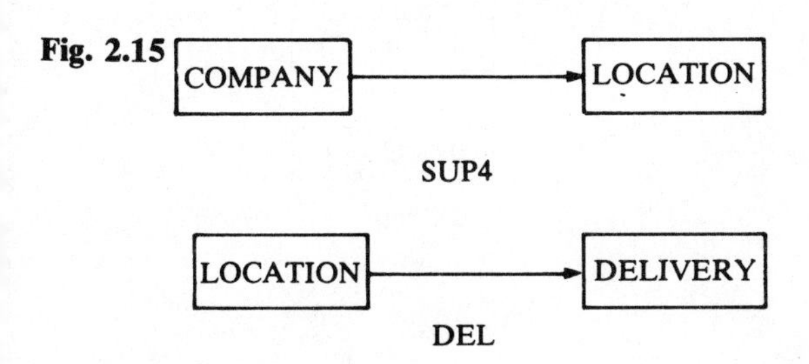

As noted in the preamble there are many other 'normal' forms but we shall not pursue the study of files any further. It is sufficient to have indicated how relational information in files is just one realization of the mathematical concept of relation. The practical use of the relations SUP4 and DEL requires the implicit association of LOCATION *of* SUP4 with LOCATION *of* DEL. This is an equivalence relation (between components in *different* files having the *same* name). Similar equivalence relations can be used to define interconnections within other data structures. As an informal illustration, we give Figure 2.16. Here, the first diagram is the usual one for a tree, the second looks like a data structure diagram, and the last two give possible implementations. Notice that the equivalences, implied by like field entries and 'array' selectors, are different but the resulting structural links are preserved; mathematically these equivalences have been 'factored out'. Hence we can define any representation of *this* tree as an element of the set $T = D/E$ (D factored

Fig. 2.16

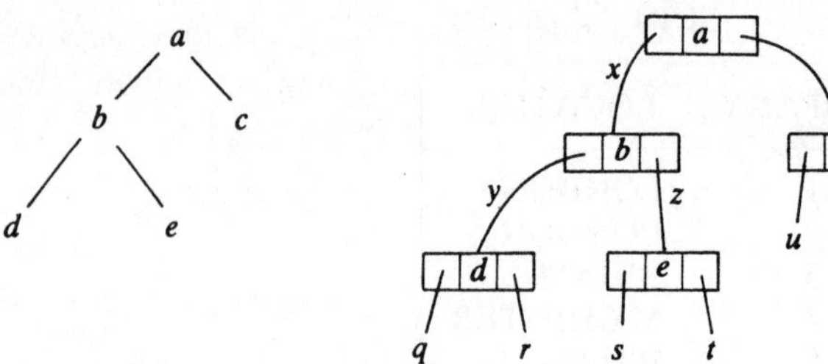

(i) (ii)

1	2	a	3
2	4	b	5
3	0	c	0
4	0	d	0
5	0	e	0

(iii)

1	0	c	0
2	5	b	3
3	0	e	0
4	2	a	1
5	0	d	0

(iv)

by E) where

$$D=\{\mathbf{a}=(x, a, p),\\ \mathbf{b}=(y, b, z),\\ \mathbf{c}=(u, c, v),\\ \mathbf{d}=(q, d, r),\\ \mathbf{e}=(s, e, t)\}$$

and

$$E=\{(x, \mathbf{b}), (y, \mathbf{d}), (p, \mathbf{c}), (z, \mathbf{e})\}.$$

We shall discuss the requirements necessary to 'factor out' equivalences in Section 3.4.

There are no exercises for this section.

2.7 Composite relations

In much the same way that we may have to consult several inter-related files to extract certain pieces of information (for instance DELIVERY *of* LOCATION *of* COMPANY 3, via files DEL and SUP4 in Example 2.6.3) we often need to link binary relations together. Guided by the preceding discussion on files we can immediately put this concept on a firm footing.

Definition. Given three sets A, B and C and relations σ between A and B and ρ between B and C we can define a relation between A and C which goes from A to B via σ and then to C via ρ – it is called the *composite relation* of ρ over σ and is written $\rho \circ \sigma$ (read 'rho of sigma') – so that

$$(\rho \circ \sigma)(a) = \rho(\sigma(a)). \quad /\!/$$

Therefore $(x, y) \in (\rho \circ \sigma)$ if there is a $z \in B$ such that

$$(x, z) \in \sigma$$

and

$$(z, y) \in \rho,$$

and hence it follows that $\mathscr{D}_{\rho\circ\sigma} = \overset{-1}{\sigma}\mathscr{D}_\rho$.

To illustrate the situation, consider Figure 2.17. The domain and range of σ and ρ are shaded ▨ and ▧ respectively. Consequently the double-hatched segments of A, B and C represent $\mathscr{D}_{\rho\circ\sigma}$, $\mathscr{D}_\rho \cap \mathscr{R}_\sigma$ and $\mathscr{R}_{\rho\circ\sigma}$ respectively.

Fig. 2.17

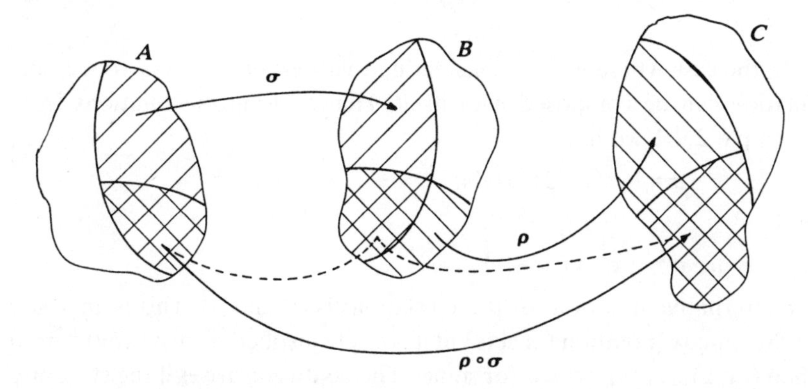

Note. Our ordering of the relations σ and ρ follows that used by analytical mathematicians and operates from right to left. Therefore $(\rho \circ \sigma)(a)$ means take a, transform by σ, then transform the result by ρ. Algebraists would write this as $a\sigma\rho$. Care should be taken when reading mathematics books to ascertain the conventions adopted by the author.

Example 2.7.1

Let σ, ρ be relations on $\mathbb{N}$ defined by

$$\sigma=\{(x, x+1): x\in\mathbb{N}\},$$

$$\rho=\{(x^2, x): x\in\mathbb{N}\}.$$

Therefore

$$\mathscr{D}_\rho=\{x^2: x\in\mathbb{N}\}$$

$$\mathscr{D}_\sigma=\{x: x, x+1\in\mathbb{N}\}=\mathbb{N}$$

$$\mathscr{D}_{\rho\circ\sigma}=\sigma^{-1}\mathscr{D}_\rho=\{x: x\in\mathbb{N} \text{ and } x+1=y^2 \text{ where } y\in\mathbb{N}\}$$

$$=\{3, 8, 15, 24, \ldots\}. \quad \text{(see Figure 2.18)} \quad //$$

Fig. 2.18

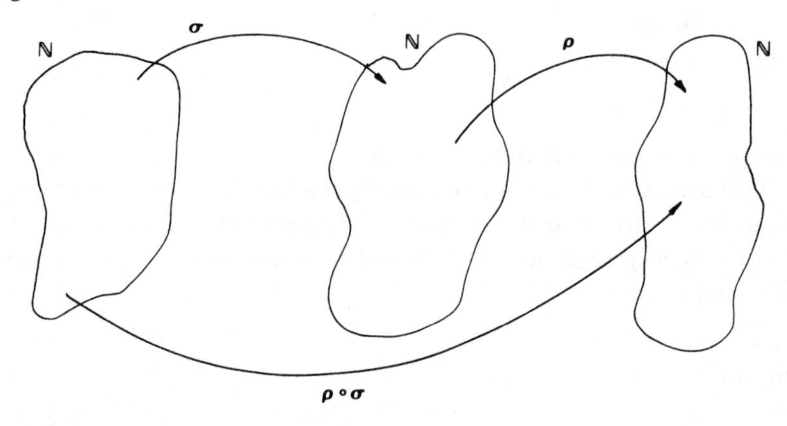

In the case where we are considering relations *on* a set (as above) any relation can be composed with itself. Hence, using the relations from Example 2.7.1 we have

$$\sigma\circ\sigma=\{(x, x+2): x\in\mathbb{N}\}$$

and

$$\rho\circ\rho=\{(x^4, x): x\in\mathbb{N}\}.$$

An alternative notation for these relations is σ^2 and ρ^2. This is an abuse of the 'square' notation for sets but it is easily justified since if $(x, y)\in\sigma\circ\sigma$ then $((x, z), (z, y))\in\sigma\times\sigma$ for some z; no confusion arises if the structure of the required result is known.

Carrying this idea further we can define σ^n for any $n\in\mathbb{N}$, $n>1$, by

$$\sigma^n=\{(x, y): x\sigma z, \text{ and } z\sigma^{n-1}y \text{ for some } z\}.$$

Again, taking the relations σ and ρ from Example 2.7.1, we get

$$\sigma^n=\{(x, x+n): x\in\mathbb{N}\}$$

and

$$\rho^n = \{(x^{2^n}, x) : x \in \mathbb{N}\}.$$

At this point it is instructive to consider how far the analogy with multiplication can be taken. Let A be a set and R a relation on A. It then follows that the three relations $I_A \circ R$, R and $R \circ I_A$ are all equal; so I_A, the identity relation on A, behaves in the same way as the number 1 with respect to multiplication of numbers. To complete the analogy we would like to be able to write $R^{-1} \circ R = I_A = R \circ R^{-1}$; however, this cannot in general be done. In order to be able to do this, further conditions must be met as we shall see in Chapter 3.

Now for some exercises. These embody 'easy to describe' situations in mathematical terms but be careful to include only the pairs that you require. You may find it useful to devise some sort of diagrammatic aid.

Exercises 2.7

1. If R and S are defined on P, the set of all people, by

   $R = \{(x, y) : x, y \in P$ and x is the father of $y\}$

   $S = \{(x, y) : x, y \in P$ and x is the daughter of $y\}$,

   describe, explicitly, the following relations.
   (*a*) R^2,
   (*b*) S^2,
   (*c*) $R \circ S$,
   (*d*) $S \circ R$,
   (*e*) $S \circ R^{-1}$,
   (*f*) $R^{-1} \circ S$,
   (*g*) $R^{-1} \circ S^{-1}$,
   (*h*) $S^{-1} \circ R$,
   (*i*) $S^{-1} \circ S^{-1}$,
   (*j*) $S^{-1} \circ R^{-1}$.

2.8 Closures of relations

The notion of closure is fundamental to mathematics and is used (implicitly) in most aspects of the subject. To illustrate the concept consider the following.

Take an object x_0 and a process p that generates a set, and define a sequence $x_1, x_2, x_3, \ldots, x_n, \ldots$ such that

$$x_1 \in p(x_0),$$
$$x_2 \in p(x_1),$$
$$\vdots$$
$$x_n \in p(x_{n-1}),$$
$$\vdots$$

The set containing all elements of all possible sequences derived using p and starting at x_0 is the closure of the process p about x_0. So, the 'answer' is in $p^n(x_0)$ for some value of n, but we do not know *in advance* the value of n. Moreover, if we take any element y of this closure and perform the process p on y then we get nothing new; the result is already included in the closure; the set cannot be extended in this way; it is closed.

Example 2.8.1

Take the square S, labelled as in Figure 2.19 and define the process r by saying that, from a given position of S, r generates (the set of) all positions obtained by rotating S clockwise through a right angle. So $r(S)$ gives the configuration shown in Figure 2.20. After performing r four times we are back where we started and hence the closure is the set of positions

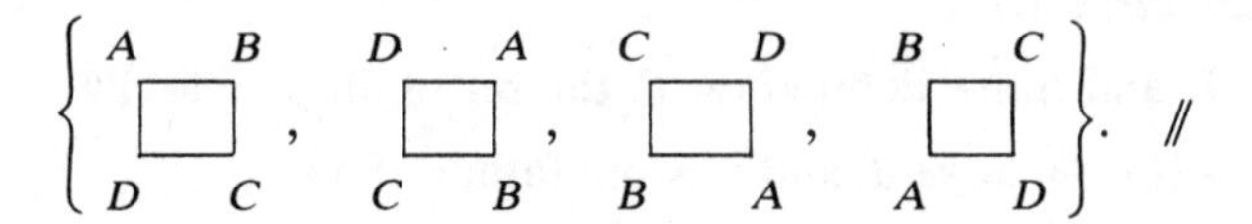

Fig. 2.19
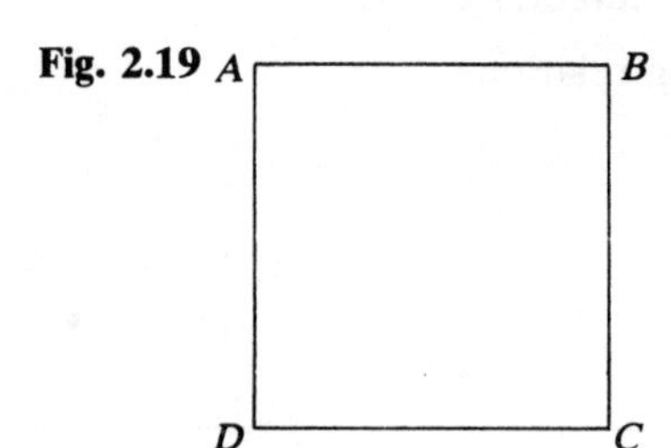

Fig. 2.20
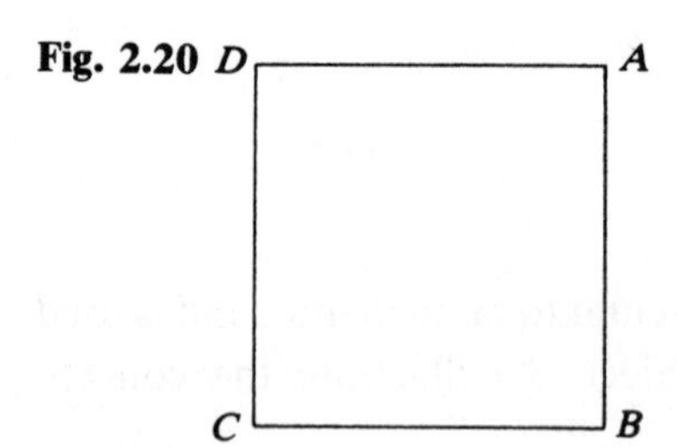

Consider now what happens when the process is defined by a relation. (In fact, this is always possible because we can specify a suitable relation by the set $\{(x, y): y \in p(x)$ where p is the process$\}$.) What we need for the closure of a relation A is all the composite relations $A, A^2, A^3, \ldots, A^n, \ldots$, which must then be combined in the usual set-theoretic way.

Definition. The (*transitive*) *closure* of a relation A on a set is the infinite union

$$\bigcup_{n=1}^{\infty} A^n = A \cup A^2 \cup A^3 \cup \ldots. \quad //$$

The transitivity of closing a relation is obvious by its very nature but the word 'transitive' is often included to stress the difference between this and a similar operation to be defined shortly. The transitive closure of a relation A is denoted by A^+.

Examples 2.8.2

1. If R is a relation defined on $\mathbb{N}$ by

   $R = \{(x, y): y = x + 1\}$

   then

   $R^+ = \{(x, y): x < y\}$.
2. If σ is a relation on $\mathbb{Q}$ defined by

   $\sigma = \{(x, y): x < y\}$

   then

   $\sigma^+ = \sigma$.
3. If ρ is a relation on $\mathbb{Q}$ defined by

   $\rho = \{(x, y): x * y = 1\}$

   then

   $\rho^+ = \{(x, x): x \neq 0\} \cup \rho$.
4. Let L be the set of stations on the London underground railway and a, b and c be consecutive stations. If N is the relation on L defined by $N = \{(x, y): x$ is the next station to $y\}$ then (a, b) and $(b, c) \in N$, and (a, a), (b, b), (c, c) and $(a, c) \in N^2$. Hence $N^+ = U_L = L \times L$. //

As can be seen from these examples, the closure of a relation is not in general reflexive. However, it is sometimes convenient to make it so. This is easily done; first we adopt the convention that the identity relation on a set X, $I = \{(x, x): x \in X\}$, is the zero power of any relation on X. Thus $A^0 = I$ for any A.

Definition. The *reflexive closure* A^*, of the relation A is defined by

$$A^* = \bigcup_{n=0}^{\infty} A^n$$

and the closure relations are associated by the obvious identity

$$A^* = A^+ \cup I. \quad /\!/$$

Examples 2.8.3

Using the relations defined in the previous set of examples we have,

$$R^* = \{(x, y): x \leqslant y\},$$
$$\sigma^* = \{(x, y): x \leqslant y\},$$
$$\rho^* = \rho^+ \cup \{(0, 0)\},$$
$$N^* = N^+. \quad /\!/$$

Practical methods of actually *computing* closure relations will be discussed in Chapters 6 and 7.

Exercises 2.8

1. Using the relations R and S of Exercises 2.7, describe the following closure relations:
   (*a*) R^+,
   (*b*) S^+,
   (*c*) R^*,
   (*d*) S^*,
   (*e*) $(S \circ S^{-1})^+$,
   (*f*) $(R^{-1} \circ R)^*$,
   (*g*) $(S^2 \circ R^2)^+$.

3 FUNCTIONS

The concept of a function may already be quite familiar to the reader but in this chapter we consider functions as a subset of binary relations. We properly define a function, and the allied concept of a mapping, and examine various properties which they may possess. We then use functions to formalize the process of counting and give a proper definition of the cardinality of a set. Finally we consider some functions of special interest and functions that are conveniently denoted by operators.

3.1. Functions and mappings

Definition. A binary relation ρ between sets A and B is a *function* if $a\rho b$ and $a\rho c$ imply that $b=c$, so for any $x \in A$ there is at most one $y \in B$ such that $x\rho y$. Another way of putting this is to say that

$$\rho(x)=\varnothing \quad \text{or} \quad \rho(x)=\{y\}. \quad //$$

The important fact to remember is that if $\rho(x)$ gives any element of B, then it gives only *one* element. In the case when $\rho(x) \neq \varnothing$ it is usual to drop the set brackets and write

$$y=\rho(x).$$

Functions are conventionally named by lower case letters $f, g, h, \ldots$ or by other names particular to a special situation, for example sin, log, $F_n, \ldots$. If the function f is between sets A and B, this fact can be denoted by

$$f: A \to B.$$

Further, if $x \in A$ and xfy then we denote this by using the symbol '$\mapsto$' thus:

$$f: x \mapsto y.$$

The latter notation is often used to specify the rule that defines the function (if such a general rule exists).

Example 3.1.1

$f: A \to A$ where $A = \{-1, 0, 1\}$

defined by

$f: x \mapsto x^3$. //

Even when f is not a mapping (see below) we often use phrases such as 'f maps x to x^3'. The definitions of domain and range hold just as for relations but you should notice that the inverse of a function may not be a function itself.

Example 3.1.2

On the set $\{-1, 0, 1\}$, $f: x \mapsto x^2$ is a function but its inverse is not since $f^{-1}(1) = \{-1, 1\}$. //

Definition. A function $f: A \to B$ is a *mapping* if its domain is the whole of A, i.e. if $\mathcal{D}_f = A$. A function that is not a mapping is called a *partial function.* A mapping *on* a set is called a *transformation.* //

Note. The terminology often varies from that given here, particularly in American texts. The terms 'function' and 'mapping' are sometimes used synonymously, and a mapping as we define it is referred to as a 'total function'.

Of course any function can be regarded as a mapping on its own domain; this is sometimes useful when forming composite functions. As we shall see in subsequent sections, manipulation of compositions becomes even easier when further conditions are met.

Any function $f: A \to \mathbb{R}$ is said to be *real-valued* and a function on $\mathbb{R}$ is called a *real function.*

We now consider restrictions that help us trace what happens to individual elements and the results of applying functions to them. Here we shall deal only with the classification of functions between sets and assume that we know very little about the sets involved; later we shall look at functions between sets on which operations such as addition are defined.

Definition. A function $f: A \to B$ is said to be *surjective* (or *onto*) if $\mathcal{R}_f = B$. This means that given $b \in B$, $f^{-1}(b) \neq \varnothing$ but it does not guarantee that $|f^{-1}(b)| = 1$. A function $f: A \to B$ is said to be *injective* (or 1 to 1, written 1–1) if when given $a_1, a_2 \in A$ such that $f(a_1) = f(a_2)$ then $a_1 = a_2$. //

So, if $f: A \to B$ and f is surjective then for any $b \in B$, $f^{-1}(b) \in \mathcal{P}(A) \backslash \varnothing$. Pictorially this could be interpreted as every point in B acting as the 'sharp end' of at least one f-arrow starting in A. Except in very simple cases, this is difficult to illustrate.

On the other hand, a pictorial characterization of injectivity is easily given in the form of a restriction or prohibition. The function f is *not* injective if the situation depicted in Figure 3.1(a) occurs.

For comparison and emphasis, we give in Figure 3.1(b) the mirror-image restriction which distinguishes a function from an arbitrary binary relation.

Fig. 3.1

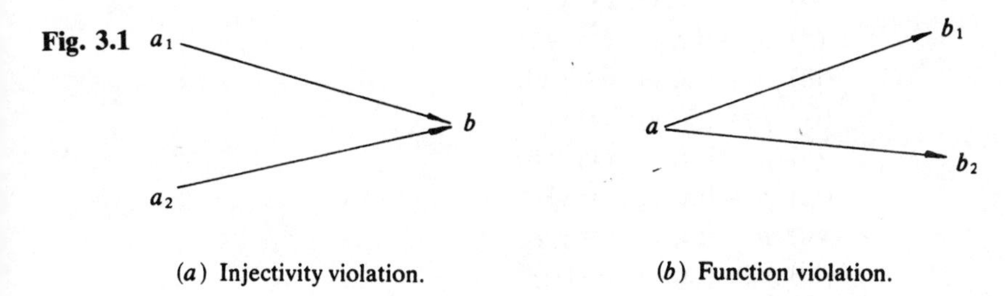

(a) Injectivity violation.

(b) Function violation.

If $f: A \to B$ is injective and $a \in \mathscr{D}_f$ then

$$a = f^{-1}(f(a)).$$

Further, it follows that if $b \in \mathscr{R}_f$ (i.e. $b \in \mathscr{D}_{f^{-1}}$, and notice that f^{-1} is now a function) then

$$b = f(f^{-1}(b)).$$

Hence, if we define the function $I_X : X \to X$ to be the *identity mapping* on X, i.e. $I_X : x \mapsto x$ for all $x \in X$, then whenever f is injective

$$f^{-1} \circ f = I_{\mathscr{D}_f}$$

and

$$f \circ f^{-1} = I_{\mathscr{R}_f}.$$

Using the function f in Example 3.1.2, we see that $f^{-1}(f(1)) \neq 1$, indicating that the first of these identities is in general *false* but calculations become much easier in cases when both identities are valid as will be seen in the next section. Before going on, we note some terminology incorporating the properties discussed above. A function f is *bijective* if it is both surjective and injective. A bijective mapping is called a *bijection.* 'Bijection' means literally something that can be 'thrown both ways'. Hence, using a bijection f between A and B, we can take elements in A, go to B via f, perform some calculations, then move back to A by f^{-1}.

The terms *injection* and *surjection* are also used to describe injective and surjective mappings, but we shall make little use of these.

Exercises 3.1

1. Which of the following relations on $\{-10, -9, \ldots, 0, \ldots, 9, 10\}$ are functions? Give a counter-example for any that you believe not to be a function. (In part (iv) the order relation $\leqslant$ is that induced by $\mathbb{Z}$. In parts (viii)–(x) $|x|$ is defined by
$|x| = x$ if $x \geqslant 0$,
$|x| = -x$ if $x < 0$.)
   (i) $\rho_1 = \{(x, y): x = y^2\}$,
   (ii) $\rho_2 = \{(x, y): x^2 = y\}$,
   (iii) $\rho_3 = \{(x, y): x = -y\}$,
   (iv) $\rho_4 = \{(x, y): x \leqslant y\}$,
   (v) $\rho_5 = \{(x, y): x * y = 6\}$,
   (vi) $\rho_6 = \{(x, y: x^3 = y\}$,
   (vii) $\rho_7 = \{(x, y): x = y^3\}$,
   (viii) $\rho_8 = \{(x, y): x = |y|\}$,
   (ix) $\rho_9 = \{(x, y): |x| = |y|\}$,
   (x) $\rho_{10} = \{(x, y): y * |y| = x * |x|\}$.

2. Construct a function $f: A \to A$ where $A = \{0, 1\}$ whose inverse is not a function.

3. Which of the following functions are mappings?
   (i) f on $\mathbb{R}$ defined by $\{(x, x^4): x \in \mathbb{R}\}$,
   (ii) f on $\mathbb{R}$ defined by $\{(x^3, x): x \in \mathbb{R}\}$,
   (iii) f on $\mathbb{R}$ defined by $\{(x, x^2): x \in \mathbb{R}\}$,
   (iv) f on $\mathbb{R}$, $f: x \mapsto \sin(x)$,
   (v) f on $\mathbb{R}$, $f: x \mapsto 1/x$,
   (vi) f on $\mathbb{Q}$, $f: x \mapsto \sin^{-1}(x)$,
   (vii) $f: A \to \mathcal{P}(A)$ defined by $f: x \mapsto \{x\}$,
   (viii) $f: \mathcal{P}(A) \to A$ defined by $f = \{(x, y): y \in x \cap \{a\}\}$ where a is a fixed element of A.

4. Let $f: A \to B$ and $g: B \to C$ be relations. What is the domain of $g \circ f$
   (i) when f and g are functions,
   (ii) when f is a function and g is a mapping,
   (iii) when f is a mapping and g is a function, and
   (iv) when f and g are both mappings?

5. Prove that if the function f is injective then so is f^{-1}.

6. If f is a surjective function does this imply that f^{-1} is a mapping?

7. Give an example to show that the function on $A = \{-1, 0, 1\}$ defined by $f: x \mapsto x^2$ is such that $f^{-1} \circ f \neq I_A$.

8. If $f:A\to B$ and $g:B\to C$ are functions prove that
   (i) if f and g are injective then $g\circ f$ is injective, and
   (ii) if f and g are surjective then $g\circ f$ is surjective.
9. If $f:A\to B$ and $g:B\to C$ are functions and g is surjective, is this sufficient to ensure that $g\circ f$ is surjective?

3.2. Inverse functions and mappings

Now we shall collect together the results derived in the previous section and investigate the implications for composite functions. Given a function $f:A\to B$ then f^{-1} is also a function iff f is injective, and is a mapping iff f is bijective. The most well-behaved case is when f is a bijection; then f^{-1} is also a bijection and the composite functions $f^{-1}\circ f$ and $f\circ f^{-1}$ *are* identity mappings.

Consider the functions $f:A\to B$ and $g:B\to C$. Then
(i) if f and g are injective then so is $g\circ f$, and
(ii) if f and g are surjective then so is $g\circ f$
(see Exercises 3.1 Question 8). The inverse *relation* of $g\circ f$ is $f^{-1}\circ g^{-1}$. (The ordering needs to be reversed as indicated in Figure 3.2.)

Fig. 3.2

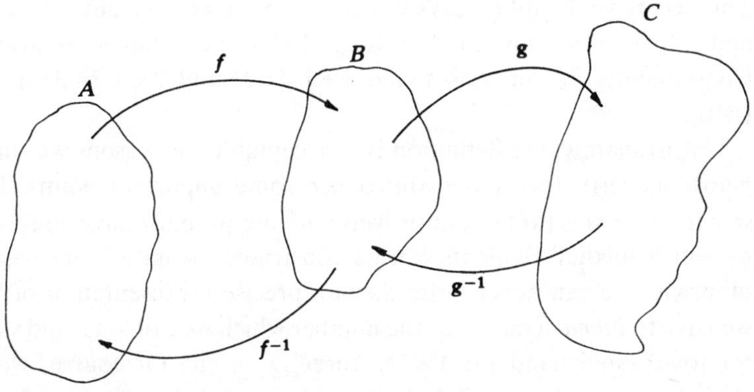

Notice that if g is a mapping, i.e. $\mathscr{D}_g = B$, then $\mathscr{R}_f \subseteq \mathscr{D}_g$ and hence $\mathscr{D}_{g\circ f} = \mathscr{D}_f$. Similarly if $\mathscr{R}_f \supseteq \mathscr{D}_g$ then $\mathscr{R}_{g\circ f} = \mathscr{R}_g$. If f and g are injective then so is $g\circ f$ hence $f^{-1}\circ g^{-1}$ is a function. To summarize: $\mathscr{R}_f = \mathscr{D}_g$ implies that $g\circ f:\mathscr{D}_f \to \mathscr{R}_g$ is a mapping; if, also, f and g are injective then $f^{-1}\circ g^{-1}:\mathscr{R}_g \to \mathscr{D}_f$ is a bijection. These criteria are trivially satisfied if f and g are both bijections.

Exercises 3.2

1. Let $f:A\to B$ and $g:C\to B$ be functions; show that
   (i) if f is surjective and g is a mapping then
   $\mathscr{R}_{g^{-1}\circ f} = C$,

(ii) if f and g are bijections then

$(g^{-1} \circ f)^{-1} = f^{-1} \circ g,$

and

(iii) if $\mathcal{R}_g \subseteq \mathcal{R}_f$ then $(f \circ f^{-1} \circ g)C = \mathcal{R}_g$.

3.3 Cardinality of sets and countability

We are almost in a position to use bijections to formalize the concept of cardinality and the process of counting. (Counting is important not only for its own sake but also because a function is *computable* iff a related set is *countable.*) However, we must first give a proper definition of the set $\mathbb{N}$. To clarify our intentions we note that any number (one, two, . . . , etc) can be used in two distinct ways: as a noun or as an adjective quantifying some other noun. Initially, we shall investigate numbers as nouns.

As a *preliminary* definition of $\mathbb{N}$ let

$$\mathbb{N} = \{1\} \cup \{n+1 : n \in \mathbb{N}\}.$$

This recursive definition says that $1 \in \mathbb{N}$ and if we take any element of $\mathbb{N}$ and add one to this number then the result obtained is also in $\mathbb{N}$. Consequently, $\mathbb{N}$ contains 1 and $1+1$ (=2) and $2+1$ (=3) and $3+1$ (=4) . . . .

Unfortunately, the definition is unacceptable for reasons we shall give below; nevertheless it does introduce some important points. For instance, since $\mathbb{N}$ is (at least intuitively) infinite we must have a mechanism by which further elements can be constructed from a finite base set – otherwise we can never write down a precise representation of $\mathbb{N}$; also we have to *invent* a name for the number which we call 'one' and similarly for 'two' (shorthand for $1+1$), 'three', . . . , etc. Of course, we could choose any names (or symbols) for this purpose but it would be rather perverse to use unconventional ones. Now for the bad news. Examination of our preliminary definition will reveal that it contains two new symbols '1' and '+'; the rest are valid set-building symbols. '1' can be explained away as above; however, '+' denotes an operation *on* the set $\mathbb{N}$ and hence cannot be used to define $\mathbb{N}$. (Operations will be defined in Section 3.6.)

To get out of this predicament we return to basic set theory. Remember, essentially all that is required is to construct a number that is one bigger than the previous largest one. The analogous process for sets, namely the construction of a superset with one more element than a given set, is easily achieved.

Example 3.3.1

Let $A=\{x, y, z\}$ and $B=\{x, y, z, A\}$. Then $A\subseteq B$ and $A\in B$ so $B\backslash A=\{A\}$ and has just a single element. //

This construction can be carried out on any set. Starting from a set X we can define the *successor set*, written $X^{\oplus}$, such that

$$X^{\oplus}=X\cup\{X\}.$$

So far so good. In order to use this process in the creation of $\mathbb{N}$ we need some initial set; we choose $\{\varnothing\}$ which has one element. (Many authors start with $\varnothing$. This generates $\{0\}\cup\mathbb{N}$. We do not regard 0 as a natural number and hence the reason for our choice. No universal convention exists with respect to 0 and $\mathbb{N}$. Always check the conventions adopted in any other text you consult.)

From $\{\varnothing\}$ we generate the sequence

$$\{\varnothing\},$$
$$\{\varnothing\}^{\oplus}=\{\varnothing,\{\varnothing\}\},$$
$$\{\varnothing\}^{\oplus\oplus}=\{\varnothing,\{\varnothing\},\{\varnothing,\{\varnothing\}\}\},$$

etc.

This leads to a progression which is more foreboding than 1, 1+1, 1+1+1, . . . , but at least its construction is well-defined. To tidy up matters we choose temporary names for these sets.

Rename $\{\varnothing\}$ as 1, $1^{\oplus}$ as 2, $2^{\oplus}$ as 3, ... Then

$$1=\{\varnothing\}$$
$$2=\{\varnothing, 1\}$$
$$3=\{\varnothing, 1, 2\}.$$

So the set 3 has 3 elements etc. To avoid confusion let us again change the notation and define

$$\mathbb{N}_m=m^{\oplus}\backslash\{\varnothing\}$$
$$=\{1, 2, \ldots, m\}$$

and

$$\mathbb{N}=\mathbb{N}_1\cup(\mathbb{N}_{m^{\oplus}}\colon m\in\mathbb{N}).$$

It then follows, *by definition*, that $|\mathbb{N}_m|=m$ (the number m) and that if $a, b\in\mathbb{N}$

$$a\leqslant b \text{ iff } \mathbb{N}_a\subseteq\mathbb{N}_b.$$

So our belief in the ordering on $\mathbb{N}$ has been formally justified. Now that the sets $\mathbb{N}$ and $\mathbb{N}_m$ (for every $m\in\mathbb{N}$) are properly defined we can use them in arguments about counting. As always we need some terminology.

Definition. Two *sets* are said to be *bijective*, written $X \sim Y$, if there exists a bijection between them. A non-empty set is said to be *finite* if it is bijective with $\mathbb{N}_m$ for some $m \in \mathbb{N}$. If $X \sim \mathbb{N}_m$ then the cardinality of X, written $|X|$, is m. (Consequently, the use of numbers as adjectives follows immediately – for example, if P is the set of all people and $X \subseteq P$ such that $X \sim \mathbb{N}_m$ then X is a set of m people.) Recall that the empty set $\varnothing$ is bijective only with itself, is finite and has cardinality zero, so $|\varnothing|=0$.

A set is said to be *denumerable* if it is bijective with $\mathbb{N}$. The symbol $\aleph_0$ (aleph nought) is commonly used to denote the cardinality of $\mathbb{N}$.

A set X is *countable* if it is finite or denumerable and can be counted by any bijection

$$f:\mathbb{N} \to X \quad \text{(if } X \text{ is denumerable)},$$

or

$$f:\mathbb{N}_m \to X \ (\text{if } |X| = m),$$

or

$$f:\varnothing \to X,$$

the ith element of X simply being the image of i under f. //

Before setting about deriving some useful results, we note one essential property of sets and bijections. The relation σ defined on a set of sets S by

$$\sigma = \{(X, Y): X \sim Y\}$$

is an equivalence relation and partitions S into classes consisting of sets having the same cardinality. Therefore all that is required in order to demonstrate that two sets are the same size is to construct a bijective mapping between the sets.

Example 3.3.2.

To show that

$$|\mathbb{N}| = |\mathbb{Z}|.$$

The mapping

$$\psi : n \mapsto \begin{cases} \dfrac{1-n}{2} \text{ if } n \text{ is odd} \\ \dfrac{n}{2} \text{ otherwise} \end{cases}$$

is a bijection between $\mathbb{N}$ and $\mathbb{Z}$. //

In the preceding example the Greek letter ψ was used for the bijection. The use of the Greek letters ϕ, ψ, χ... to denote arbitrary bijections is common practice in texts on logic and will be used here in this context (amongst others). However, in an attempt to avoid unnecessary confusion with the empty set $\varnothing$ we shall avoid the use of phi within this section.

Example 3.3.3

To show that

$$|\mathbb{N}| = |\mathbb{Q}|.$$

This is a little more involved. First we consider denumerably many copies of $\mathbb{N}$, one copy for each $n \in \mathbb{N}$. We can write this set as $\mathbb{N} \times \mathbb{N}$ and set out its elements as in Figure 3.3.

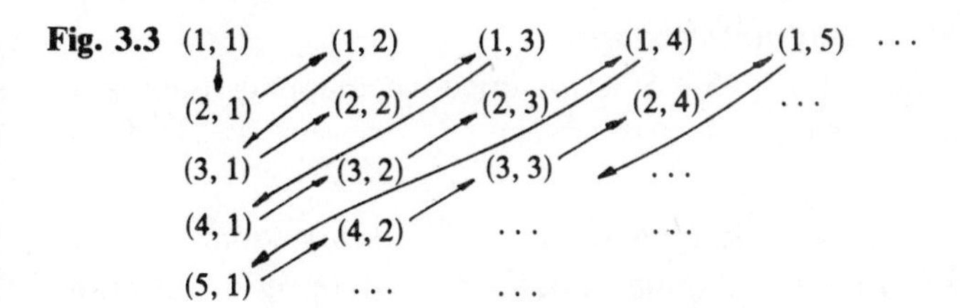

Fig. 3.3

The ordering displayed is a bijection

$$\mathbb{N} \times \mathbb{N} \to \mathbb{N}$$

given by

$$(x, y) \mapsto \frac{(x+y-1)(x+y-2)}{2} + y.$$

Now any positive element of $\mathbb{Q}$ can be identified with p/q, where p and q have no common factors, and associated with the element (p, q) of $\mathbb{N} \times \mathbb{N}$ by the natural injection, so writing

$$T = \{x : x \in \mathbb{Q}, x > 0\},$$

we must have

$$|T| \leqslant |\mathbb{N} \times \mathbb{N}| = |\mathbb{N}|.$$

(The use of '$\leqslant$' between these infinite numbers has not been justified. $|A| \leqslant |B|$ should read as 'A is bijective with a subset of B'. Proof of the non-obvious fact that this is an order relation lies outside the scope of this text.) But by the same injection, any $p \in \mathbb{N}$ can be associated with $p/1$ and hence with $(p, 1)$, so

$$|\mathbb{N}| \leqslant |T| \leqslant |\mathbb{N}|,$$

therefore

$$|T| = |\mathbb{N}|.$$

The chain of argument here essentially says: take the diagonal ordering of $\mathbb{N} \times \mathbb{N}$ and leave some elements out (those pairs having non-trivial common factors). Intuitively this gives a method of enumerating the elements of T, but it is difficult to give the formula which identifies the elements of T with $\mathbb{N}$. All we need now is a way of repeating the method for all rationals. There are many ways of doing this. In order to broaden the discussion we choose the following. Extend the injections already used between T and $\mathbb{N}\times\mathbb{N}$ to operate between $\mathbb{Q}$ and $\mathbb{Z}\times\mathbb{N}$ then

$$|\mathbb{N}|=|\mathbb{Z}|\leq|\mathbb{Q}|\leq|\mathbb{Z}\times\mathbb{N}|=|\mathbb{N}\times\mathbb{N}|=|\mathbb{N}|$$

which gives the result. (The bijection between $\mathbb{Z}\times\mathbb{N}$ and $\mathbb{N}\times\mathbb{N}$ is left to the inventiveness of the reader.) //

The previous example was somewhat lengthy but raises several important points which we shall now itemize.

(1) If S is finite and $\chi: S\to S$ is an injective mapping then χ is bijective.

Proof. Suppose S is finite and $\chi: S\to S$ is an injection. If $S=\varnothing$ the result follows trivially. For $S\neq\varnothing$ there is a bijection $\psi:\mathbb{N}_m\to S$ for some $m\in\mathbb{N}$ and $\psi^{-1}\circ\chi\circ\psi$ is an injection $\mathbb{N}_m\to\mathbb{N}_m$ and hence is a bijection. (The proof of this fact is left as an exercise for the reader.) The basic idea is simply that of rearranging m objects and is often referred to as the pigeonhole principle; given m pigeons, each in its own box, any rehousing scheme that does not put two or more pigeons in the same box must use all m boxes, i.e. $\psi^{-1}\circ\chi\circ\psi$ is a rearrangement of $\mathbb{N}_m$. (See Figure 3.4.) But ψ is a bijection therefore so is $\chi\circ\psi$ and hence χ is also a bijection. (Conversely, it follows if $\chi: S\to S$ is a non-surjective injection then S must be infinite.)

Fig. 3.4

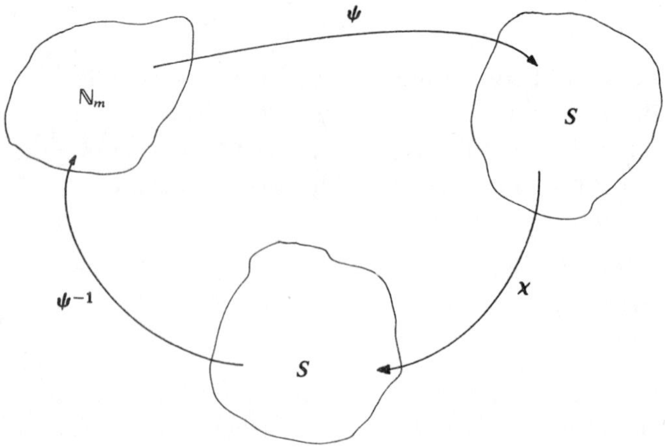

(2) The set $\mathbb{N}$ is infinite, since the mapping on $\mathbb{N}$ defined by $n \mapsto n+1$ is injective but not bijective (no element is mapped to 1) hence by the converse of the previous result $\mathbb{N}$ cannot be finite.

More definitions: a subset A of $\mathbb{R}$ is *bounded above* (*below*) if it has an *upper* (*lower*) *bound*. A is *bounded* if it is bounded above *and* below.

(3) A bounded subset of $\mathbb{N}$ is finite.

Proof. Any subset of $\mathbb{N}$ is bounded below by 0; additionally let A ($\subseteq \mathbb{N}$) be bounded above by some $m \in \mathbb{N}$. Define a mapping $\chi: A \to \mathbb{N}$ such that if we write $A = \{a_1, a_2, \ldots, a_i \ldots\}$ with $a_1 < a_2 < a_3 \ldots < m$ (such an ordering is possible since $A \subseteq \mathbb{N}$) then $\chi(a_i) = i$.

A consequence of this definition is that $\chi(a_i) \leqslant a_i$ and χ is clearly injective. It must also be a bijection:

$\chi: A \to \mathbb{N}_n$ for some $n \leqslant m$.

If this were not so then there would be an $a_p \in A$ such that

$\chi(a_p) > m$

and thus

$a_p \geqslant \chi(a_p) > m.$

But A is bounded by m so this is a contradiction, hence χ is a bijection onto $\mathbb{N}_n$ and A is thus finite.

(4) Any subset of a finite set is finite.

Proof. Suppose $A \subseteq B$ and B is finite. If $B = \varnothing$ then $A = \varnothing$ and the result follows, otherwise $B \sim \mathbb{N}_m$ for some $m \in \mathbb{N}$. Thus there is a bijection $\chi: B \to \mathbb{N}_m$. Applying χ to A yields a subset of $\mathbb{N}_m$ and so $\chi(A)$ is bounded and by (3) is finite. Since $\chi(A)$ is bijective with A then A is finite.

(5) A direct consequence of (4) is the deduction that any set having an infinite subset is itself infinite.

Having demonstrated some rather unbelievable facts about the sizes of the sets $\mathbb{Z}$, $\mathbb{Q}$, $\mathbb{N}$, $\mathbb{N} \times \mathbb{N}, \ldots$ it is reasonable to ask whether there exist any sets that are bigger than $\mathbb{N}$. The answer is yes. In fact given *any* set we can (by using Cantor's diagonal argument, below) devise a strictly bigger set.

We shall not consider the general case but examine a fairly straightforward example which serves to illustrate the fundamental idea and is adequate for our purposes.

Example 3.3.4
To show

$$|[0, 1[| > |\mathbb{N}| \quad ([0, 1[= \{x: x \in \mathbb{R}, 0 \leq x < 1\}).$$

Proof. (After Cantor.) Each number between 0 and 1 can be written as an infinite decimal of the form

$$0.d_{n1}d_{n2}d_{n3}d_{n4} \ldots .$$

Assume that these numbers can be counted using $\mathbb{N}$ and that the nth number has the value given above. All that is required to show that the set [0, 1[cannot be counted is to construct an element of [0, 1[that is not included in the enumeration. The construction is as follows: take the nth digit from the decimal form of the nth number. This gives the number

$$0.d_{11}d_{22}d_{33} \ldots .$$

We now construct a new number

$$0.\delta_{11}\delta_{22}\delta_{33} \ldots ,$$

where each digit δ_{ii} is different from the corresponding d_{ii}. This new number then differs from every number in the original list; explicitly it differs from the nth number in its nth digit. Hence, [0, 1[is strictly larger than $\mathbb{N}$ and no counting bijection exists. //

The principle involved in the construction of

$$0.\delta_{11}\delta_{22}\delta_{33} \ldots$$

in the previous proof is basically sound, though, as the astute reader will observe, there is a problem when the decimal representation of a number is not unique. This occurs when a representation terminates in an infinite progression of 0s or 9s. (For example, 0.39999 . . . and 0.40000 . . . .) To avoid the possibility of the number $0.\delta_{11}\delta_{22}\delta_{33} \ldots$ being merely a different representation of a number already in the list we can stipulate that δ_{ii} is different from 0, d_{ii} and 9. Corresponding difficulties occur in similar constructions but, in order to restrict attention to the main points, these will generally be ignored.

In fact it can be shown that $[0, 1[\sim \mathbb{R}$. Obviously $\mathbb{R}$ is a very important set and its cardinality is denoted by the special symbol $\aleph_1$ (aleph one). So much for very large sets. Before concluding this section we investigate ways in which sets can be combined and the relationships between the cardinals of the individual sets and the cardinal of the resultant set. The first of these is very straightforward.

Theorem. If $A \sim B$ and $C \sim D$ then $(A \times C) \sim (B \times D)$.

Proof. Let $\chi: A \to B$ and $\psi: C \to D$ be bijections. Then

$$(a, c) \mapsto (\chi(a), \psi(c))$$

is a bijection between $A \times C$ and $B \times D$. //

Theorem. If Z is a finite set and $\{X, Y\}$ is a partition of Z, then $|Z| = |X| + |Y|$.

Proof. Since Z is finite, $Z \sim \mathbb{N}_m$ for some $m \in \mathbb{N}$ and there is a bijection $\chi: Z \to \mathbb{N}_m$. (See Figure 3.5.) Moreover, since $X \subseteq Z$ and $Y \subseteq Z$, then

$$\chi(X) \subseteq \mathbb{N}_m$$

and

$$\chi(Y) \subseteq \mathbb{N}_m.$$

Let ψ_1 be a bijection from $\chi(X)$ to $\mathbb{N}_{p_1}$ for some $p_1 \leq m$, so $(\psi_1 \circ \chi)(X) = \mathbb{N}_{p_1}$. Similarly, let ψ_2 be a bijection from $\chi(Y)$ to $\mathbb{N}_{p_2}$ for some $p_2 \leq m$ where $(\psi_2 \circ \chi)(Y) = \mathbb{N}_{p_2}$.

Now, if $\sigma: \mathbb{N} \to \mathbb{N}$ is defined by

$$\sigma: x \mapsto x + p_1,$$

then σ is a bijection between $\mathbb{N}_{p_2}$ and $\mathbb{N}_{p_1+p_2} \backslash \mathbb{N}_{p_1}$ and hence the mapping

$$z \mapsto \begin{cases} \psi_1 \circ \chi(z) & \text{if } z \in X \\ \sigma \circ \psi_2 \circ \chi(z) & \text{if } z \in Y \end{cases}$$

is injective, and (since Z is finite) bijective between Z and $\mathbb{N}_{p_1+p_2}$, and $Z \sim \mathbb{N}_{p_1+p_2}$. Therefore

$$m = p_1 + p_2$$

and

$$|Z| = |X| + |Y|. \quad //$$

Example 3.3.5

To illustrate the mechanics of the previous theorem, examine the case when

$$Z = \{a, b, c, d, e, f\},$$

$$X = \{b, e\},$$

$$Y = \{a, c, d, f\}.$$

Then a bijection between Z and $\mathbb{N}_6$ is given by

$$\chi = \{(a, 5), (b, 1), (c, 6), (d, 3), (e, 4), (f, 2)\},$$

so

$$\chi(X) = \{1, 4\}$$

and

$$\chi(Y)=\{2, 3, 5, 6\}.$$

Fig. 3.5

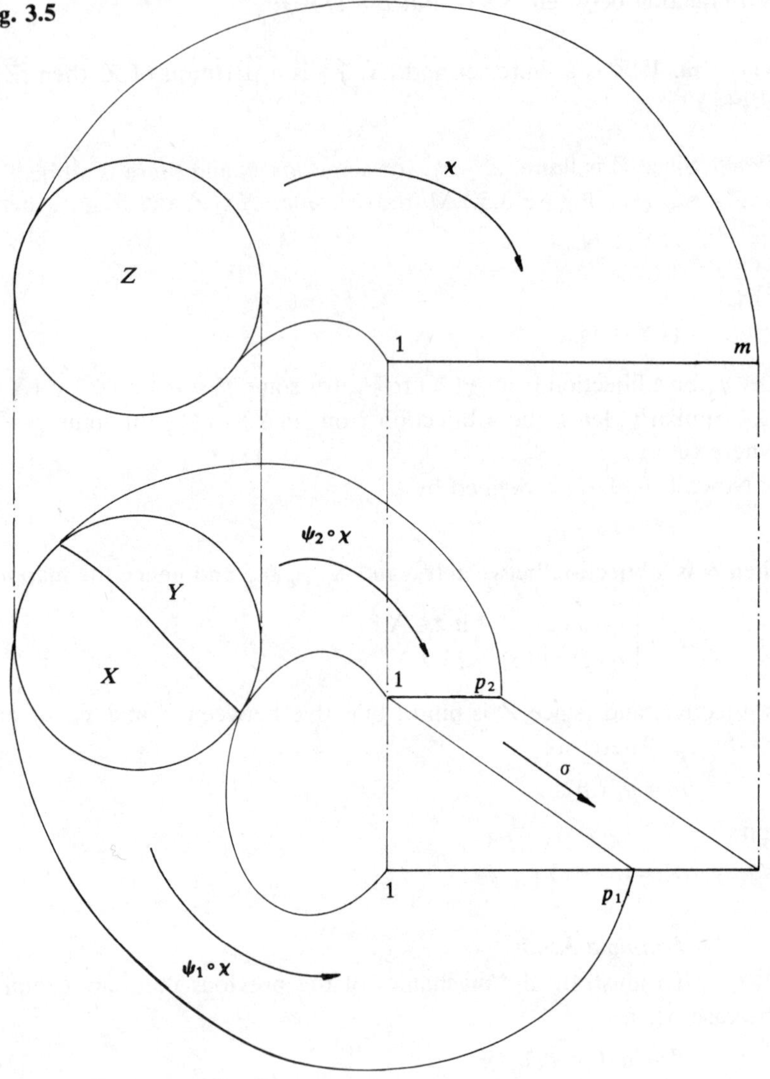

Two suitable ψs are

$$\psi_1=\{(1, 2), (4, 1)\},$$

$$\psi_2=\{(3, 1), (5, 2), (6, 3), (2, 4)\}.$$

So if σ is defined by $x \mapsto x+2$ then

$$\psi_1 \circ \chi(X)=\{1, 2\},$$

$$\psi_2 \circ \chi(Y)=\{1, 2, 3, 4\}$$

and

$$\sigma \circ \psi_2 \circ \chi(Y) = \{3, 4, 5, 6\}.$$

Combining $\psi_1 \circ \chi$ and $\sigma \circ \psi_2 \circ \chi$ thus gives the 'obvious' result. //

The previous example is uncharacteristic in that it requires very detailed manipulation of bijections. This is seldom necessary since we can nearly always refer to well-known established results.

We conclude our discussion on countability by giving one further basic result, proving that for finite sets A and B, $A \times B$ is finite and $|A \times B| = |A| * |B|$. This is not only of value in its own right but also provides a vehicle for the introduction of *proof by induction.*

Inductive proofs utilize the two fundamental concepts present in the definition of $\mathbb{N}$. These are (i) that there is some initial element (in $\mathbb{N}$ this is the number 1) and (ii) given an intermediate state there is a method for moving on to the next (in $\mathbb{N}$ this involves creating a successor of the largest number so far included in $\mathbb{N}$). More properly, if (i) for some $n_0 \in \mathbb{N}$ – usually, but not necessarily, 1 – we can show that the statement $P(n_0)$ about n_0 is true, and (ii) for any $n \in \mathbb{N}$ $(n \geqslant n_0)$ the validity of $P(n)$ implies the validity of $P(n+1)$ – here $n+1$ is the successor of n, no addition is implied – then we deduce that $P(m)$ is true for all $m \geqslant n_0$. Step (i) is the *basis of the induction* and step (ii) is the *induction step.*

Of course all we are really doing is condensing the direct proof

$$P(n_0) \Rightarrow \ldots \Rightarrow P(m)$$

by obtaining a shorthand verification of the intermediate deductions. Let us look at the promised example.

Example 3.3.6

If A and B are finite sets then

$$|A \times B| = |A| * |B|.$$

Proof. Since A and B are finite,

$$A \sim \mathbb{N}_m \quad \text{with bijection } \tau : A \to \mathbb{N}_m$$

and

$$B \sim \mathbb{N}_n \quad \text{for some } m, n \in \mathbb{N}.$$

We shall use induction on n, the size of B. Notice, however, that the size of A is also required to be finite because we assume familiarity only with multiplication of finite quantities.

Basis of induction. If $B = \varnothing$ then $A \times B = \varnothing$ so, trivially,

$$|A \times B| = 0 = |A| * 0 = |A| * |B|.$$

If $B=\{b\}$ (say) then the mapping

$$A \to A \times B$$

defined by $a \mapsto (a, b)$ is clearly bijective, so

$$|A \times B| = |A| = |A| * 1 = |A| * |B|.$$

Induction step. Assume that

$$|A \times B_k| = |A| * |B_k|,$$

where B_k is any subset of B and $|B_k| = k \in \mathbb{N}$. Then

$$|A| * |B_k| = m * k \in \mathbb{N}$$

so there is a bijection

$$\psi : A \times B_k \to \mathbb{N}_{m*k}.$$

Now if $k < n$ we can take a subset of B which has $j = k+1$ elements. Let $B_j = B_k \cup \{x\}$, where B_k is a set of size k, and let χ be a mapping

$$\chi : A \times B_j \to \mathbb{N}$$

defined by

$$\chi : (a, b) \mapsto \psi(a, b) \quad \text{if } b \in B_k$$

and

$$\chi : A \times \{x\} \to \{m * k + 1, \ldots, m * k + m\}$$

defined by

$$\chi : (a, x) \mapsto \tau(a) + m * k.$$

Clearly χ is a bijection onto $\mathbb{N}_{m*k+m}$ and

$$\begin{aligned} m * k + m &= m * (k+1) \\ &= m * j. \end{aligned}$$

So

$$\begin{aligned} |A \times B_j| &= m * j \\ &= |A| * |B_j|. \end{aligned}$$

Hence the identity holds for all subsets of B, up to and including B, so

$$|A \times B| = |A| * |B|. \quad //$$

Exercises 3.3

1. Prove that the relation defined between sets by

   $\{(A, B) : A \sim B\}$

   is an equivalence relation.
2. Construct a bijection between the sets $\mathbb{Z} \times \mathbb{N}$ and $\mathbb{N} \times \mathbb{N}$.
3. Prove that if A and B are sets and $A \cup B$ is finite then

   $|A \cup B| + |A \cap B| = |A| + |B|.$

4. Prove (by contradiction) that any injection

   $\mathbb{N}_m \to \mathbb{N}_m$, for any $m \in \mathbb{N}$

   is a bijection.
5. *Without* using the fact that $|\mathcal{P}(A)| = 2^{|A|}$ show that for any finite set A such that $|A| \geq 2$

   $|\mathcal{P}(A)| > |A|$.
6. Repeat Question 5 in the case when A is not necessarily finite.
7. By considering the set $\{0, 1\}^{\mathbb{N}}$ to be the set of sequences

   $a_1, a_2, a_3, \ldots, a_n, \ldots$ with $a_i \in \{0, 1\}$,

   show that $|\{0, 1\}^{\mathbb{N}}| > |\mathbb{N}|$.
8. Prove, by constructing suitable bijections and using induction, that if $A_1, \ldots, A_n$ are finite sets then

   $|A_1 \times \ldots \times A_n| = |A_1| * \ldots * |A_n|$.

3.4. Some special kinds of functions

At this point in the text we deviate from our main development of mathematical concepts to include a brief discussion of four important kinds of functions; namely permutations, sequences, functionals and equivalence preserving mappings. These functions have many uses but of particular note is their application to graph theory, to tracing computations, to the definition of programming languages and translation of languages, and to graphics.

Definition. A *permutation* of (the elements of) a set A is a bijection on A. //

Permutations of finite sets are of special interest in computing, and when A is finite we are able to calculate the *number* of different permutations on A.

If $|A| = n \in \mathbb{N}$ then we denote by ${}_nP_n$ the number of such permutations. Moreover, the value of ${}_nP_n$ is easy to calculate. We may regard the problem of constructing a bijection on A as filling boxes labelled 1 to n (as in Figure 3.6) with the objects a_1 to a_n.

The order in which the boxes are filled is irrelevant (we can get any other order by shuffling the boxes by yet a further bijection) so we shall fill them from left to right. Box 1 can be filled in n ways since we have a free choice from the whole of A. Removing this element from A gives a set of size $n-1$. Hence box 2 can be filled in $n-1$ ways; box 3 in $n-2$ ways and so on. Continuing in this way box $n-1$ can be filled in 2 ways and finally there is no choice for box n, it must be filled with the sole remaining element of A. Hence the number of different permutations

Fig. 3.6 1 2 $n-1$ n

		....		

of A is

$$n*(n-1)*(n-2)*\ldots*3*2*1.$$

This product is the *factorial* of n, usually written $n!$ Hence

$$_nP_n = n!$$

Since $A \sim \mathbb{N}_n$ we can limit our consideration to $\mathbb{N}_n$. Any permutation of $\mathbb{N}_n$ must specify the image of each element in $\mathbb{N}_n$ (which must of course be unique and distinct). Let a permutation on $\mathbb{N}_n$ be called ψ. Then ψ could be specified by a set of n pairs thus

$$\psi = \{(1, x_1), (2, x_2), \ldots, (n, x_n)\}$$

where

$$\{x_1, x_2, \ldots, x_n\} = \mathbb{N}_n,$$

but, of course, it does not necessarily follow that $x_1 = 1$, etc. Another way to represent ψ is by the array

$$\psi = \begin{pmatrix} 1 & 2 & 3 & 4 & \ldots & n \\ x_1 & x_2 & x_3 & x_4 & \ldots & x_n \end{pmatrix}.$$

Example 3.4.1

Let σ be a permutation on $\mathbb{N}_6$ defined by

$$\sigma = \begin{pmatrix} 1 & 2 & 3 & 4 & 5 & 6 \\ 5 & 6 & 3 & 1 & 4 & 2 \end{pmatrix},$$

then $\sigma(1) = 5$, $\sigma(3) = 3$, etc. //

A consequence of this notation is the simplicity with which composite permutations can be computed. Suppose that ψ is a permutation on $\mathbb{N}_n$, as above, and that χ is a second permutation on the same set, then χ could be written as a collection of pairs in the order dictated by x_1, x_2, etc., rather than 1, 2, etc. If the two arrays are then written one above the other (the first permutation to be applied being written first) then the top and bottom rows give the resulting permutation.

Example 3.4.2

Let σ be as defined in Example 3.4.1 and

$$\rho = \begin{pmatrix} 1 & 2 & 3 & 4 & 5 & 6 \\ 3 & 2 & 6 & 1 & 4 & 5 \end{pmatrix}.$$

Equivalently, we can write

$$\rho = \begin{pmatrix} 5 & 6 & 3 & 1 & 4 & 2 \\ 4 & 5 & 6 & 3 & 1 & 2 \end{pmatrix},$$

so $\rho \circ \sigma$ can be computed thus:

$$\sigma = \begin{pmatrix} 1 & 2 & 3 & 4 & 5 & 6 \\ 5 & 6 & 3 & 1 & 4 & 2 \end{pmatrix}$$

$$\rho = \begin{pmatrix} 5 & 6 & 3 & 1 & 4 & 2 \\ 4 & 5 & 6 & 3 & 1 & 2 \end{pmatrix}$$

$$\rho \circ \sigma = \begin{pmatrix} 1 & 2 & 3 & 4 & 5 & 6 \\ 4 & 5 & 6 & 3 & 1 & 2 \end{pmatrix}.$$

(The bottom row of σ and the top row of ρ are marked "identical".)

Consequently, for example,

$$\rho \circ \sigma(2) \; (= \rho(\sigma(2)) = \rho(6)) = 5, \text{ etc.} \quad //$$

It also follows that the representation of the inverse of a (finite) permutation is obtained by interchanging the two rows of the representation of the original permutation. Although the array representation is useful in computations it may be wasteful of space, particularly in cases when many of the elements are not changed by the permutation. There is a simpler notation which can be applied directly to certain simple permutations, and indirectly to all finite permutations.

Definition. A permutation ρ on a finite set A is called a *cycle* if $A = \{a_1, \ldots, a_n\}$, where the labelling of the elements of A has been suitably chosen, and

$$\rho = \begin{pmatrix} a_1 & a_2 & a_3 & \ldots & a_{n-1} & a_n \\ a_2 & a_3 & a_4 & \ldots & a_n & a_1 \end{pmatrix}.$$

Now suppose that $A \subseteq B$ and B is finite. Extending ρ to act on B, we can define σ such that

$$\sigma : x \mapsto \begin{cases} \rho(x) & \text{if } x \in A, \\ x & \text{if } x \in B \backslash A. \end{cases}$$

Thus σ behaves like ρ in all cases which do not leave an element of B unchanged. Applying σ to A moves the elements round in a cyclic fashion and in both circumstances, provided that the domain is known, we can denote the permutation by

$$(a_1, a_2, a_3, \ldots, a_n).$$

This is a cycle of *length n.* //

Example 3.4.3

Again consider

$$\rho = \begin{pmatrix} 1 & 2 & 3 & 4 & 5 & 6 \\ 3 & 2 & 6 & 1 & 4 & 5 \end{pmatrix}.$$

The permutation is a cycle of length 5, and can be denoted by

$$(1, 3, 6, 5, 4). \quad //$$

Not all permutations are cycles. For example, the permutation σ in Example 3.4.1 is not a cycle. Recall that

$$\sigma = \begin{pmatrix} 1 & 2 & 3 & 4 & 5 & 6 \\ 5 & 6 & 3 & 1 & 4 & 2 \end{pmatrix}.$$

So $\sigma(1) = 5$, $\sigma(5) = 4$, $\sigma(4) = 1$, which implies that σ contains the cycle $(1, 5, 4)$. Starting at 2 we also obtain the cycle $(2, 6)$. In fact we have

$$\sigma = (1, 5, 4) \circ (2, 6)$$

and

$$\sigma = (2, 6) \circ (1, 5, 4).$$

Indeed any finite permutation can be expressed as the composition of cycles and these cycles may be composed in any order. By their construction it must also follow that no element can occur in more than one cycle, viz. the cycles are *disjoint.*

Theorem. Every permutation ρ of a finite set A is expressible as the composition of disjoint cycles.

Proof. Since $|A| = n \in \mathbb{N}$ then $A \sim \mathbb{N}_n$ so without loss of generality we can again limit consideration to a permutation ρ of $\mathbb{N}_n$.

We claim that $\rho = \sigma_1 \circ \sigma_2 \circ \sigma_3 \circ \ldots \circ \sigma_r$ where each σ_i is a cycle and the cycles are disjoint. The validity of this result is demonstrated by constructing the required cycles. First find the lowest $x_1 \in \mathbb{N}_n$ such that $\rho(x_1) \neq x_1$ and all x: $1 \leq x < x_1$ are such that $\rho(x) = x$. If no such x_1 can be found then $\rho = I$ which is trivially an empty composition of cycles. If such an x_1 can be found then compute

$$x_1, \rho(x_1), \rho^2(x_1), \rho^3(x_1), \ldots, \text{etc.}$$

All these elements are in $\mathbb{N}_n$ and hence this infinite progression must include repeats. Suppose that $\rho^k(x_1)$ is the first term that has already appeared in the progression. We then claim that $\rho^k(x_1) = x_1$. If this were not so then we would have

$$\rho^l(x_1) = \rho^k(x_1) \quad \text{for some } 0 < l < k,$$

hence

$$\rho^{l-1}(x_1) = \rho^{-1} \circ \rho^l(x_1) = \rho^{-1} \circ \rho^k(x_1) = \rho^{k-1}(x_1), \text{ etc.}$$

and so

$$\rho^{l-l}(x_1) = \rho^{k-l}(x_1),$$

i.e.

$$\rho^{k-l}(x_1) = \rho^0(x_1) = x_1.$$

But $k-l<k$ so ρ^{k-l} gives a repeated value and hence we have a cycle within ρ given by

$$\sigma_1 = (x_1, \rho(x_1), \rho^2(x_1), \ldots, \rho^{k-1}(x_1)).$$

If all the non-stationary elements of $\mathbb{N}_n$ under ρ are included in σ_1 then $\rho = \sigma_1$ a single cycle which is necessarily disjoint. Otherwise find the next highest element, x_2 of $\mathbb{N}_n$ such that x_2 is non-stationary and x_2 does not occur in σ_1. From x_2 construct the set of distinct powers under ρ to give

$$\sigma_2 = (x_2, \rho(x_2), \rho^2(x_2), \ldots, \rho^m(x_2)).$$

This cycle is of length not less than two and is disjoint from σ_1. (The proof is left as an exercise.) If all non-stationary elements have now been included we have

$$\rho = \sigma_1 \circ \sigma_2 = \sigma_2 \circ \sigma_1.$$

Obviously the set of non-stationary elements not included in these cycles is being reduced and hence must eventually be reduced to $\varnothing$. Consequently,

$$\rho = \sigma_1 \circ \sigma_2 \circ \sigma_3 \circ \ldots \circ \sigma_r \quad \text{for some } r \in \mathbb{N}. \quad //$$

So much for permutations of sets; now to consider a slightly different situation. Take a set A: $|A| = n$ and $B \subseteq A$ with $|B| = r \leqslant n$. How many bijective *functions* are there from A to B? Or, equivalently, how many injective mappings from B to A? The answer, *the number of permutations of size r from a set of size n*, is denoted by ${}_nP_r$ and is calculated in the same way as ${}_nP_n$ except that we stop after having filled r boxes, thus

$${}_nP_r = n*(n-1)*\ldots*(n-r+1).$$

If we continued filling n boxes then it is clear that the last $n-r$ boxes could be filled in ${}_{n-r}P_{n-r}$ ways from the $n-r$ elements that remain, so

$${}_nP_n = {}_nP_r * {}_{n-r}P_{n-r}$$

and again we obtain

$${}_nP_r = \frac{{}_nP_n}{{}_{n-r}P_{n-r}} = \frac{n!}{(n-r)!}.$$

In the calculation of ${}_nP_r$ we are effectively counting the number of bijective functions from A to B. We now concern ourselves only with the size of the sets involved and investigate how many such functions exist.

Definition. Given a finite set A with $B \subseteq A$ and $|A| = n \geqslant r = |B|$ then B is said to be a *combination* of size r from a set of size n. The number of such combinations is written ${}_nC_r$. //

The calculation of ${}_nC_r$ is as follows. Assume $|A| = n$. Take any subset B of A such that $|B| = r$; B is the range of a permutation of size r of the set A. The number of injective functions on A having B as their range is ${}_nP_r$. If f is any one such function and g is another having the same domain and range, then g can be related to f by $g = \phi \circ f$ where ϕ is a *permutation on B*. Both g and f define the same combination and in fact the number of functions defining that combination is equal to the number of permutations ϕ on B. Consequently,

$$ {}_nP_r = {}_nC_r * {}_rP_r $$

and therefore

$$ {}_nC_r = \frac{{}_nP_r}{{}_rP_r} = \frac{n!}{r!\,(n-r)!}. $$

Since relative complements are unique and $|A \backslash B| = n - r$, it follows both logically and arithmetically that

$$ {}_nC_r = {}_nC_{n-r}. $$

So much for permutations and combinations. We now turn to mathematical objects which might be familiar to you but which you will probably not regard as functions.

Definition. A *sequence* in a set S is a mapping $\mathbb{N} \to S$.

If $\sigma : \mathbb{N} \to S$ is a given sequence and $\sigma(n) = s_n$ then it is usual to denote the sequence not by σ but by (s_n) or by

$$ (s_1, s_2, \ldots, s_n, \ldots). $$

In such cases s_n is called the nth term of the sequence. //

Some of the most important questions that arise in connection with sequences concern the 'distance' between successive elements in a sequence (s_n and s_{n+1}, say), and distances between elements s_n for all $n \geqslant n_0$ (where n_0 is some special element of $\mathbb{N}$) and a fixed element of S. However, in the general case we have no concept of distance and hence we shall return to such questions later.

Of special interest in problems associated with the translation of (computer) languages are functionals. These are functions whose domains are not sets of simple objects such as numbers, but are sets of functions.

Definition. Given sets A, B and C, denote by $[B \to C]$ the set of all functions from B to C. Any function $f : A \to [B \to C]$ is a *functional.* Consequently,

$$a \in \mathscr{D}_f \Rightarrow f(a) \text{ is a function}$$

and

$$f(a) : B \to C.$$

Further,

$$b \in \mathscr{D}_{f(a)} \Rightarrow f(a)(b) \in C. \quad //$$

At first sight combinations such as $f(a)(b)$ appear so unusual as to suggest that they may be illegal. This is basically because we are used to regarding functions as special objects, in some way different from the kinds of element to be found in domains and ranges of functions. Of course, a set of functions is just as valid as any other set.

With reference to advanced programming languages names of integer variables are no different from names of function variables and both can be treated in similar ways. Although the functions involved are very complex, programming languages readily give us an example of the importance of functionals.

Example 3.4.4

Let P be a set of programs, i.e. the textual form of the programs – the string of characters that would be submitted to a compiler. Similarly, let I and O be the sets of input values and output values that would be available for a program to read and to write.

A compiler (for the relevant language) is then a functional of type $P \to [I \to O]$, since given $p \in P$ it attempts to produce runnable code which when executed will read $i \in I$ and produce $o \in O$. //

As will be readily apparent, the proper definition of functions (or functionals) of the complexity of a compiler is long and detailed. Nevertheless, the concepts involved can be studied in simpler situations.

Example 3.4.5

Over the sets $\mathbb{R}$, $\mathbb{R} \to \mathbb{R}$, etc., if

$f : a \mapsto [x \mapsto a + x]$,

then

$$f(2): x \mapsto 2 + x$$

and

$$f(2)(3) = 5.$$

Whereas

$$f(3): x \mapsto 3 + x \quad \text{and} \quad f(3)(3) = 6, \text{etc.} \quad //$$

Manipulation of functionals causes no great practical difficulty provided that reference is always made to the underlying functionality (e.g. $A \to B$ or $A \to [B \to C]$). Hence we shall in future regard them merely as functions having non-trivial ranges and will refer to them as such.

Finally in this section, we define functions that preserve some of the structure of their domains. Subsequently we shall see that in certain situations it is desirable to preserve many of the algebraic properties that sets may possess but initially we shall limit consideration to a particularly simple case.

Definition. Given that X is a set on which is defined an equivalence relation ρ then X *factored by* ρ is the set of ρ-equivalence classes in X and is denoted by X/ρ. //

Definition. Given that X and Y are sets on which are defined equivalence relations ρ_X and ρ_Y and $f: X \to Y$ is a *mapping*, let $\hat{f}$ be a *relation*

$$\hat{f}: X/\rho_X \to Y/\rho_Y,$$

such that

$$\hat{f} = \{([x], [f(x)]): x \in X\},$$

where $[x]$ denotes the equivalence class of x. If $\hat{f}$ is a function then

$$x_1 \rho_X x_2 \Rightarrow \hat{f}([x_1]) = \hat{f}([x_2])$$

and f is an *equivalence preserving mapping.* In this case we say that $f: X \to Y$ *induces* the mapping $\hat{f}: X/\rho_X \to Y/\rho_Y$. //

A useful way to visualize the workings of such an equivalence preserving mapping is by the diagram in Figure 3.7.

Fig. 3.7

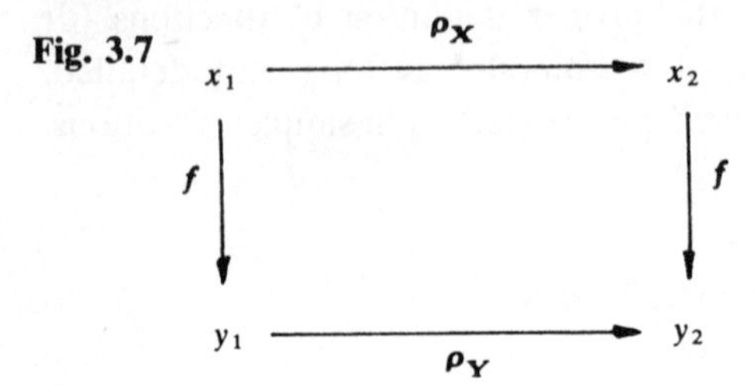

Providing that f behaves in the desired way with respect to the equivalence relations then we can go from x_1 to y_2 either via x_2 using $y_2 = f(x_2)$ and $x_2 \rho_X x_1$ or via y_1 using $y_1 = f(x_1)$ and $y_2 \rho_Y y_1$.

Example 3.4.6

Let $X = \{1, 2, 3\}$, $Y = \{1, 4, 9\}$, ρ_X and ρ_Y be such that

$$X/\rho_X = \{\{1\}, \{2, 3\}\},$$
$$Y/\rho_Y = \{\{1\}, \{4, 9\}\},$$

and $f: X \to Y$ such that $x \mapsto x^2$. Then

$$\hat{f}([1]) = [f(1)] = [1] = \{1\},$$
$$\hat{f}([2]) = [4] = \{4, 9\}$$
$$\hat{f}([3]) = [9] = \{4, 9\}.$$

Now

$$\{2, 3\} \in X/\rho_X \Rightarrow 2\rho_X 3$$
$$\Rightarrow [2] = [3]$$

and $\hat{f}([2]) = \hat{f}([3])$ so $\hat{f}$ is a function and f preserves equivalences. //

Example 3.4.7

Let X, Y and f be as before but with equivalence relations σ_X and σ_Y inducing partitions $\{\{1\}, \{2, 3\}\}$ and $\{\{1, 4\}, \{9\}\}$ respectively. In this case the induced relation gives

$$\hat{f}([2]) = [f(2)] = [4] = \{1, 4\},$$
$$\hat{f}([3]) = [f(3)] = [9] = \{9\}.$$

Again $2\sigma_X 3$ so $[2] = [3]$ in X/σ_X, but $(4, 9) \notin \sigma_Y$ so $[4] \neq [9]$ in Y/σ_Y. Comparing with Figure 3.7 this example gives rise to the relationships shown in Figure 3.8. Since we cannot join the sides of the rectangle in all cases equivalences are *not* preserved. //

Fig. 3.8

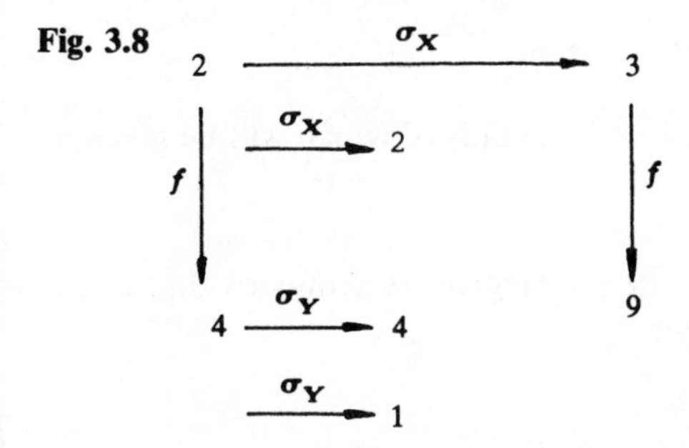

In Section 5.1 and later sections we shall show how these diagrams can be used to *define* operations in such a way that the corners of the rectangle *always* fit together. Once this has been done we can put diagrams together just like building blocks.

There are no exercises for this section.

3.5. Analytical properties of real functions

This part consists of bridging material that utilizes the set theory of Chapter 1 to give a sound foundation to the study of calculus. Our presentation of this material is aimed not at developing computational techniques but at making precise such statements as

'The limit of $f(x)$ as x tends to zero is y',

'The slope of the graph of f at a is b'

and

'f has a smoothly turning graph', etc.

(These last two concepts are of obvious relevance in graphics.) The basic definitions will be given and used to derive selected results. This is sufficient for the purpose of illustrating how the major theorems are obtained.

3.5.1 Sequences

A *real sequence* is a mapping with domain $\mathbb{N}$ and range $\mathbb{R}$. Sequences are written (a_n). If, as n increases, a_n becomes increasingly 'close' to some fixed value $a \in \mathbb{R}$ we say that the sequence (a_n) has limit a, or a_n tends to a as n tends to infinity. More precisely we make the following definition.

Definition. If (a_n) is a real sequence and for each $\varepsilon > 0$ there is an $N_\varepsilon \in \mathbb{N}$ such that

$$N > N_\varepsilon \Rightarrow |a_N - a| < \varepsilon,$$

then we say that the sequence (a_n) has *limit* a and write

$$\operatorname*{limit}_{n\to\infty} a_n = a \quad \text{or} \quad a_n \to a \text{ as } n \to \infty.$$

(Here $|x|$ denotes the modulus of $x \in \mathbb{R}$, details of which will be given in Section 5.3.4.) //

If (a_n) has a limit it is said to be *convergent*. A sequence that is not convergent is said to be *divergent*.

Example 3.5.1

(i) The sequence (a_n) defined by $a_n = 1/n$ has limit 0, for given $\varepsilon > 0$ we choose N_ε to be any natural number greater than $1/\varepsilon$.

Now

$$N > N_\varepsilon \Rightarrow |a_N - 0| = \frac{1}{N} < \frac{1}{N_\varepsilon} < \varepsilon,$$

hence

$$\operatorname*{limit}_{n\to\infty} \frac{1}{n} = 0.$$

(ii) The sequence (a_n) defined by $a_n = (-1)^n$ is divergent. //

Proposition. If (s_n) and (t_n) are sequences and $\lambda \in \mathbb{R}$ then $(s_n + t_n)$, $(s_n t_n)$ and (λs_n) are sequences, and if $\text{limit}_{n\to\infty} s_n = s$ and $\text{limit}_{n\to\infty} t_n = t$, then

(i) $\text{limit}_{n\to\infty} (s_n + t_n) = s + t$,
(ii) $\text{limit}_{n\to\infty} (s_n t_n) = st$,
(iii) $\text{limit}_{n\to\infty} (\lambda s_n) = \lambda s$, and if $t \neq 0$, then
(iv) $s_n / t_n \to s/t$ as $n \to \infty$.

Proof. (i) If $\varepsilon > 0$ then we can find $N_\varepsilon \in \mathbb{N}$ such that

$$|s_N - s| < \varepsilon/2 \text{ and } |t_N - t| < \varepsilon/2 \quad \text{whenever } N > N_\varepsilon$$

but

$$\begin{aligned} |s_N + t_N - (s+t)| &= |s_N - s + t_N - t| \\ &\leq |s_N - s| + |t_N - t| \\ &< \varepsilon/2 + \varepsilon/2 = \varepsilon \quad \text{whenever } N > N_\varepsilon, \end{aligned}$$

hence

$$\operatorname*{limit}_{n\to\infty} (s_n + t_n) = s + t.$$

Similarly, for (ii)

$$\begin{aligned} |s_N t_N - st| &= |s_N t_N - s_N t + s_N t - st| \\ &\leq |s_N t_N - s_N t| + |s_N t - st| \\ &\leq |s_N||t_N - t| + |s_N - s||t|. \end{aligned}$$

Now, given $\varepsilon > 0$, there is an $N_\varepsilon \in \mathbb{N}$ such that whenever $N > N_\varepsilon$

$$|s_N - s| < \frac{1}{2} \frac{\varepsilon}{|t| + 1}$$

and

$$|t_N - t| < \frac{1}{2} \frac{\varepsilon}{|s| + 1}$$

and

$$|s_N| < |s| + 1.$$

Hence

$$\begin{aligned}|s_N||t_N-t|+|s_N-s||t| &\leq (|s|+1)|t_N-t|+|s_N-s||t| \\ &< \frac{1}{2}\varepsilon+\frac{1}{2}\frac{\varepsilon}{|t|+1}|t| \\ &< \tfrac{1}{2}\varepsilon+\tfrac{1}{2}\varepsilon=\varepsilon,\end{aligned}$$

and therefore $(s_n t_n) \to st$.

Parts (iii) and (iv) are left as an exercise. //

Definition. Let (a_n) be a sequence in $\mathbb{R}$. The *series* $\sum a_n$ is defined to be the sequence (s_n):

$$s_n = \sum_{r=1}^{n} a_r.$$

The number s_n is called the nth *partial sum* of the series. If (s_n) is convergent the series $\sum a_n$ is said to be *convergent* and the number $\text{limit}_{n\to\infty} s_n$ is called the *sum* of the series. This is denoted by

$$\sum_{n=1}^{\infty} a_n. \quad //$$

3.5.2 Continuity

The concept of continuity is almost completely ignored in the study of elementary calculus. It suffers from a common characteristic of informal mathematics in that it is 'obvious' but, in a school context, quite difficult to define in a useful way. In fact a proper definition is based on the concept of a limit. Throughout this section I denotes an interval of the real line $\mathbb{R}$. If $f: I \to \mathbb{R}$ and $f(x)$ becomes increasingly 'close' to some point $b \in \mathbb{R}$ as x 'approaches' a point $a \in I$ we say that the limit of $f(x)$ as x tends to a is b. Formally we make the following definition.

Definition. A function $f: I \to \mathbb{R}$ has *limit* b at a if for all $\varepsilon > 0$ there is a $\delta_\varepsilon > 0$ such that

$$0 < |x-a| < \delta_\varepsilon \Rightarrow |f(x)-b| < \varepsilon.$$

We write this as

$$\operatorname*{limit}_{x\to a} f(x) = b$$

or

$$f(x) \to b \text{ as } x \to a. \quad //$$

We note that this makes no assumption about f at $x = a$, in fact it need not even be defined.

Example 3.5.2

(i) $\text{limit}_{x\to a}\, x = a$, choose $\delta_\varepsilon = \varepsilon$.

(ii) $\text{limit}_{x\to 2}\, x^2 = 4$, since

$|x-2| < \delta_\varepsilon$ iff $-\delta_\varepsilon < x-2 < \delta_\varepsilon$ iff $4-\delta_\varepsilon < x+2 < 4+\delta_\varepsilon$,

therefore

$$\begin{aligned}|x^2-4| &= |x+2|\,|x-2|,\\ &< \delta_\varepsilon |x+2|,\\ &< \delta_\varepsilon(4+\delta_\varepsilon),\end{aligned}$$

therefore if we choose $\delta_\varepsilon = \min(1, \varepsilon/5)$, then

$|x^2-4| < 5\delta_\varepsilon \leqslant \varepsilon$. //

It is easy to show that the following are all equivalent,

$$\lim_{x\to a} f(x) = b, \quad \lim_{x\to a} (f(x)-b) = 0,$$

$$\lim_{h\to 0} f(a+h) = b, \quad \lim_{h\to 0} (f(a+h)-b) = 0.$$

We are now in a position to treat the concept of continuity for real-valued functions. Roughly speaking $f: I \to \mathbb{R}$ is continuous at $a \in I$ if points 'close' to a are mapped to points 'close' to $f(a)$. The following makes this idea precise.

Definition. A function $f: I \to \mathbb{R}$ is *continuous at* $a \in I$ if

$$\lim_{x\to a} f(x) = f(a);$$

f is said to be *continuous* if it is continuous at each point of its domain. //

It is implicit in this definition that f is *defined* at $x = a$. This definition of continuity corresponds to what we would expect intuitively. We make this clear with the aid of pictures.

Figure 3.9(i) is a pictorial representation, or graph, of the continuous function $f_1: [-2, 2] \to \mathbb{R}$ defined by $f_1(x) = |x|$. Figure 3.9(ii) represents the function $f_2: [0, 4] \to \mathbb{R}$ defined by

$$\begin{aligned} f_2(x) &= x &&\text{for } 0 \leqslant x \leqslant 2,\\ &= x+1 &&\text{for } 2 < x \leqslant 4;\end{aligned}$$

f_2 is continuous at each point of $[0, 4]$ *except* at $x = 2$, for there are no intervals of the form $2-\delta < x < 2+\delta$ for which $|f_2(x) - f_2(2)| < 1$.

Before leaving the topic of limits and continuity we state some results informally; their proofs can be found in most texts on real analysis.

Fig. 3.9

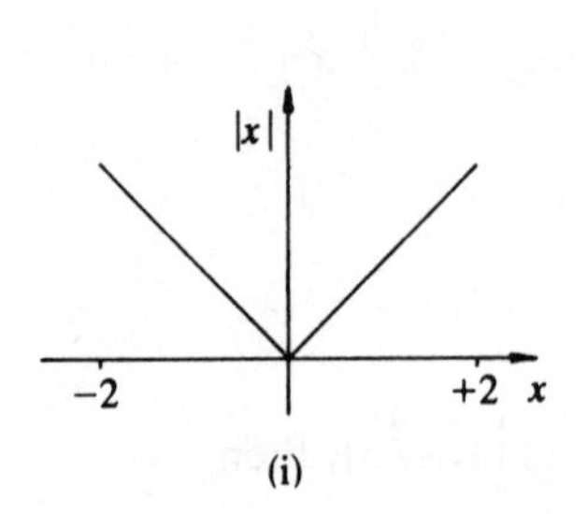

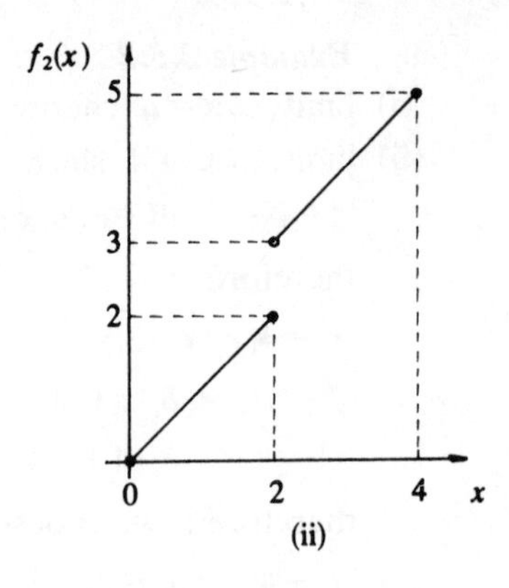

If $f(x) \to l$ and $g(x) \to m$ as $x \to a$, then

$$(f+g)(x) = f(x) + g(x) \to l + m,$$
$$(fg)(x) = f(x)g(x) \to lm,$$
$$(f/g)(x) = f(x)/g(x) \to l/m \quad \text{provided } m \neq 0,$$
$$(\lambda f)(x) = \lambda f(x) \to \lambda l \quad \text{for all } \lambda \in \mathbb{R}.$$

It follows that if f and g are continuous at a point a then so are the functions λf, $f+g$, gh and f/g, provided a is in the domain of each of these 'new' functions.

3.5.3 Differentiability

The pictorial representation of a function $f: I \to \mathbb{R}$, discussed in the previous section, suggests that such functions determine a second function $f': I \to \mathbb{R}$, where

$$f'(a) = \text{'slope' of the graph of } f \text{ at } a.$$

In general $\mathcal{D}_{f'} \subseteq \mathcal{D}_f$, for the slope may not be defined at each point of $\mathcal{D}_f$ – in fact it may not be defined anywhere, in which case f' does not exist. We now define the derived function, f', precisely.

Definition. $f: I \to \mathbb{R}$ is *differentiable* at $a \in I$ if

$$\lim_{h \to 0} \frac{f(a+h) - f(a)}{h}$$

exists. The set of points for which this limit exists constitutes the domain $\mathcal{D}_{f'}$ of the *derived* function $f': I \to \mathbb{R}$ and

$$f'(x) = \lim_{h \to 0} \frac{f(x+h) - f(x)}{h};$$

f' is also called the *derivative* of f (with respect to x) and is sometimes written $\mathrm{d}f/\mathrm{d}x$. The ratio

$$\frac{f(x+h)-f(x)}{h}$$

is often written $\delta f/\delta x$ where δ is read 'a small change in'. In this notation

$$\operatorname*{limit}_{\delta x\to 0}\frac{\delta f}{\delta x}=\frac{\mathrm{d}f}{\mathrm{d}x}. \quad //$$

If f is differentiable at a then f is continuous at a, for

$$f(a+h)-f(a)=\left(\frac{f(a+h)-f(a)}{h}\right)h,$$

hence

$$\begin{aligned}\operatorname*{limit}_{h\to 0}(f(a+h)-f(a)) &= \operatorname*{limit}_{h\to 0}\left(\frac{f(a+h)-f(a)}{h}\right)\operatorname*{limit}_{h\to 0} h\\ &= f'(a)\cdot 0\\ &= 0.\end{aligned}$$

In other words $\text{limit}_{h\to 0}\, f(a+h)=f(a)$ and thus f is continuous at a by definition. Continuity is thus a necessary condition for differentiability but it is not a sufficient condition as the following example shows.

Example 3.5.3

$f(x)=|x|$ is not differentiable at $x=0$ for

$$\frac{f(0+h)-f(0)}{h}=\frac{|h|}{h}$$

and

$$\begin{aligned}\frac{|h|}{h} &= +1 \quad \text{if } h>0,\\ &= -1 \quad \text{if } h<0,\end{aligned}$$

hence in any interval $]-h, h[$, however small, $|h|/h$ assumes both the values ± 1 and a limit does not exist at $h=0$. //

Example 3.5.4

(i) If $f:\mathbb{R}\to\mathbb{R}$ is a constant function

$f(x)=c \quad$ for all $x\in\mathbb{R}$

then

$$\begin{aligned}\frac{f(x+h)-f(x)}{h} &= \frac{c-c}{h}\\ &= 0\end{aligned}$$

and

$$\lim_{h\to 0} 0 = 0,$$

thus $f'(x) = 0$ for all $x \in \mathbb{R}$. Conversely, if $f'(x) = 0$ for all $x \in \mathbb{R}$ then f is a constant function.

(ii) If $f : \mathbb{R} \to \mathbb{R}$ is defined by

$$f(x) = x^2 \quad \text{for all } x \in \mathbb{R}$$

then

$$\begin{aligned}\frac{f(x+h)-f(x)}{h} &= \frac{(x+h)^2 - x^2}{h} \\ &= \frac{x^2 + 2xh + h^2 - x^2}{h} \\ &= 2x + h\end{aligned}$$

and

$$\begin{aligned}\lim_{h\to 0} (2x+h) &= \lim_{h\to 0} 2x + \lim_{h\to 0} h, \\ &= 2x + 0,\end{aligned}$$

hence $f'(x) = 2x$ for all $x \in \mathbb{R}$. //

Proposition. If f is differentiable at x and $\lambda \in \mathbb{R}$ then λf is differentiable at x and

$$(\lambda f)'(x) = \lambda f'(x).$$

Proof.

$$\begin{aligned}(\lambda f)'(x) &= \lim_{h\to 0} \frac{\lambda f(x+h) - \lambda f(x)}{h} \\ &= \lambda \lim_{h\to 0} \frac{f(x+h)-f(x)}{h} = \lambda f'(x). \quad //\end{aligned}$$

The following results are useful for differentiating functions defined in terms of other functions.

Proposition. If f and g are both differentiable at x, then

(i) $f+g$ is differentiable at x and

$$(f+g)'(x) = f'(x) + g'(x),$$

(ii) fg is differentiable at x and

$$(fg)'(x) = (fg')(x) + (f'g)(x),$$

(iii) if $g(x) \neq 0$, f/g is differentiable at x and

$$(f/g)'(x) = \frac{(gf')(x) - (fg')(x)}{g^2(x)}.$$

Proof. This is left as an exercise. //

These formulae may be used to prove some further, possibly familiar, simple results.

Example 3.5.5

(i) If $f:\mathbb{R}\to\mathbb{R}$ is defined by

$$f(x)=\frac{1}{x}$$

then

$$f'(x)=-1/x^2,$$

where

$$\mathscr{D}_f=\mathscr{D}_{f'}=\mathbb{R}\backslash\{0\}.$$

(ii) If $f:\mathbb{R}\to\mathbb{R}$ is given by

$$f(x)=x^n \quad \text{for } n\in\mathbb{N}$$

then

$$f'(x)=nx^{n-1}. \quad //$$

Proposition. (The chain rule). If f is differentiable at x and g is differentiable at $y=f(x)$ then $g\circ f$ is differentiable at x and

$$(g\circ f)'(x)=g'(y)f'(x).$$

Proof. Let $w=g(y)=g(f(x))=g\circ f(x)$. Then

$$\begin{aligned}\frac{\mathrm{d}(g\circ f)}{\mathrm{d}x}&=\operatorname*{limit}_{\delta x\to 0}\frac{\delta w}{\delta x}\\&=\operatorname*{limit}_{\delta x\to 0}\frac{\delta w}{\delta y}\frac{\delta y}{\delta x}\quad(\text{provided }\delta y\neq 0)\\&=\left(\operatorname*{limit}_{\delta x\to 0}\frac{\delta w}{\delta y}\right)\left(\operatorname*{limit}_{\delta x\to 0}\frac{\delta y}{\delta x}\right)\quad(\text{by Section 3.5.1})\end{aligned}$$

but f is differentiable at x so $\delta y\to 0$ as $\delta x\to 0$.

$$\operatorname*{limit}_{\delta x\to 0}\frac{\delta y}{\delta x}=\frac{\mathrm{d}f}{\mathrm{d}x},$$

similarly

$$\operatorname*{limit}_{\delta y\to 0}\frac{\delta w}{\delta y}=\frac{\mathrm{d}g}{\mathrm{d}y}$$

so

$$\frac{\mathrm{d}(g\circ f)}{\mathrm{d}x}=\frac{\mathrm{d}w}{\mathrm{d}y}\frac{\mathrm{d}y}{\mathrm{d}x}$$

and

$$(g\circ f)'(x) = g'(y)f'(x). \quad /\!/$$

The derivative of f' is written f'' or d^2f/dx^2 and called the second derivative of f. Similarly the derivative of $d^{n-1}f/dx^{n-1}$ ($n \geqslant 3$) is written $f^{(n)}$ or d^nf/dx^n and called the nth derivative of f. If f' exists and is continuous, f is said to be of class C^1; f is of class C^n if $f^{(n)}$ exists and is continuous, and of class C^∞ if $f^{(n)}$ exists for all $n \in \mathbb{N}$.

3.5.4 Integration

If $f:[a, b] \to \mathbb{R}$, $n \in \mathbb{N}$, $h = (b-a)/n$ and $x_k = a + kh$ for $0 \leqslant k < n$ then we can define a real sequence $(s_n(f))$ by

$$s_n(f) = \sum_{k=0}^{n-1} f(x_k)h.$$

If $(s_n(f))$ has a limit we say that f is *integrable* on $[a, b]$ and write

$$\operatorname*{limit}_{n\to\infty} s_n(f) = \int_a^b f(x)\,dx.$$

$\int_a^b f(x)\,dx$ is called the *Riemann integral* of f on $[a, b]$.

The shaded area of Figure 3.10 is a pictorial representation of $s_5(f)$ for a continuous function f on $[a, b]$. For increasingly large values of n we would intuitively expect the shaded area to be an increasingly good approximation to the 'area' under the graph between $x = a$ and $x = b$, and the limiting value (if it exists) to be precisely this area.

Fig. 3.10

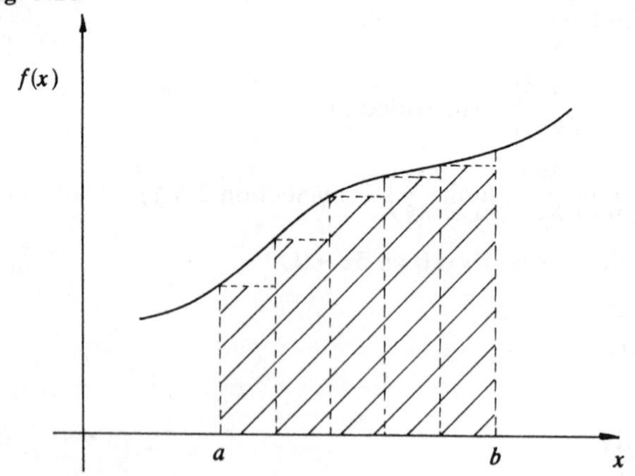

If $\mathcal{F}$ denotes the set of all real-valued functions on $[a, b]$ then integration may be regarded as a function

$$\mathcal{F} \to \mathbb{R},$$

the domain of which is

$$\left\{f \in \mathscr{F}: \int_a^b f(x)\,\mathrm{d}x \text{ exists}\right\}.$$

Some important properties of the integral are summarized below.

Proposition.
(i) If f is continuous on $[a, b]$ then f is integrable on $[a, b]$.
(ii) If f is integrable on $[a, b]$ and $x \in [a, b]$ then f is integrable on $[a, x]$ and $[x, b]$, and

$$\int_a^b f(x)\,\mathrm{d}x = \int_a^x f(x)\,\mathrm{d}x + \int_x^b f(x)\,\mathrm{d}x.$$

(iii) If f is integrable on $[a, b]$ and $\lambda \in \mathbb{R}$ then λf is integrable on $[a, b]$ and

$$\int_a^b (\lambda f)(x)\,\mathrm{d}x = \lambda \int_a^b f(x)\,\mathrm{d}x.$$

(iv) If f and g are integrable on $[a, b]$ then $f+g$ is integrable on $[a, b]$ and

$$\int_a^b (f+g)(x)\,\mathrm{d}x = \int_a^b f(x)\,\mathrm{d}x + \int_a^b g(x)\,\mathrm{d}x.$$

Proof. No formal proof of (i) is given but we remark that for a continuous function f it is intuitively clear that the area under f is a well-defined concept and hence we would expect the integral of f to exist.

The proofs of (ii), (iii) and (iv) follow from the properties of sequences. Consider (iv); if f and g are integrable on $[a, b]$ then the sequences

$$s_n(f) = \sum_{k=0}^{n-1} f(x_k)h \quad \text{and} \quad s_n(g) = \sum_{k=0}^{n-1} g(x_k)h$$

both have limits, hence so does the sequence $s_n(f) + s_n(g)$ and

$$\operatorname*{limit}_{n\to\infty} (s_n(f) + s_n(g)) = \operatorname*{limit}_{n\to\infty} s_n(f) + \operatorname*{limit}_{n\to\infty} s_n(g). \quad //$$

To evaluate an integral we rarely work from the definition by evaluating limits. The following theorem is fundamental; it states that integration and differentiation are essentially inverse processes.

Theorem. If $f: [a, b] \to \mathbb{R}$ is continuous and we define $F: [a, b] \to \mathbb{R}$ by

$$F(t) = \int_a^t f(x)\,\mathrm{d}x,$$

then F is differentiable on $[a, b]$ and $F' = f$.

Sketch proof. We have

$$\frac{F(t+h)-F(t)}{h}=\frac{\int_a^{t+h} f(x)\,\mathrm{d}x-\int_a^t f(x)\,\mathrm{d}x}{h}$$

$$=\frac{\int_a^t f(x)\,\mathrm{d}x+\int_t^{t+h} f(x)\,\mathrm{d}x-\int_a^t f(x)\,\mathrm{d}x}{h}$$

(using (ii) of the previous proposition)

$$=\frac{1}{h}\int_t^{t+h} f(x)\,\mathrm{d}x.$$

Consider

$$\lim_{h\to 0}\frac{1}{h}\int_t^{t+h} f(x)\,\mathrm{d}x.$$

From our definition of integral and its interpretation as area it is clear that for increasingly small values of h, $\int_t^{t+h} f(x)\,\mathrm{d}x$ approaches $f(t)h$ and

$$\lim_{h\to 0}\frac{1}{h}\int_t^{t+h} f(x)\,\mathrm{d}x=f(t),$$

hence

$$F'(t)=f(t)\quad \text{for all } t\in[a, b]. \quad //$$

In the notation of the theorem, if Φ is any function for which

$$\Phi'=f,$$

then $F-\Phi$ is a constant function, for

$$F'(t)=f(t)$$
$$=\Phi'(t),$$

hence

$$(F-\Phi)'(t)=0\quad \text{for all } t\in[a, b]$$

and from Section 3.5.3 we conclude that

$$F-\Phi=\lambda$$

for some $\lambda\in\mathbb{R}$. Clearly, we have

$$\Phi(t)=\int_a^t f(x)\,\mathrm{d}x+\lambda.$$

Φ is called an *indefinite* integral of f, and written

$$\int f(t)\,\mathrm{d}t;$$

$\int f(t)\,\mathrm{d}t$ is only determined up to a constant, and strictly denotes the equivalence class of functions $[\Phi]$, where $\Phi_1\sim\Phi_2$ iff Φ_1 and Φ_2 are indefinite integrals of f.

Proposition. If Φ is an indefinite integral of f, then

$$\int_a^b f(x)\,\mathrm{d}x = \Phi(b) - \Phi(a).$$

Proof.

$$\begin{aligned}\Phi(b)-\Phi(a) &= (F(b)+\lambda)-(F(a)+\lambda)\\ &= F(b)-F(a)\\ &= \int_a^b f(x)\,\mathrm{d}x - \int_a^a f(x)\,\mathrm{d}x\\ &= \int_a^b f(x)\,\mathrm{d}x. \quad //\end{aligned}$$

As with differentials, these results may be used to derive many integrals some examples of which are given below.

Example 3.5.6

(i) If

$$F(t) = \int_0^t x\,\mathrm{d}x,$$

then

$$F'(t) = t,$$

but the indefinite integrals of the function $x \mapsto x$ are

$$\Phi(t) = \frac{t^2}{2} + \lambda \quad \text{for } \lambda \in \mathbb{R}$$

and thus

$$\begin{aligned}F(t) &= \Phi(t) - \Phi(0)\\ &= \frac{t^2}{2}.\end{aligned}$$

(ii) More generally, if $f: x \mapsto x^n$ for $n \in \mathbb{Z}\backslash\{-1\}$ and

$$F(t) = \int_a^t x^n\,\mathrm{d}x,$$

then

$$F'(t) = t^n$$

and the indefinite integrals are

$$\Phi(t) = \frac{t^{n+1}}{(n+1)} + \mu \quad \text{for } \mu \in \mathbb{R},$$

so that

$$F(t) = \Phi(t) - \Phi(a)$$
$$= \frac{t^{n+1}}{(n+1)} - \frac{a^{n+1}}{(n+1)}.$$

Things clearly go wrong at $n = -1$; this case will be considered in the next section. //

3.5.5 Some special real functions

We assume that the reader is familiar with the geometric definition of the sine and cosine functions from which it follows that

$$\sin : \mathbb{R} \to [-1, 1],$$
$$\cos : \mathbb{R} \to [-1, 1],$$

where $\mathscr{D}_{\sin} = \mathscr{D}_{\cos} = \mathbb{R}$ and $\mathscr{R}_{\sin} = \mathscr{R}_{\cos} = [-1, 1]$. We also assume familiarity with the periodic nature of these functions. Some other elementary properties are summarized in the following proposition.

Proposition. For all $x, y \in \mathbb{R}$ we have

(i) $\sin(x+y) = \sin(x)\cos(y) + \cos(x)\sin(y)$,
(ii) $\sin(x-y) = \sin(x)\cos(y) - \cos(x)\sin(y)$,
(iii) $\cos(x+y) = \cos(x)\cos(y) - \sin(x)\sin(y)$,
(iv) $\cos(x-y) = \cos(x)\cos(y) + \sin(x)\sin(y)$,
(v) $\sin^2 x + \cos^2 x = 1$,
(vi) $\frac{\mathrm{d}}{\mathrm{d}x}(\sin(x)) = \cos(x)$, $\frac{\mathrm{d}}{\mathrm{d}x}(\cos(x)) = -\sin(x)$. //

These results follow quickly from the definitions and are left as an exercise for the reader. We shall not concern ourselves with their justification and you may regard them as assumptions.

In Section 3.5.4 we determined the integrals $\int x^n \,\mathrm{d}x$ for $n \neq -1$. In fact the integral $\int_1^t 1/x \,\mathrm{d}x$ exists for all $t > 0$ and is denoted by $\ln(t)$; ln is a mapping $]0, \infty[\to \mathbb{R}$ and has the property

$$\ln(xy) = \ln(x) + \ln(y) \quad \text{for all } x, y \in]0, \infty[,$$

for

$$\frac{\mathrm{d}}{\mathrm{d}x}\ln(xy) = \frac{y}{xy} = \frac{1}{x} = \frac{\mathrm{d}}{\mathrm{d}x}\ln(x).$$

Hence by results in Section 3.5.4

$$\ln(xy) - \ln(x) = \lambda \quad \text{for some } \lambda \in \mathbb{R} \text{ and all } x \in]0, \infty[.$$

In particular, when $x = 1$

$$\ln(y) - \ln(1) = \lambda$$

and $\ln(1)=0$, so

$$\ln(y)=\lambda.$$

Therefore we have

$$\ln(xy)=\ln(x)+\ln(y).$$

It can also be shown that ln is bijective, and hence there is a function

$$\exp:\mathbb{R}\to\,]0,\infty[$$

such that

$$\ln(\exp(p))=p \quad \text{for all } p\in\mathbb{R}$$

and

$$\exp(\ln(q))=q \quad \text{for all } q\in\,]0,\infty[.$$

It follows from the properties of the function ln that

$$\exp(x+y)=\exp(x)\exp(y) \quad \text{for all } x, y\in\mathbb{R},$$

$$\exp(0)=1$$

and

$$\frac{\mathrm{d}}{\mathrm{d}x}(\exp(x))=\exp(x).$$

It is conventional to denote $\ln(x)$ by $\log_e x$ and $\exp(x)$ by e^x: $\log_e x$ is called the *natural logarithm* of x, and the function $x\mapsto e^x$ is called the *exponential* function.

If $a>0$, $f:\,]-a,a[\,\to\mathbb{R}$ is C^∞ and $x\in\,]-a,a[$, then

$$\sum_{k=0}^{\infty} f^{(k)}(0)\frac{x^k}{k!}$$

is called the *Maclaurin series* for f at x. For certain functions it can be shown that the Maclaurin series is convergent to the function value at x; in other words for certain f we have

$$f(x)=\operatorname*{limit}_{N\to\infty}\sum_{k=0}^{N} f^{(k)}(0)\frac{x^k}{k!}.$$

In particular this is true for the sine, cosine and exponential functions for which

$$\sin(x)=\sum_{k=0}^{\infty}(-1)^k\frac{x^{2k+1}}{(2k+1)!}=x-\frac{x^3}{3!}+\frac{x^5}{5!}-\frac{x^7}{7!}+\ldots,$$

$$\cos(x)=\sum_{k=0}^{\infty}(-1)^k\frac{x^{2k}}{(2k)!}=1-\frac{x^2}{2!}+\frac{x^4}{4!}-\frac{x^6}{6!}+\ldots,$$

$$\exp(x)=\sum_{k=0}^{\infty}\frac{x^k}{k!}=1+x+\frac{x^2}{2!}+\frac{x^3}{3!}+\ldots,$$

for all $x\in\mathbb{R}$.

Exercises 3.5

1. Show that the sequences defined below are convergent and determine their limits
   (i) $s_n = 1/n^2$,
   (ii) $s_n = 3n/(n+3)$,
   (iii) $s_n = 1 + 1/2^n$.
2. If (s_n) and (t_n) are sequences with $\text{limit}_{n\to\infty} s_n = s$ and $|t_n| < |s_n|$ for all $n \in \mathbb{N}$, show that $\text{limit}_{n\to\infty} t_n = 0$.
3. Prove that if (s_n) and (t_n) are convergent sequences with limits s and t respectively, then the sequence (p_n) defined by $p_n = \lambda s_n$ is convergent with limit λs. Show also that if $t \neq 0$ then $\text{limit}_{n\to\infty} s_n/t_n = s/t$.
4. Obtain the derivatives of the following functions from first principles. In each case specify the domain of the derived function precisely.
   (i) $f:\mathbb{R}\to\mathbb{R}$ defined by
   $f(x) = x^{1/2}$,
   (ii) $f:\mathbb{R}\backslash\{0\}\to\mathbb{R}$ defined by
   $f(x) = 1/x$,
   (iii) $f:\mathbb{R}\to\mathbb{R}$ defined by
   $f(x) = |x|^2$.
5. Show that if f and g are both differentiable at x then
   (i) $f+g$ is differentiable at x and
   $(f+g)'(x) = f'(x) + g'(x)$,
   (ii) fg is differentiable at x and
   $(fg)'(x) = (fg')(x) + (f'g)(x)$,
   (iii) if $g(x) \neq 0$, f/g is differentiable at x and
   $$(f/g)'(x) = \frac{(gf')(x) - (fg')(x)}{g^2(x)}.$$
6. Show that if $a_k \in \mathbb{R}$ for $0 \leqslant k \leqslant N$ and $p:\mathbb{R}\to\mathbb{R}$ is defined by
   $$p(x) = \sum_{k=0}^{N} a_k x^k$$
   then
   $$p'(x) = \sum_{k=1}^{N} a_k k x^{k-1}.$$
7. Determine the derivatives of the following functions.
   (i) $f:\mathbb{R}\to\mathbb{R}$ defined by
   $f(x) = x\sin(x)/(1+\cos(x))$

(ii) $g:\mathbb{R}\rightarrow\mathbb{R}$ defined by

$g(x)=\sin(x^2)+x\cos^2(x)$.

8. Evaluate the following integrals

(i) $\displaystyle\int_1^2 x^3\,dx$,

(ii) $\displaystyle\int_{-1}^1 x^{1/3}\,dx$,

(iii) $\displaystyle\int_0^{\pi/2} \cos(x)\,dx$.

9. Determine the indefinite integrals defined by

(i) $\displaystyle\int x\,e^{x^2}\,dx$,

(ii) $\displaystyle\int \left(\frac{x^3+2xe^x+1}{x}\right)dx$,

(iii) $\displaystyle\int \sin(x)\cos(x)\,dx$.

3.6 Operations

Some functions, such as integer addition, are so commonly used as to justify the introduction of a simpler notation which can be written more conveniently. This can then be used to describe the underlying concepts of earlier sections, thus enabling us to make proofs more concise and at the same time isolate exactly the properties that give rise to the deductions. After preliminary definitions we consider some consequences that follow when these functions possess certain simple properties, and this serves as an introduction to the more structured treatment to follow in Chapters 5 to 8.

Definition. An *operation over a set S* is a function $f:S^n\rightarrow S$ for some $n\in\mathbb{N}$. Implicit within this definition are two important points which deserve special mention. Firstly, since an operation is also a function, the result of performing an operation is *well-defined,* so given an ordered n-tuple of elements from S, f generates at most one element of S. Secondly, since the range of the operation lies within the set S on which the operation acts we say that it is *closed* on S.

An operation $S^n\rightarrow S$ is said to be of *order n.* We shall usually restrict consideration to situations where the order is one or two, in which case the operations are called *monadic* (or *unary*), and *dyadic* (or *binary*) respectively. The individual elements of the n-tuples in the domain of the operation are called *operands*. Operations are normally denoted by symbols called *operators*. For monadic operations it is usual to place the operator symbol before the operand. //

The most common example of this is the negation operation on $\mathbb{R}$. Here, assuming that the addition operation has already been defined,

$$-x$$

represents the operation

$$x \mapsto y \colon x + y = 0$$

(x is mapped to y where $x + y = 0$).

Definition. Binary operations can be denoted in three standard ways. These are called *infix* (where the operator is placed between the operands), *prefix* (the operator is placed before the operands) and *postfix* (the operator is placed after the operands). //

Example 3.6.1

$a + b$ infix,

$+ab$ prefix,

$ab+$ postfix. //

The transition from one form to another is not difficult but is best described in terms of directed graphs which are discussed in Section 7.6.

In line with most mathematical texts (except some on advanced algebra and formal logic) we shall use infix notation. The other notations do have the advantage of not requiring brackets to dictate the order of evaluation of complicated expressions and this makes them eminently suitable for automatic manipulation. The reader might like to examine the correspondence between the following pairs of expressions written in infix and postfix forms respectively.

(1) $a + b * c + (d + e * (f + g))$,

$abc * + defg + * + +$,

(2) $(a + b) * c + d + e * f + g$,

$ab + c * d + ef * + g +$,

(3) $a + (b * (c + d) + e) * f + g$,

$abcd + * e + f * + g +$.

Example 3.6.2

Consider further the algebraic expression

$a + b * c + (d + e * (f + g))$

and the representation given in Figure 3.11 which is called a tree structure.

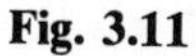
Fig. 3.11

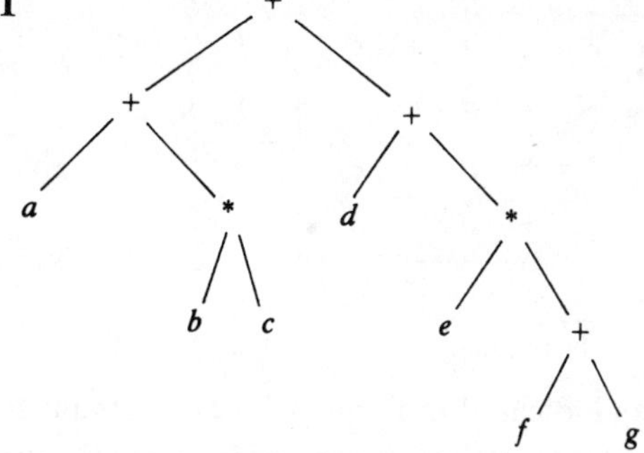

We know, because of the properties possessed by the arithmetic operators, that we can evaluate this in many ways. However, working from left to right and from the bottom we get

$$\alpha \leftarrow b * c,$$

$$\beta \leftarrow a + \alpha,$$

$$\gamma \leftarrow f + g,$$

$$\delta \leftarrow e * \gamma,$$

$$\pi \leftarrow d + \delta,$$

$$\rho \leftarrow \beta + \pi,$$

where Greek symbols denote intermediate results and ρ is the result. //

The evaluation of this expression, from the tree, was quite straightforward; however, it would have been a different story if we had to work directly on the original expression. Indeed the usual (infix) expression as given in the example is irregular in that some subexpressions are enclosed in brackets while others are not, a peculiar state of affairs if one ignores the knowledge, which we have but the tree does not, about the various symbols in the tree. Obviously the prefix and postfix versions of this expression are more uniform in their behaviour. Evaluation of the postfix version goes as follows:

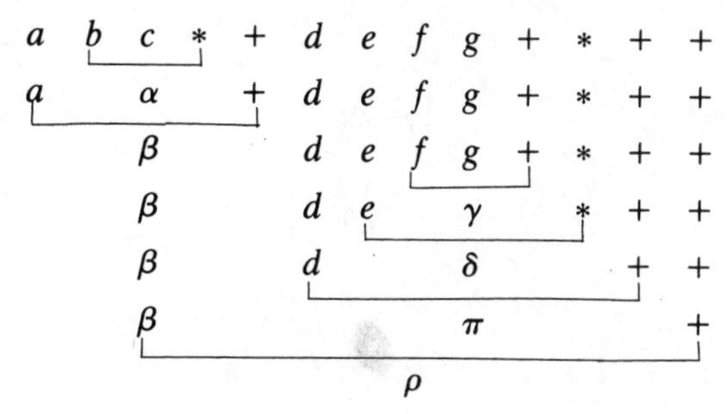

Similarly, the prefix version:

```
+  +  a  *  b  c   +  d  *  e  +  f  g
         └‾‾‾‾‾┘
+  +  a     α      +  d  *  e  +  f  g
   └‾‾‾‾‾‾‾‾┘
+        β         +  d  *  e  +  f  g
                               └‾‾‾‾┘
+        β         +  d  *  e     γ
                         └‾‾‾‾‾‾‾‾┘
+        β         +  d     δ
                   └‾‾‾‾‾‾‾‾┘
+        β             π
└‾‾‾‾‾‾‾‾‾‾‾‾‾‾‾‾‾‾‾‾‾‾┘
            ρ
```

'Traversing' the tree as indicated in Figures 3.12 we obtain from (a) the prefix version, from (b) the postfix version and from (c) the bracketed

Fig. 3.12

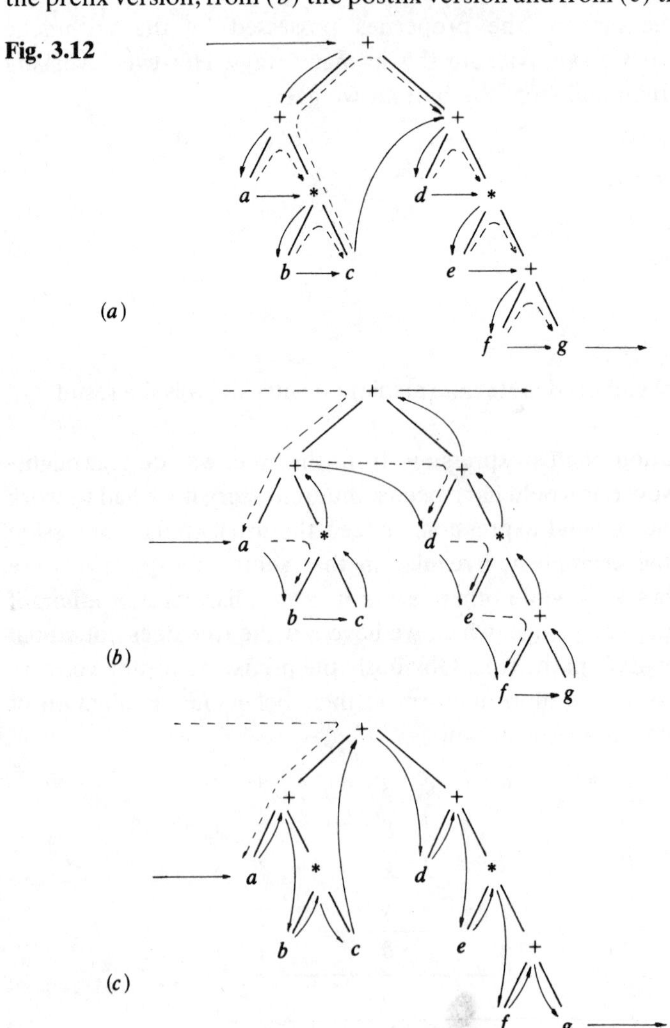

infix version

$$((a+(b*c))+(d+(e*(f+g)))).$$

We shall return to these matters later.

Of course we are already familiar with many binary operations, for example, the common arithmetic operations denoted by +, $*$, − and /, and the set operations of union ($\cup$) and intersection ($\cap$).

Operations defined by their effect on finite sets can often best be specified by a table of the type traditionally used when learning multiplication.

Example 3.6.3

An operation, denoted by $\circledast$, may be defined on the set $\{a, b, c\}$ by the table in Figure 3.13

Fig. 3.13

$\circledast$	a	b	c
a	a	a	b
b	b	a	c
c	a	b	b

Consequently,

$$a \circledast b = a,$$
$$b \circledast b = a,$$
$$c \circledast b = b. \ldots \quad //$$

(Symbols such as $\oplus$ and $\circledast$ will be used to denote different operations as required throughout the text.) As some of the operations encountered in the mathematics of computing do not lend themselves to wordy definitions the use of tables is of considerable importance.

We now turn our attention to properties of operations which, together with their consequences, provide the framework for all algebraic aspects of mathematics in that they specify how objects can be manipulated.

Definition. The (binary) operation $\circledast$ on a set A is said to be *commutative* if

$$a \circledast b = b \circledast a \quad \text{for all } a, b \in A. \quad //$$

Consequently, the usual addition operation on $\mathbb{Z}$ is commutative but subtraction is not.

Definition. The operation $\circledast$ on A is said to be *associative* if

$$(a \circledast b) \circledast c = a \circledast (b \circledast c) \quad \text{for all } a, b, c \in A. \quad //$$

Notice that in the definition of associativity the order of the operands a, b and c is preserved (the operation may not be commutative!) and the parentheses are included to indicate the order of evaluation.

Thus $(a \circledast b) \circledast c$ requires that $a \circledast b$ should be calculated first and the result, x say, should be composed with c to give $x \circledast c$. If an operation is associative then the order of calculation is irrelevant and hence the brackets are not required.

Example 3.6.4

Over $\mathbb{Z}$

$$(1+2)+3=1+2+3=1+(2+3),$$

but

$$(1-2)-3=-4$$

and

$$1-(2-3)=2,$$

so subtraction is not associative. //

Commutativity and associativity are the two major properties that can be attributed to single operations. Before describing a property that connects two operations we define some terms relating to special elements of the set on which an operation acts.

Definition. If $\circledast$ is a binary operation on a set A and $l \in A$ such that

$$l \circledast a = a \quad \text{for all } a \in A,$$

then l is called a *left identity* with respect to $\circledast$ (on A). Similarly, if there is an element r of A such that

$$a \circledast r = a \quad \text{for all } a \in A,$$

then r is a *right identity* wrt (with respect to) $\circledast$. Further, if there is an element, e say, which is both a left and a right identity, so

$$e \circledast a = a \circledast e = a \quad \text{for all } a \in A,$$

then e is a (two-sided) *identity* wrt $\circledast$. //

Example 3.6.5

Over $\mathbb{R}$, 0 is a right identity of the subtraction operation and a (two-sided) identity wrt addition, since

$$a-0=a$$

but

$$0-a \neq a \quad \text{unless } a=0,$$

while

$$a + 0 = a$$

and

$$0 + a = a \quad \text{for all } a. \quad //$$

Definition. If $\circledast$ is an operation on A with identity e and $x \circledast y = e$, then x is said to be a *left inverse* of y and y is a *right inverse* of x. Further, if x and y are such that

$$x \circledast y = e = y \circledast x$$

then y is an *inverse* of x wrt $\circledast$ and x is said to be *invertible*. //

Note. Some texts relate left (right) inverses to left (right) identities but, as we shall soon see, in the more common cases identities are two-sided and hence no distinction needs to be made.

The existence and uniqueness of identities and inverses are necessary in solving equations. A less common property, but one which is of importance in the algebra associated with logic, is idempotence.

Definition. Given an operation $\circledast$ over a set A, any $x \in A$ such that $x \circledast x = x$ is said to be *idempotent* wrt $\circledast$. //

Trivially, any subset is idempotent wrt the set operations of intersection and union. Finally, returning to operations:

Definition. Given a set A on which there are defined two operations $\circledast$ and $\oplus$ then if

$$a \circledast (b \oplus c) = (a \circledast b) \oplus (a \circledast c) \quad \text{for all } a, b, c \in A,$$

we say that $\circledast$ *distributes over* $\oplus$. //

If this seems strange, notice the similarity between the above identity and the usual arithmetic on $\mathbb{R}$ whence

$$3 * (1 + 2) = (3 * 1) + (3 * 2).$$

The reader may be surprised that only a few special properties were discussed in the previous section, and could well believe that little or nothing can be derived from the knowledge that a set and associated operations have certain of these properties. In fact most common algebra can be derived from a relatively small set of basic rules as will be seen in subsequent sections. For the moment we shall concern ourselves with demonstrating how some simple deductions can be made from our elementary assumptions. Most of the work is relegated to the exercises but we will illustrate with two examples.

Example 3.6.6

If $\circledast$ is an operation on the set A and there is an identity wrt $\circledast$, then the identity element is unique.

Proof. Assume that x and y are both (two-sided) identities wrt $\circledast$, so

$$x \circledast a = a \circledast x = a$$

and

$$y \circledast a = a \circledast y = a \quad \text{for all } a \in A.$$

Then

$$x = x \circledast y \quad \text{since } y \text{ is an identity}$$

and

$$x \circledast y = y \quad \text{since } x \text{ is an identity.}$$

Therefore $x = y$. //

Example 3.6.7

If $\circledast$ is an associative operation on the set A and e is the (two-sided) identity wrt $\circledast$ then, if $x \in A$ and x is invertible, the inverse of x wrt $\circledast$ is unique.

Proof. Assume that x' and x'' are inverses of x so

$$x \circledast x' = x' \circledast x = e$$

and

$$x \circledast x'' = x'' \circledast x = e.$$

Then

$$\begin{aligned} x' &= x' \circledast e && \text{property of } e \\ &= x' \circledast (x \circledast x'') && \text{property of } x'' \\ &= (x' \circledast x) \circledast x'' && \text{associativity} \\ &= e \circledast x'' && \text{property of } x' \\ &= x'' && \text{property of } e. \quad // \end{aligned}$$

Exercises 3.6

1. Consider the following 'definitions' of $\circledast$. Decide whether or not each is the definition of a valid binary operation and if so whether the operation is (a) commutative, and (b) associative.

Also, if applicable, find (c) the two-sided identity, and (d) the two-sided inverse of a typical element x.

Assume the standard properties of real arithmetic.

(i) $x \circledast y = x - y$, on $\mathbb{N}$,

(ii) $x \circledast y = (x * y) - 1$, on $\mathbb{Z}$,

(iii) $x \circledast y =$ the largest of x and y, on $\mathbb{N}$,

(iv) $x \circledast y = +\sqrt{(x^2 + y^2)}$, on $\{x: 0 \leqslant x, x \in \mathbb{R}\}$,

(v) $x \circledast y = x/y$, on $\{x: 0 < x, x \in \mathbb{R}\}$.

2. The operation ϕ is defined on the set $\{a, b, c\}$ in Figure 3.14. Verify that ϕ is associative and commutative and find the identity element.

Fig. 3.14

ϕ	a	b	c
a	b	c	a
c	a	b	c
b	c	a	b

3. Assuming the usual properties of $+$, $-$, $*$ and $/$ on $\mathbb{R}$ show that the operation ψ defined over $[1, \infty[$ by

$$a \psi b = \frac{(a * b) + 1}{a + b}$$

is associative. Give precise reasons for each step in your answer. Hint: do not think too much about the domain.

4. If $\circledast$ is an associative operation on the set A, with identity e and such that each $a \in A$ is invertible and the inverse is denoted by a' show that

$(a \circledast b)' = b' \circledast a'$.

5. Show that if $\circledast$ is an associative operation on the set A, with identity e and such that $a \circledast a = e$ for every $a \in A$, then $\circledast$ is commutative.

6. Let $\circledast$ be an associative operation on the set A such that for any $a, b \in A$ if $a \circledast b = b \circledast a$ then $a = b$. Show that every element of A is idempotent wrt $\circledast$. What can you say about A if the operation has an identity?

4 BASIC CONCEPTS OF ARITHMETIC

Having properly defined operations and described some of their classifying properties we are now in a position to consider what can usefully be done with a collection of operations acting on a set.

A set together with associated operations is called an *algebraic structure.* Some of the more common algebraic structures will be examined later but before proceeding with this we take an informal look at arithmetic. For the most part, formal definitions will be omitted, the emphasis being on 'following the rules' (which we invent) even when this leads to strange ways of combining familiar symbols such as those normally used to represent decimal digits.

4.1 'Small' finite arithmetics

Arithmetic, which we may regard as a set with two operations that act like addition and multiplication, can be studied in many ways. In order to clarify the requirements of an arithmetic system we shall adopt a constructive approach and regard the integers $(0, 1, 2, \ldots)$ merely as symbols. Furthermore, we shall concern ourselves only with finite arithmetics where the set of 'numbers' involved will always be finite, and initially this set will be small. By this we mean that if the set $A \sim \mathbb{N}_m$ then we demand m separate symbols, no combinations are allowed. If we use only denary numbers this means that $m < 10$; and, since all sets of a given size are bijective we need consider only the sets $\mathbb{N}_m$.

To simplify matters still further we, arbitrarily, choose to base our discussion on the set $\mathbb{N}_6$. What needs to be done is to construct an addition table and a multiplication table for the arithmetic on $\mathbb{N}_6$. Choice of a specific set makes the discussion explicit. Moreover, this set is big enough to identify the underlying structure; this would not be true of $\mathbb{N}_2$ which may be thought by some to be more relevant. We begin with addition.

The addition operation requires an identity, usually written 0. However, $0 \notin \mathbb{N}_6$ so we use $\mathbb{Z}_6 = \{0, 1, 2, 3, 4, 5\}$. Obviously, $\mathbb{Z}_6 \sim \mathbb{N}_6$ so we can work with $\mathbb{Z}_6$ without any great loss. Thus so far we have the table in Figure 4.1.

Fig. 4.1

+	0	1	2	3	4	5
0	0	1	2	3	4	5
1	1					
2	2					
3	3					
4	4					
5	5					

For the operation to be commutative it must have a symmetric table. Associativity is more difficult to deal with but if the operation is to be associative and, as is usual, we require additive inverses, then entries in each row and column must be unique. This claim is substantiated by the following argument.

If

$$a+b=a+c,$$

then

$$-a+(a+b)=-a+(a+c),$$
$$(-a+a)+b=(-a+a)+c,$$
$$0+b=0+c,$$
$$b=c.$$

Consider now the operations defined in Figure 4.2. Of these three candidates for an addition operation on $\mathbb{Z}_6$ only C satisfies all the conditions, and even this looks rather strange. The operation A is not commutative and B violates the 'unique result' criterion.

Having discussed the properties that are desirable how do we construct a suitable operation? As will already be apparent the most difficult property to validate is associativity. In the procedure that follows we use associativity as the major construction step and hence this property holds automatically as do the others mentioned above.

Fig. 4.2

A	0	1	2	3	4	5
0	0	1	2	3	4	5
1	1	2	0	4	5	3
2	2	0	1	5	3	4
3	3	5	4	0	2	1
4	4	3	5	1	0	2
5	5	4	3	2	1	0

B	0	1	2	3	4	5
0	0	1	2	3	4	5
1	1	1	2	3	4	5
2	2	2	2	3	4	5
3	3	3	3	3	4	5
4	4	4	4	4	4	5
5	5	5	5	5	5	5

C	0	1	2	3	4	5
0	0	1	2	3	4	5
1	1	5	3	4	2	0
2	2	3	1	5	0	4
3	3	4	5	0	1	2
4	4	2	0	1	5	3
5	5	0	4	2	3	1

Step 1. The number 0 is the additive identity so we have the situation given in Figure 4.3.

Fig. 4.3

+	0	1	2	3	4	5
0	0	1	2	3	4	5
1	1					
2	2					
3	3					
4	4					
5	5					

Step 2. 'Invent' another row of the table consistent with the 'unique result' criterion. (In order to stress the technique used we purposely choose results which are different from those that the reader would normally expect.) Hence we take

+	0	1	2	3	4	5
1	1	3	0	5	2	4.

To satisfy the commutativity requirement we also fill the corresponding column, hence we obtain Figure 4.4.

Fig. 4.4

+	0	1	2	3	4	5
0	0	1	2	3	4	5
1	1	3	0	5	2	4
2	2	0				
3	3	5				
4	4	2				
5	5	4				

Step 3. Fill other entries in the table using associativity. We follow this part of the construction in minute detail.

$$\begin{aligned} 2+2 &= 2+(1+4) \\ &= (2+1)+4 \\ &= 0+4 \\ &= 4. \end{aligned}$$

$$\begin{aligned} 2+3 &= 2+(1+1) \\ &= (2+1)+1 \\ &= 0+1 \\ &= 1. \end{aligned}$$

$$\begin{aligned} 2+4 &= (2+1)+5 \\ &= 5 \end{aligned}$$

$$\begin{aligned} 2+5 &= (2+1)+3 \\ &= 3. \end{aligned}$$

Here we are using the facts that $2+1=0$ and $0+x=x$.

Again,

$$\begin{aligned}3+3&=(1+1)+3\\&=1+(1+3)\\&=1+5\\&=4, \quad \text{etc.}\end{aligned}$$

Hence, from the assumed values of $1+x$, we have the operation in Figure 4.5.

Fig. 4.5

+	0	1	2	3	4	5
0	0	1	2	3	4	5
1	1	3	0	5	2	4
2	2	0	4	1	5	3
3	3	5	1	4	0	2
4	4	2	5	0	3	1
5	5	4	3	2	1	0

Having worked through the complete process we must note an extra constraint on Step 2. The values in the non-zero row must be chosen so as to 'generate' the whole of $\mathbb{Z}_6$. For example starting from 1, as we did,

$$\begin{aligned}1+1&=3,\\3+1&=5,\\5+1&=4,\\4+1&=2,\\2+1&=0,\\0+1&=1.\end{aligned}$$

Therefore by adding enough 1s we can get all of $\mathbb{Z}_6$.

Now we turn to multiplication. First note that the multiplicative identity must be different from zero; otherwise, for any x and y we have

$$\begin{aligned}x&=0*x=x*0,\\y&=0+y=y+0\end{aligned}$$

and so

$$\begin{aligned}x*y&=x*(0+y)\\&=(x*0)+(x*y)\\&=x+(x*y),\end{aligned}$$

which implies $x=0$. So 0 will not do. In fact, we need a number that will additively generate $\mathbb{Z}_6$. We *could* therefore proceed in a similar way from the partial table given in Figure 4.6.

Fig. 4.6

*	0	1	2	3	4	5
0		0				
1	0	1	2	3	4	5
2		2				
3		3				
4		4				
5		5				

However, in this case we need not insist on the 'unique result' criterion. (In normal arithmetic there is no non-zero integer which multiplied by two gives the answer 1! So in a finite set there *may* be repeats.) Instead of repeating the construction procedure for multiplication we turn to the problem of linking the two operations so that the multiplication distributes over addition. This is tackled in a similar fashion to associativity. Take the addition operation already derived and notice that $1+1=3$. So, assuming the desired distributivity, we obtain

$$\begin{aligned}3*0&=(1+1)*0\\&=(1*0)+(1*0)=0+0=0,\\3*2&=(1*2)+(1*2)=2+2=4,\\3*3&=(1*3)+(1*3)=3+3=4,\\3*4&=(1*4)+(1*4)=4+4=3,\\3*5&=(1*5)+(1*5)=5+5=0.\end{aligned}$$

Now $3+3=4$, $3+1=5$, $1+4=2$ and $1+2=0$ hence, proceeding as above, we obtain the operation set out in Figure 4.7.

Fig. 4.7

*	0	1	2	3	4	5
0	0	0	0	0	0	0
1	0	1	2	3	4	5
2	0	2	1	4	3	5
3	0	3	4	4	3	0
4	0	4	3	3	4	0
5	0	5	5	0	0	5

Therefore, starting from an almost arbitrary choice for entries in the row of a non-identity element in the addition table and obeying a handful of simple restrictions, we arrive at a respectable arithmetic system. Now all that is required to confirm our faith in such a system is to repeat the exercise with initial choices (and hopefully the consequences) that do not clash with our preconceived ideas of what $1+1$ actually is. In short if, in normal (infinite) arithmetic, $a+b=c$ and $c\in\mathbb{Z}_6$ then we would like our arithmetic to give the answer c. Hence we are led to the choice

+	0	1	2	3	4	5
1	1	2	3	4	5	?.

By default 5+1 must be 0 since $6 \notin \mathbb{Z}_6$ and 0 is the only element of $\mathbb{Z}_6$ not in the row. Following through the construction from this choice gives Figure 4.8. (This is known as modulo 6 arithmetic; it works exactly as normal integer arithmetic except that all integers are replaced by the remainder obtained on dividing them by 6.)

Fig. 4.8

+	0	1	2	3	4	5
0	0	1	2	3	4	5
1	1	2	3	4	5	0
2	2	3	4	5	0	1
3	3	4	5	0	1	2
4	4	5	0	1	2	3
5	5	0	1	2	3	4

*	0	1	2	3	4	5
0	0	0	0	0	0	0
1	0	1	2	3	4	5
2	0	2	4	0	2	4
3	0	3	0	3	0	3
4	0	4	2	0	4	2
5	0	5	4	3	2	1

Exercises 4.1

1. By analogy with the 'natural' arithmetic derived for $\mathbb{Z}_6$ construct a similar arithmetic for $\mathbb{Z}_{16}$ using the hexadecimal characters $\{0, 1, 2, 3, 4, 5, 6, 7, 8, 9, \mathrm{A}, \mathrm{B}, \mathrm{C}, \mathrm{D}, \mathrm{E}, \mathrm{F}\}$.

2. Derive an arithmetic for $\mathbb{Z}_6$ which is compatible with

+	0	1	2	3	4	5
2	2	3	4	0	5	1.

3. By considering 1+3, or otherwise, show that the following partial table leads to conflicts.

+	0	1	2	3	4	5
0	0	1	2	3	4	5
1	1	0	3	4	5	2.

4.2 'Larger' finite arithmetics

We have just constructed an arithmetic for $\mathbb{Z}_6$. How can this system be extended to enable us to count past 5? What is required is a set of n-place numbers, n-tuples from $\mathbb{Z}_6$ with an arithmetic inherited from $\mathbb{Z}_6$. To illustrate, consider ordered triples from $\mathbb{Z}_6$, i.e. elements of $\mathbb{Z}_6 \times \mathbb{Z}_6 \times \mathbb{Z}_6$. If we order $\mathbb{Z}_6$ in the usual way so that $0<1<2<3<4<5$ and thus define an ordering on $\mathbb{Z}_6 \times \mathbb{Z}_6 \times \mathbb{Z}_6$ by the rule

$$(a, b, c)<(x, y, z)$$

if $a<x$ or $a=x$ and $b<y$ or $a=x$, $b=y$ and $c<z$,

then, in ascending order, the elements of $\mathbb{Z}_6^3$ ($=\mathbb{Z}_6\times\mathbb{Z}_6\times\mathbb{Z}_6$) are

$$\begin{array}{ll}
(0,0,0),(0,0,1),\ldots,(0,0,5), & \\
(0,1,0),\ldots, & (0,1,5),\\
\vdots & \vdots\\
(0,5,0),\ldots, & (0,5,5),\\
(1,0,0),\ldots, & (1,0,5),\\
\vdots & \vdots\\
(5,5,0),\ldots, & (5,5,5).
\end{array}$$

Altogether there are 6^3 ($=216$) different triples so we should be able to arrange matters such that we can count from 0 to 215 and perform arithmetic on $\mathbb{Z}_6^3$ provided we stay within these limits.

In the ordering given above,

$$\begin{array}{ll}
(0,0,5) \text{ is immediately before } & (0,1,0)\\
(0,1,5)\ldots & \ldots(0,2,0)\\
\vdots & \\
(0,5,5)\ldots & \ldots(1,0,0).
\end{array}$$

We should therefore like to be able to write

$$(0,0,5)+1=(0,1,0);$$

however, so far we do not have an interpretation of 1 in $\mathbb{Z}_6^3$ and although it may be obvious what is coming we must be careful not to use any undefined concepts. To facilitate description of the processes involved let us write $\mathbb{Z}_6^3$ as $A_2\times A_1\times A_0$ and consider the sum of (a_2, a_1, a_0) and (b_2, b_1, b_0). The componentwise addition gives $(a_2+b_2, a_1+b_1, a_0+b_0)$ where the additions are in $\mathbb{Z}_6$ and this would seem to be quite a good start.

Examples 4.2.1

$$(0,1,3)+(4,2,1)=(4,3,4)$$

which looks more convincing if it is written

$$\begin{array}{r}
0,1,3\\
+\underline{4,2,1}\\
=4,3,4.
\end{array}$$

But

$$\begin{array}{r}
1,2,5\\
+\underline{2,3,1}\\
=3,5,0
\end{array}
\quad\text{and}\quad
\begin{array}{r}
0,0,5\\
+\underline{0,0,1}\ (=1?)\\
=0,0,0\ (=0?).
\end{array}\quad //$$

The addition in A_0 is self-contained and is correct; however, in order that the sum of sufficiently large integers (such as $5+1$) should indicate a result strictly outside A_0 we need to invoke some effect in A_1 and also possibly in A_2. To show how this works two tables (Figure 4.9) are required.

Fig. 4.9

$+_s$	0	1	2	3	4	5
0	0	1	2	3	4	5
1	1	2	3	4	5	0
2	2	3	4	5	0	1
3	3	4	5	0	1	2
4	4	5	0	1	2	3
5	5	0	1	2	3	4

$+_c$	0	1	2	3	4	5
0	0	0	0	0	0	0
1	0	0	0	0	0	1
2	0	0	0	0	1	1
3	0	0	0	1	1	1
4	0	0	1	1	1	1
5	0	1	1	1	1	1

If we now take any two numbers a and b in $\mathbb{Z}_6$ their sum (in $\mathbb{Z}$) is

$$6*(a+_c b)+(a+_s b).$$

Example 4.2.2

4 plus 4 gives

$$6*(1)+(2)=8 \text{ in } \mathbb{Z}.$$

The $+_s$ table gives the 'single column' sum of two elements of $\mathbb{Z}_6$ while the $+_c$ table tells us when a 'carry' into another copy of $\mathbb{Z}_6$ is required, and contains only 0s and 1s. The values in $+_c$ are limited in this way because if

$$0 \leq x < n \quad \text{and} \quad 0 \leq y < n,$$

then

$$0 \leq x \leq x+y < x+n < n+n = 2n \quad (\text{and } n=6 \text{ in } \mathbb{Z}_6). \quad //$$

Indeed we can make an even better estimate since

$$0 \leq x \leq n-1 \quad \text{and} \quad 0 \leq y \leq n-1,$$

and hence

$$0 \leq x+y \leq 2n-2 < 2n-1.$$

Returning to the summation of (a_2, a_1, a_0) and (b_2, b_1, b_0), if we denote the answer by (d_2, d_1, d_0) then

$$d_0 = a_0 +_s b_0$$

$$x_0 = a_0 +_c b_0$$

$$d_1 = a_1 +_s b_1 +_s x_0$$

$$x_1 = \textbf{if } a_1 +_c b_1 = 1 \textbf{ then } 1$$
$$\qquad \textbf{else } (a_1 +_s b_1) +_c x_0$$

$$d_2 = a_2 +_s b_2 +_s x_1$$

$$x_2 = \textbf{if } a_2 +_c b_2 = 1 \textbf{ then } 1$$
$$\qquad \textbf{else } (a_2 +_s b_2) +_c x_1.$$

Since $0 \leq a_i + b_i < 2n-1$ and $x_i = 0$ or 1 the resultant carry from $a_i + b_i + x_{i-1}$ can never be greater than 1, hence the use of the rather awkward definitions of x_1 and x_2.

Notice that as a consequence of these definitions the numbers $(0, 0, 0)$ and $(0, 0, 1)$ in $\mathbb{Z}_6^3$ *do* act as 0 and 1 in the new arithmetic. Notice also that if $x_2 = 1$ then the result of the addition is too large for $\mathbb{Z}_6^3$ and the calculation is said to have *overflowed.* We shall say more about overflow in Section 4.3 but for the remainder of the current section it will usually be ignored.

Similarly, one can use the operations $*_p$ and $*_c$ (single column product and carry) given in Figure 4.10 to derive a multiplication on $\mathbb{Z}_6^3$ but we shall not pursue this.

Fig. 4.10

$*_p$	0	1	2	3	4	5
0	0	0	0	0	0	0
1	0	1	2	3	4	5
2	0	2	4	0	2	4
3	0	3	0	3	0	3
4	0	4	2	0	4	2
5	0	5	4	3	2	1

$*_c$	0	1	2	3	4	5
0	0	0	0	0	0	0
1	0	0	0	0	0	0
2	0	0	0	1	1	1
3	0	0	1	1	2	2
4	0	0	1	2	2	3
5	0	0	1	2	3	4

So far we have discussed only symbols that look like positive numbers. Of course, the symbols $0, 1, 2, \ldots$ have been manipulated in ways that we regard as 'natural' and hence we could interpret them as (non-negative) numbers. The arithmetic on $\mathbb{Z}_6^3$ had numbers in the range $0 = (0, 0, 0)$ to $215 = (5, 5, 5)$ which was derived from the range (0 to 5) in $\mathbb{Z}_6$. If now we take $\{-3, -2, -1, 0, 1, 2\}$ instead of $\mathbb{Z}_6$ we have a system that includes negative numbers, but behaves in a strange way.

It turns out that if we take two copies of $\mathbb{Z}_6$ and one copy of $\{-3, -2, -1, 0, 1, 2\}$, which we shall call $\mathbb{Z}_6^-$, so as to form

$$\mathbb{Z}_6^- \times \mathbb{Z}_6 \times \mathbb{Z}_6$$

then an arithmetic with the range -108 to 107 can easily be derived. In fact the arithmetic is the same except that values 3, 4 and 5 in A_2 are now interpreted as $-3, -2, -1$ respectively, and consequently $(-2, 4, 2)$, for example, is evaluated as

$$(-2 * 36) + (4 * 6) + 2 = -46 \text{ in } \mathbb{Z}.$$

The bijection between the two systems, defined by

$$3 \mapsto -3$$

$$4 \mapsto -2$$

$$5 \mapsto -1$$

$$x \mapsto x \quad \text{otherwise,}$$

can be applied at any time, provided that at the end of the calculation no 3s, 4s or 5s remain in A_2. No fixed convention will be adopted as to when the bijection or its inverse should be applied, the notation being chosen for convenience. At this stage the reader is strongly advised to ignore any apparent discrepancies related to overflow from A_2; this will be properly considered in a simpler system in subsequent sections, but notice that the new system 'rolls over' from positive to negative hence

$$\begin{aligned}(2,5,5)+(0,0,1) &= (3,0,0) \text{ in the old system}\\ &= (-3,0,0) \text{ in the new system.}\end{aligned}$$

(In $\mathbb{Z}$ this is $107+1=-108$!)

Calculations involving addition and subtraction in the new arithmetic are very simple and hinge on two identities. First,

$$(5,5,5)+(0,0,1)=(0,0,0)$$

($(5,5,5)$ being equivalent to $(-1,5,5)$), and second,

$$(a_2, a_1, a_0)+(b_2, b_1, b_0)=(5,5,5)$$

iff

$$a_2+b_2=5,\ a_1+b_1=5 \text{ and } a_0+b_0=5.$$

Thus, to compute the additive inverse of (a_2, a_1, a_0) we first find (b_2, b_1, b_0), this is called the *fives complement*, and then add $1=(0,0,1)$; this gives the *sixes complement*. The process is illustrated in the following example.

Example 4.2.3

The inverse of $(-3,4,1)$, or $(3,4,1)$.

From	3, 4, 1	
obtain	2, 1, 4	5s complement
	+ 1	
then	2, 1, 5	6s complement.
Check	3, 4, 1	
	+2, 1, 5	
	=0, 0, 0	

so $-(-3,4,1)=(2,1,5)$. //

Subtraction is thus reduced to the addition of a suitable (sixes) complement.

Example 4.2.4

$(1, 3, 4)-(2, 1, 5)$

Take	$2, 1, 5$	
obtain	$3, 4, 0$	5s complement
then	$3, 4, 1$	6s complement
add $(1, 3, 4)$	$1, 3, 4$	
	$5, 1, 5=(-1, 1, 5).$	

A quick check, in $\mathbb{Z}$, verifies the calculation to be $58-83=-25$. //

Of course, the reason for forming the so-called 5s complement and 6s complement is due to the fact that we are essentially working in $\mathbb{N}_6$ (or $\mathbb{Z}_6$); in general if we were computing in $\mathbb{N}_m$ we would use $m-1$ and m complements respectively.

Finally, it must be emphasized that in computing we very rarely, if ever, have dealings with the set $\mathbb{Z}$ but with $\mathbb{Z}_m^n$ for some fixed m and n. The range is always bounded and even though the bounds may be very large we must not forget that they exist. At the risk of doing just this it is usual to omit the commas and all leading zeros (except the 0th). Hence

$(1, 3, 4)$ would be written as 134,

$(0, 0, 6)$ as 6

and

$(0, 0, 0)$ as 0.

Exercises 4.2

1. Denote $\mathbb{Z}_m^- \times \mathbb{Z}_m^{n-1}$ by $\mathbb{Z}_m^{n-}$ and let

$$\mathbb{Z}_m^- = \left\{\frac{-m}{2}, \ldots, 0, \ldots, \frac{m}{2}-1\right\} \text{if } m \text{ is even}$$

and

$$\mathbb{Z}_m^- = \left\{-\frac{m-1}{2}, \ldots, 0, \ldots, \frac{m-1}{2}\right\} \text{ if } m \text{ is odd.}$$

Perform the calculations

(a) $10-7$,

(b) $17-23$ and

(c) $(-8)+(-21)$

in each of the 'natural' arithmetics

$\mathbb{Z}_7^{4-}, \mathbb{Z}_{10}^{3-}, \mathbb{Z}_5^{5-}$ and $\mathbb{Z}_{12}^{2-}$.

(Note: the calculations (a), (b) and (c) are given in $\mathbb{Z}$ and must be translated into the relevant finite notation before performing the required computation.)

4.3 Binary arithmetic

From the arithmetics over $\mathbb{Z}_m^n$ and $\mathbb{Z}_m^{n-}$ just constructed, it is a simple matter to extract the base 2 arithmetics. There are two so-called binary arithmetics; the first, the *sign and modulus* form, is defined on

$$\{-,+\}\times\mathbb{Z}_2^n$$

and is $\mathbb{Z}_2^n$ (as defined in the previous section) together with an appended sign to extend the range. The sign is then encoded in binary form with 0 for '+' and 1 for '−'. The second arithmetic, *twos complement arithmetic*, is $\mathbb{Z}_2^{n-}$ with $\{0,1\}$ used in all n positions. This is the kind of binary arithmetic used by most computers and hence we restrict discussion to $\mathbb{Z}_2^{n-}$. To make matters more specific we consider $\mathbb{Z}_2^{5-}$, the range of which is from

$$1\ 0\ 0\ 0\ 0\ (=-16)$$

to

$$0\ 1\ 1\ 1\ 1\ (=+15)$$

(so $32=2^5$ different numbers are represented). Moreover -1 is represented by 11111 and the calculation of twos complements is trivial. All we need do is 'flip' all the binary digits, popularly called *bits* (*bi*nary digi*ts*), to give the 1s complement, and then add 1 to give the twos complement.

Examples 4.3.1

```
−(0 1 0 1 1)     (=11)
  1 0 1 0 0      (1s complement)
+         1
 ----------
  1 0 1 0 1      (2s complement)
```

$(=-2^4+2^2+2^0=-11)$

```
−(1 0 1 1 0)     (=−10)
  0 1 0 0 1      (1s complement)
+         1
 ----------
  0 1 0 1 0      (2s complement)
```

$(=2^3+2^1=10)$ //

As should be evident, there are going to be problems caused by the range restrictions. We cannot avoid these but we need to know when an 'error' has occurred. The complement form makes the checking of such overflow conditions relatively easy requiring knowledge of only the highest valued bit. (In $\mathbb{Z}_2^{5-}$ this is the 2^4 bit.) This bit indicates the sign of the number represented and hence is called the *sign bit*. Before examining how the sign bit is involved, recall that adding 1 to the maximum positive number in $\mathbb{Z}_m^{n-}$ gives the maximum negative number (this is the largest negative number, the negative number furthest away from zero), in other words the numbers repeat in a cyclic fashion. In $\mathbb{Z}_2^{5-}$ we obtain the situation depicted in Figure 4.11.

Fig. 4.11 In $\mathbb{Z}_2^{5-}$

15 −16 0 15 −16 0 15 −16 0 15 −16

−49 −48 −32 −17 −16 0 15 16 32 47 48

In $\mathbb{Z}$

What can happen when we add two numbers x and y where

$$-a \leqslant x < a \quad \text{and} \quad -a \leqslant y < a$$

(in $\mathbb{Z}_2^{5-}$, $a = 16$)?

Trivially the sum is bounded thus:

$$-2a \leqslant x + y \leqslant 2a - 2 < 2a - 1.$$

This in itself is not particularly helpful so we consider three subcases.

(i) **if** $-a \leqslant x < 0$ and $-a \leqslant y < 0$
then $-2a \leqslant x + y < 0$,

(ii) **if** $0 \leqslant x < a$ and $0 \leqslant y < a$
then $0 \leqslant x + y \leqslant 2a - 2 < 2a - 1$,

(iii) **if** $-a \leqslant x < 0$ and $0 \leqslant y < a$
then $-a \leqslant x + y < a$.

First notice that the result in case (iii) is within the required range and thus is always correct. To appreciate how errors can arise in cases (i) and (ii) it is necessary to realize that if $z \in \mathbb{Z}$ and $-2a \leqslant z < -a$ then z is represented in the finite arithmetic by z' where $z' = z + 2a$ and $0 \leqslant z' < a$. Similarly, if $a \leqslant z < 2a - 1$ then z is represented by z'' in the finite arithmetic where $z'' = z - 2a$ and $-a \leqslant z'' < 0$. Hence the answer is wrong if in (i) it is positive, or if in (ii) it is negative.

To interpret these conclusions in terms of the properties of the sign bit consider the different possibilities when adding two numbers.

(a) Both negative

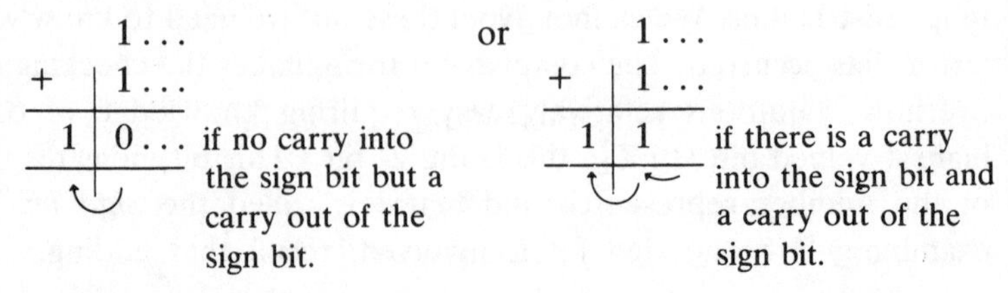

(*b*) Both positive

```
  0...          or      0...
 +0...                 +0...   if there is a carry
 -----                 -----   into sign bit.
  0... if no carry into  1
 -----  the sign bit.  -----
```

(*c*) Different signs

```
  0...          or        0...
 +1...              +  |  1...
 -----              -------------
  1... if no carry into  1|  0...  if there is a carry
 -----  sign bit.    -------------  into the sign bit and
                                    consequently also a
                                    carry out of the sign
                                    bit.
```

Rationalizing these cases, an *overflow* (overflow error) occurs iff there is either a carry into the sign bit or a carry out of the sign bit *but not both.* To illustrate we give some examples in $\mathbb{Z}_2^{5-}$; try to match these with the cases (i)–(iii) and (*a*)–(*c*) above.

Examples 4.3.2

In $\mathbb{Z}_2^{5-}$

```
     1 0 1 0 1  (−11)           1 1 1 0 0  (−4)
 + | 1 1 0 1 0  (−6)        + | 1 0 1 1 1  (−9)
 ---------------            ---------------
   1|0 1 1 1 1  (15)          1|1 0 0 1 1  (−13)

     1 1 1 0 1  (−3)            0 0 1 0 1  (5)
 + | 0 0 1 1 0  (6)         +   0 0 1 1 1  (7)
 ---------------            ---------------
   1|0 0 0 1 1  (3)             0 1 1 0 0  (12)

            0 1 1 0 0   (12)
        +   0 1 0 1 0   (10)
        -------------
            1 0 1 1 0   (−10). //
```

So much for addition and subtraction; now we turn our attention to multiplication and division. First multiplication: recall that in $\mathbb{Z}$ (or, more properly, $\mathbb{Z}_{10}^n$ for sufficiently large n) multiplication by 10 can be realized by 'moving the integer point' one place to the right; equivalently, and realistically in a computing context, shift all the numbers one place to the left and fill the zeroth position with 0. (In fact, multiplication by m in $\mathbb{Z}_m^n$ can always be achieved by a left shift.)

Therefore we have an easy way of multiplying by a non-negative power of 2 in $\mathbb{Z}_2^n$ simply by shifting one place to the left the required number of times.

Examples 4.3.3

In $\mathbb{Z}_2^{5-}$

0 0 0 1 1	(3)	1 1 1 1 0	(−2)
0 0 1 1 0	(∗2)	1 1 1 0 0	(∗2)
0 1 1 0 0	(∗2=12)	1 1 0 0 0	(∗2=−8)

0 0 1 0 1	(5)
0 1 0 1 0	(∗2)
1 0 1 0 0	(∗2)
0 1 0 0 0	(∗2=8). //

As demonstrated by these examples the method works equally well for negative numbers but will result in an error (overflow) if at any stage the sign bit is changed, even if it is subsequently changed back. To multiply by *any* natural number (an element of $\mathbb{N}$) we use the distributivity of multiplication over addition and express the multiplier as a sum of powers of 2.

Example 4.3.4

In $\mathbb{Z}_2^{5-}$

$3*5=3*2^2+3*2^0 \quad (2^0=1)$

so

	0 0 0 1 1	$(3*2^0)$
+	0 1 1 0 0	$(3*2^2)$
	0 1 1 1 1	(=15)

$(-5)*3=(-5)*2^1+(-5)*2^0$

so

	1 1 0 1 1	$(-5*2^0)$
+	1 0 1 1 0	$(-5*2^1)$
1	1 0 0 0 1	(=−15). //

In much the same way that multiplication is carried out by left shifts, division by positive powers of two is achieved by right shifts. (Division by other integers must be performed by what amounts to repeated subtraction of powers of two. We shall not discuss this process.) However, special care needs to be taken with negative numbers. Also, notice that in general the expected result (i.e. the arithmetically expected result in $\mathbb{R}$) will not be integer but fractional.

Examples 4.3.5

In $\mathbb{Z}_2^{5-}$ we attempt to calculate 12/4, (−6)/2 and 7/4 by right shifts of 2, 1 and 2 places respectively.

$$\begin{array}{lll} 0\ 1\ 1\ 0\ 0 & & (12) \\ 0\ 0\ 0\ 1\ 1 & |\ 0\ 0 & (3=12/4) \end{array} \qquad \begin{array}{lll} 1\ 1\ 0\ 1\ 0 & & (-6) \\ 0\ 1\ 1\ 0\ 1 & |\ 0 & (13\neq -6/2) \end{array}$$

$$\begin{array}{lll} 0\ 0\ 1\ 1\ 1 & & (7) \\ 0\ 0\ 0\ 0\ 1 & |\ 1\ 1 & (1\simeq 7/4). \end{array} \quad //$$

A single-place shift to the right automatically converts any negative number to a positive one. In $\mathbb{Z}_2^{5-}$ it converts −16 to +8. To correct this we need to subtract 16 to give −8 (i.e. (−16/2)) and this can be obtained by setting the sign bit to 1. Consequently, the correct result is achieved by propagating the sign bit, be it 0 or 1, so as to fill 'gaps' created as the right shifts take place.

Therefore (−6)/2 becomes

$$1\ 1\ 1\ 0\ 1 \quad (=-3).$$

The effect of bits (with value 1) 'falling off' the right of the number is to *truncate* the result; hence 7/4 is 1. It is common practice to *round* the number (upwards, regardless of sign) by adding back into the number the last bit to be lost. This is in line with usual arithmetic practice since a (notional) 1 in the first bit position past the integer point represents 0.5. Hence 7/4 becomes

$$\begin{array}{l|l} 0\ 0\ 0\ 0\ 1 & 1\ 1 \\ 1 & \\ \hline 0\ 0\ 0\ 1\ 0 & \end{array} \qquad (=2).$$

Exercises 4.3

1. A quick technique for computing the twos complement of a given bit pattern in $\mathbb{Z}_2^{n-}$ is as follows:

   Starting at the right end, copy all trailing 0s and the first 1 encountered. All remaining bits are then flipped.

   Ascertain that the technique works in most cases and specify exactly the cases where it fails.

2. A computation is performed in $\mathbb{Z}_2^{5-}$ thus:

   Two numbers x and y are added; call their sum z. The value y is subtracted from z giving c, and c is then subtracted from z giving d.

   What can be said about c and d? How would the results differ if the calculation were to be carried out in $\mathbb{Z}_2^{n-}$?

4.4 Logical arithmetic

Strictly speaking Boolean arithmetics operate on the sets $\mathbb{Z}_2$ and $\mathbb{Z}_2^n$ and hence involve only the numbers 0 and 1. Therefore, in order to stress the structure involved, we begin by considering a logical arithmetic on a larger set, namely $\mathbb{Z}_5$. This forms the basis of a multi-valued logic and can easily be 'shrunk' to give the more common $\mathbb{Z}_2$ case. Take the set $\mathbb{Z}_5=\{0, 1, 2, 3, 4\}$ and the operations $\vee$ and $\wedge$ defined by the tables

Fig. 4.12

$\vee$	0	1	2	3	4
0	0	1	2	3	4
1	1	1	2	3	4
2	2	2	2	3	4
3	3	3	3	3	4
4	4	4	4	4	4

$\wedge$	0	1	2	3	4
0	0	0	0	0	0
1	0	1	1	1	1
2	0	1	2	2	2
3	0	1	2	3	3
4	0	1	2	3	4

in Figure 4.12. With the usual ordering on $\mathbb{Z}_5$ (that which is induced by $\mathbb{Z}$ and $\mathbb{R}$) we see that

$$a \vee b = \text{the maximum of } a \text{ and } b$$

$$a \wedge b = \text{the minimum of } a \text{ and } b.$$

Both operations are commutative and associative, 0 is an identity for $\vee$ and 4 is an identity for $\wedge$, and $\wedge$ distributes over $\vee$ (but there are no inverses).

Distributivity can be shown, in the general case, as follows.

Example 4.4.1

Take any $\mathbb{Z}_m$ with $\wedge$ and $\vee$ defined in terms of the natural ordering and consider the six possible orderings for arbitrary elements of $\mathbb{Z}_m$ denoted by a, b and c, namely

(i) $a \leqslant b \leqslant c$,
(ii) $a \leqslant c \leqslant b$,
(iii) $b \leqslant a \leqslant c$,
(iv) $b \leqslant c \leqslant a$,
(v) $c \leqslant a \leqslant b$,
(vi) $c \leqslant b \leqslant a$.

The use of $\leqslant$ is intuitive but can be justified by defining

$$a \leqslant b \quad \text{iff } a \vee b = b.$$

It is required to show that

$$a \wedge (b \vee c) = (a \wedge b) \vee (a \wedge c).$$

This is done by examining both sides of the would-be identity for each of the orderings given above. We present the evaluations of corresponding expressions side-by-side.

(i) $a \wedge (b \vee c)$ $\quad$ $(a \wedge b) \vee (a \wedge c)$

$= a \wedge c$ $\quad$ $= a \vee a$

$= a,$ $\quad$ $= a,$

(ii) $a \wedge (b \vee c)$ $\quad$ $(a \wedge b) \vee (a \wedge c)$

$= a \wedge b$ $\quad$ $= a \vee a$

$= a,$ $\quad$ $= a,$

(iii) $a \wedge (b \vee c)$ $\quad$ $(a \wedge b) \vee (a \wedge c)$

$= a \wedge c$ $\quad$ $= b \vee a$

$= a,$ $\quad$ $= a,$

(iv) $a \wedge (b \vee c)$ $\quad$ $(a \wedge b) \vee (a \wedge c)$

$= a \wedge c$ $\quad$ $= b \vee c$

$= c,$ $\quad$ $= c,$

(v) $a \wedge (b \vee c)$ $\quad$ $(a \wedge b) \vee (a \wedge c)$

$= a \wedge b$ $\quad$ $= a \vee c$

$= a,$ $\quad$ $= a,$

(vi) $a \wedge (b \vee c)$ $\quad$ $(a \wedge b) \vee (a \wedge c)$

$= a \wedge b$ $\quad$ $= b \vee c$

$= b,$ $\quad$ $= b.$

Therefore $\wedge$ distributes over $\vee$. //

It can also be shown, and this is where we deviate from expected arithmetic behaviour, that $\vee$ distributes over $\wedge$, i.e.

$$a \vee (b \wedge c) = (a \vee b) \wedge (a \vee c).$$

Verification of this property is left as an exercise. Before we leave the general case, let us return to the tables defining $\vee$ and $\wedge$. The patterns created by equal values in the tables radiate from the identity elements as shown in Figure 4.13. In fact each operation is a 'reflection' of the other and an association defined, in $\mathbb{Z}_5$, by the pairs $(0, 4)$, $(1, 3)$, $(2, 2)$, $(3, 1)$ and $(4, 0)$ can be set up which allows such a system to be 'turned inside out'. In essence this is the principle of duality which we shall discuss in Chapter 5.

Fig. 4.13

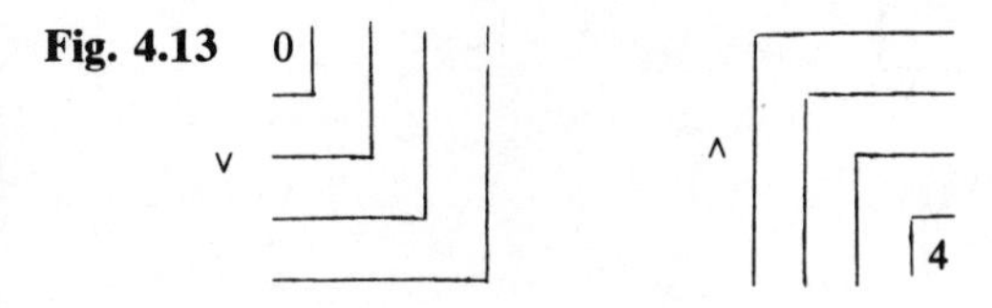

Turning to $\mathbb{Z}_2$ we have

$\vee$	0	1
0	0	1
1	1	1

$\wedge$	0	1
0	0	0
1	0	1

In $\mathbb{Z}_2$ the operation $\vee$ is usually interpreted as **or** (the result being 1 if either the first operand is 1 or the second operand is 1, inclusive of the case when both are 1). Similarly, $\wedge$ is read as **and**.

0 is the identity element wrt **or**, and

1 is the identity element wrt **and**.

Extending these processes to higher dimensions, passing from $\mathbb{Z}_2$ to $\mathbb{Z}_2^n$, is done componentwise (that is columnwise), there being no carry from one copy of $\mathbb{Z}_2$ to the next.

Example 4.4.2

$$\begin{array}{r} 0\ 1\ 1\ 1\ 0\ 1\ 0\ 1\ 0\ 0\ 1 \\ \wedge\ 0\ 0\ 1\ 1\ 1\ 1\ 0\ 0\ 1\ 0\ 1 \\ \hline 0\ 0\ 1\ 1\ 0\ 1\ 0\ 0\ 0\ 0\ 1. \end{array}$$ //

Exercise 4.4

1. From the minimum/maximum definition of $\wedge$ and $\vee$ for an arbitrary $\mathbb{Z}_n$, show that
   $a \vee (b \wedge c) = (a \vee b) \wedge (a \vee c)$.

5 ALGEBRAIC STRUCTURES

We have seen in the previous chapter some of the ways in which operations can be defined over a set and how we might wish them to interact in order to be able to perform meaningful computations. There are, of course, many different operations that can be defined on a set and consequently even more algebraic structures than sets. However, it so happens that most of the structures that people have found to be useful (by this we mean ones which describe naturally arising phenomena *and* lend themselves to computation) can be classified into a small number of types. In this chapter we first introduce terminology that is relevant to all algebraic structures and then deal with certain specific structures which are of particular relevance to computing. These will allow us to tie many loose ends which were previously 'argued away' in order that we should be able to begin our study, and also to give a firm algebraic foundation for the remainder of the book.

Central to the discussion presented here are fields, linear algebras and Boolean algebras; fields form the proper basis of simple arithmetic, linear algebra provides a framework for geometry and numerical computation and Boolean algebras embody the concepts of elementary logic. In order to obtain greater benefit from our consideration of fields we shall precede our study of this structure with a brief look at more fundamental structures which may be regarded as 'parts of a field'. Subsequently, fields are extended to vector spaces.

Similarly, we expand our treatment of Boolean algebras to include lattices and free-semirings. Several other structures will be briefly mentioned within the chapter exercises.

5.1. Algebraic structures and substructures

Definition. An *algebraic structure* is a set together with (closed) operations defined over the set. //

Typically, these operations have certain characteristic properties about which theorems can be proved and with which computations are performed. (A structure plus all associated theorems, calculations and deductions is sometimes called an *algebraic system*.)

Allied to each structure there is the notion of a substructure. To illustrate, consider a *hypothetical* structure called a *fing*, and let A be a fing. Suppose further that only one operation, $\circledast$, is defined on the fing A, hence we may more properly write $(A, \circledast)$ as the fing consisting of the set A and the operation $\circledast$. Now if $B \subseteq A$ and $(B, \circledast)$ is also a fing – in particular $\circledast$ must be closed on B – then $(B, \circledast)$ is a *subfing*. (Trivially sets are algebraic structures which have no extra, non-set, structure.)

Further, take another structure consisting of the set C and the operation $\oplus$. ($\oplus$ and $\circledast$ must have the same order. For example if one is binary then so must be the other. There may be other operations on C but these are currently of no concern to us.) Now if we can construct a mapping, $\phi : A \to C$ such that

$$\phi(x \circledast y) = \phi(x) \oplus \phi(y)$$

for any x and y in A, then ϕ is called a (*fing*) *homomorphism.*

If there is a homomorphism ϕ between A and C then we know that, in some sense, the (*homomorphic*) *image* $(\phi(A), \oplus)$ of $(A, \circledast)$ also behaves like a fing since we can either perform the operation $\circledast$ on A and then move across to C (via ϕ) or, map across to C first and then use the operation $\oplus$. In either case the result will be the same and so we can choose to do whichever we find the easier.

The situation can be characterized by the *commutative diagrams* in Figure 5.1. Diagram (*a*) indicates the sets or structures involved, and diagram (*b*) associates specific elements. The alternative forms in the bottom right entry in Figure 5.1(*b*) indicate the equality of the two different ways of performing the calculations. The commutativity of the diagram is derived from the functional compositions involved. Properly, in this case, we have $\phi \circ \circledast = \oplus \circ \phi$ which is not true commutativity since $\circledast$ and $\oplus$ are strictly different; however, they both indicate combination

Fig. 5.1 (*a*)

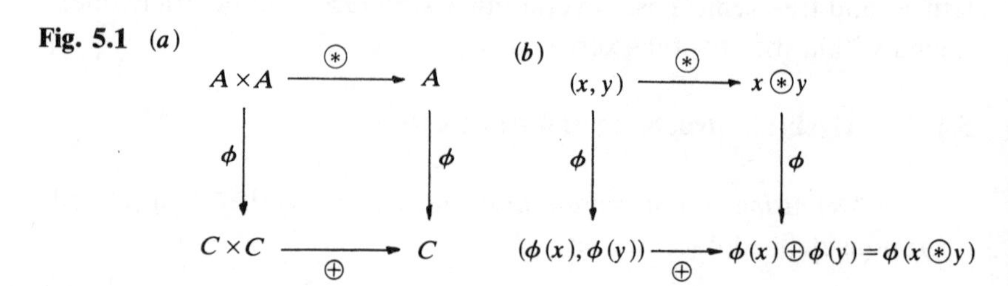

operations of the same order and therefore fit the general pattern of

$$convert \circ combine = combine \circ convert.$$

Now for an example.

Example 5.1.1

$\theta : \mathbb{Z} \to \mathbb{Z}_{10}$ defined by $\theta(n)$ is the remainder obtained by dividing n by 10. Thus

$$\theta(20) = 0,$$
$$\theta(17) = 7 \ldots .$$

If we consider the simple systems $(\mathbb{Z}, +)$ and $(\mathbb{Z}_{10}, +)$ with $+$ defined in the usual way on $\mathbb{Z}$ and defined on the 'units column' for $\mathbb{Z}_{10}$, then it is easy to see that θ is a homomorphism. For example

$$\theta(24+38) = \theta(62) = 2$$

and

$$\theta(24) + \theta(38) = 4 + 8 = 2 \text{ in } \mathbb{Z}_{10}.$$

In this case the underlying structure diagram is as shown in Figure 5.2. //

Fig. 5.2

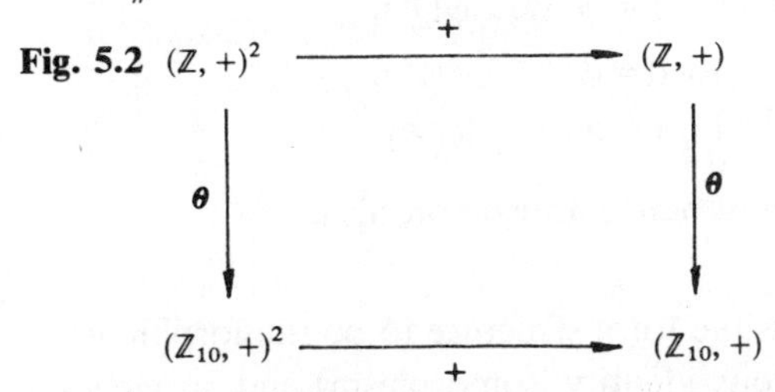

So, a homomorphism from one structure to another is a mapping between sets with extra conditions which preserve the structure. Just as mappings could be restricted by further demands on their range (to get, for example, surjections or injections) so can homomorphisms and, as would be hoped, this provides us with the ultimate machinery for stepping from structure to structure (and back!) without any loss of information.

Definitions. A homomorphism which is injective (one-to-one) is called a *monomorphism*, a homomorphism which is surjective (onto) is called an *epimorphism* and when the homomorphism is a bijection then it is called an *isomorphism*. If there exists an isomorphism between two structures then they are said to be *isomorphic*. //

The word isomorphic means 'same shape' and so it seems reasonable to expect that isomorphisms should be able to partition the set of all algebraic structures into equivalence classes (see Exercises 5.1, Question 2).

Example 5.1.2

The two structures

$$(\{\varnothing, \mathscr{E}\}, \cap, \cup)$$

and

$$(\{0, 1\}, \wedge, \vee) \quad \text{(as defined in Section 4.4)}$$

are isomorphic.

Proof. Let $\phi(\varnothing) = 0$ and $\phi(\mathscr{E}) = 1$. Clearly ϕ is a bijection. Also

$$\phi(\varnothing \cap \varnothing) = \phi(\varnothing) = 0 = 0 \wedge 0 = \phi(\varnothing) \wedge \phi(\varnothing),$$
$$\phi(\varnothing \cap \mathscr{E}) = \phi(\varnothing) = 0 = 0 \wedge 1 = \phi(\varnothing) \wedge \phi(\mathscr{E}),$$
$$\phi(\mathscr{E} \cap \varnothing) = \phi(\varnothing) = 0 = 1 \wedge 0 = \phi(\mathscr{E}) \wedge \phi(\varnothing),$$
$$\phi(\mathscr{E} \cap \mathscr{E}) = \phi(\mathscr{E}) = 1 = 1 \wedge 1 = \phi(\mathscr{E}) \wedge \phi(\mathscr{E})$$

and

$$\phi(\varnothing \cup \varnothing) = \phi(\varnothing) = 0 = 0 \vee 0 = \phi(\varnothing) \vee \phi(\varnothing),$$
$$\phi(\varnothing \cup \mathscr{E}) = \phi(\mathscr{E}) = 1 = 0 \vee 1 = \phi(\varnothing) \vee \phi(\mathscr{E}),$$
$$\phi(\mathscr{E} \cup \varnothing) = \phi(\mathscr{E}) = 1 = 1 \vee 0 = \phi(\mathscr{E}) \vee \phi(\varnothing),$$
$$\phi(\mathscr{E} \cup \mathscr{E}) = \phi(\mathscr{E}) = 1 = 1 \vee 1 = \phi(\mathscr{E}) \vee \phi(\mathscr{E}).$$

Thus ϕ is a homomorphism and hence an isomorphism. //

Finally, notice that it is possible for a structure to be isomorphic with itself (other than by the obvious identity isomorphism) and, in view of the strange phenomena associated with infinite sets, we may also have isomorphisms between a structure and one of its own substructures.

Definitions. In cases where the domain and range of the mapping are the same set, a homomorphism is called an *endomorphism* and an isomorphism is called an *automorphism.* //

Example 5.1.3

For a given set A, the structure $(\mathscr{P}(A), \cap, \cup)$ is isomorphic to $(\mathscr{P}(A), \cup, \cap)$ under the mapping $\phi: X \mapsto X'$.

Proof. Trivially ϕ is injective and surjective. If $B, C \in \mathscr{P}(A)$, then

$$\begin{aligned}\phi(B \cap C) &= (B \cap C)' \\ &= B' \cup C' \\ &= \phi(B) \cup \phi(C)\end{aligned}$$

and

$$\begin{aligned}\phi(B \cup C) &= (B \cup C)' \\ &= B' \cap C' \\ &= \phi(B) \cap \phi(C).\end{aligned}$$

We shall see later that this explicitly demonstrates the self-duality of the Boolean algebra of sets; ϕ is an automorphism. //

Exercises 5.1

1. Show that the two structures $(\mathbb{Z}_6, *)$ derived in the solution to Question 2 of Exercises 4.1 are isomorphic.
2. Let $(A, \circledast)$, $(B, \oplus)$ and $(C, \odot)$ be fings and $\phi : A \to B$ and $\theta : B \to C$ be fing isomorphisms. Show that
   $\theta \circ \phi : A \to C$
   and
   $\phi^{-1} : B \to A$
   are fing isomorphisms.

5.2 Single operation structures

We begin our detailed study of *actual* algebraic structures by considering those having only a single binary operation. Where possible, in this and subsequent sections, we shall present the structures in approximately ascending order of 'strength'. (Informally we regard structure A to be weaker than structure B if A can be viewed as B 'with some structure stripped away'. As will be seen later, some of the structures are obtained by 'bolting together' two weaker structures, and hence a strict ordering is not possible.) Each structure will usually be defined in terms of basic properties and not just in terms of more primitive structures.

Definition. A *semigroup* is a set S together with a (closed) binary operation $\circledast$ that satisfies the single requirement of being associative,

$$x \circledast (y \circledast z) = (x \circledast y) \circledast z \quad \text{for all } x, y, z \in S. \quad //$$

Definition. A *monoid* is a set M together with a binary operation $\circledast$ such that

(1) $\circledast$ is associative

$$x \circledast (y \circledast z) = (x \circledast y) \circledast z \quad \text{for all } x, y, z \in M,$$

(2) there exists $u \in M$ such that

$$u \circledast x = x = x \circledast u \quad \text{for all } x \in M;$$

u is called the *identity* wrt $\circledast$. //

Semigroups and monoids have special significance in string processing and language theory.

Example 5.2.1

Let $A = \{x, y, z\}$. Regarding x, y and z merely as symbols, not as names of objects or as 'variables', then A is an *alphabet* and we define A^* as the set of all strings of symbols taken from A. So A^* will include $x, y, z, xx, xy, yx, xxyz, zyx, \ldots$ and so on. A^* is infinite.

On A^* we can define the operation of *concatenation* $\odot$ such that if $\alpha, \beta \in A^*$ then $\alpha \odot \beta = \alpha\beta$; i.e. we write down α and follow it immediately by β. So we have

$$xyz \odot z = xyzz, \quad xz \odot yx = xzyx, \quad \text{etc.}$$

Now any string α has a finite integer length which is denoted by $|\alpha|$ and is defined to be the number of symbols in α (allowing repeats). Thus

$$|x| = 1, \quad |xy| = 2, \quad |xxxzy| = 5.$$

This is similar to, but distinct from, the notion of the size of a set. Notice in particular that

$$\begin{aligned} |x| &= 1, & |\{x\}| &= 1, \\ |xy| &= 2, & |\{x, y\}| &= 2, \\ |yx| &= 2, & |\{y, x\}| &= 2, \end{aligned}$$

but

$$|xyx| = 3, \quad |\{x, y, x\}| = |\{x, y\}| = 2.$$

So the similarity does not carry over in all aspects, though there *is* an analogue of the empty set. As a string this is invisible – it is everywhere but cannot be seen. Hence we use a special symbol for this string, Λ (capital lambda), and $\Lambda \in A^*$. So $|\Lambda| = 0$ and $\Lambda \odot \alpha = \alpha \odot \Lambda = \alpha$ for all strings α.

Therefore, for any given alphabet A, the structure $(A^*, \odot)$ is a monoid and Λ is the identity wrt $\odot$. //

(The apparent misuse of the $*$ notation in A^* is illusory; $A^* = R^*(\Lambda)$ where $R = \{(\alpha, \beta): \beta = \alpha \odot a \text{ and } a \in A\}$.)

The above example is extremely important. For, given a suitably large alphabet A, containing for instance all the symbols available on a VDU or other similar computer peripheral, all languages definable for use on the underlying computer system are specified as subsets of A^*. This concept forms the starting point of our formal study of languages in Chapter 8.

The third and final single operation structure is a straightforward and natural extension to the monoid.

Definition. A *group* is a set G together with a binary operation $\circledast$ such that

(1) $\circledast$ is associative,

(2) there exists $u \in G$, an identity wrt $\circledast$ such that

$$u \circledast x = x = x \circledast u \quad \text{for all } x \in G,$$

(3) to each element $x \in G$ there corresponds an element $y \in G$ such that

$$x \circledast y = u = y \circledast x;$$

y is called the *inverse* of x wrt $\circledast$. //

In cases where the group operation is written as $*$, u is written as 1 and the inverse of x as x^{-1}. When $+$ is used, u is written as 0 and the inverse of x as $-x$.

The most important factor that makes a group more useful than the first two structures is that within a group, $(G, \circledast)$, we can solve the equation

$$a \circledast x = b.$$

Moreover, the solution is very easy to extract (but notice that we need all the group axioms). If

$$a \circledast x = b,$$

then

$$a^{-1} \circledast (a \circledast x) = a^{-1} \circledast b \quad (a \in G \Rightarrow a^{-1} \in G),$$

$$(a^{-1} \circledast a) \circledast x = a^{-1} \circledast b \quad (\circledast \text{ associative}),$$

$$u \circledast x = a^{-1} \circledast b \quad (\text{property of inverses}),$$

so

$$x = a^{-1} \circledast b \quad (\text{property of identity}).$$

The terms group and monoid are often prefixed by the qualifier 'commutative'. This simply means that the operation in the structure concerned satisfies the extra property of commutativity, viz.

$$x \circledast y = y \circledast x \quad \text{for all } x, y \in M \text{ or } G.$$

Many other useful deductions can be made from the group axioms; we give a simple example.

Example 5.2.2

In the group $(G, *)$,

$(a * b)^{-1} = b^{-1} * a^{-1}$.

Proof.
$$\begin{aligned}(a*b)*(b^{-1}*a^{-1}) &= a*(b*b^{-1})*a^{-1} && \text{(associativity)},\\ &= a*1*a^{-1} && \text{(definition of 1)},\\ &= a*a^{-1} && \text{(property of 1)},\\ &= 1 && \text{(definition of 1)}.\end{aligned}$$

Hence $b^{-1}*a^{-1}$ is a right inverse of $a*b$. Similarly we can show it to be a left inverse and the result follows. //

Groups provide our first example of widely used isomorphisms. These are the isomorphisms commonly called *logarithms* between the groups $(\mathbb{R},+)$ and $(]0,\infty[,*)$ which enable multiplication to be performed by addition via the identity

$$a*b=\phi^{-1}(\phi(a)+\phi(b)),$$

where

$$\phi: x\mapsto \log_p(x) \quad \text{for some } p\in]1,\infty[.$$

Exercises 5.2

1. Prove the uniqueness of the identity element and of inverses in the group $(G, \circledast)$.
2. In the group $(G, \circledast)$ show that if
   $$a\circledast b=a\circledast c$$
   then
   $$b=c$$
   and if
   $$x\circledast a=y\circledast a,$$
   then
   $$x=y.$$
3. Verify that the set of permutations of a finite set forms a group under functional composition.

5.3 Rings and fields

The single operator structures of the previous section will probably have seemed unfamiliar to the reader. However we are now in a position to utilize the properties of groups to give a proper description of the arithmetic structures discussed in Chapter 4. Our main interest is in fields (in the current context, the most ideal structure arithmetically) and their classification in terms of size. However, we begin with a brief look at some structures that just ‘fall short’ of being fields, namely rings.

5.3.1 Rings

Many mathematical constructs that arise naturally in linear algebra (and especially matrices) are rings or include rings as substructures. Hence examples of rings will appear at frequent intervals throughout the remainder of this chapter and Chapter 6. However, we have already studied one collection of rings, albeit informally, and we shall refer back to these after the axiomatic formalities.

Definition. A *ring* is a set R together with two binary operations $\circledast$ and $\oplus$ such that

(1) $\circledast$ is associative,
(2) $\oplus$ is associative,
(3) $\oplus$ is commutative,
(4) $\oplus$ has an identity, written 0 and called *zero*,
(5) there are inverses wrt $\oplus$,
(6) $\circledast$ distributes over $\oplus$, so

$$x \circledast (y \oplus z) = (x \circledast y) \oplus (x \circledast z)$$

and

$$(x \oplus y) \circledast z = (x \circledast z) \oplus (y \circledast z) \quad \text{for all } x, y, z \in R. \quad //$$

Hence the modulo systems $(\mathbb{Z}_n, *, +)$ for all $n \in \mathbb{N}$ are rings.

Additionally a ring is said to be *commutative* if the multiplication ($\circledast$) is commutative and to be a *ring with unity* if there is a multiplicative identity. As usual this is denoted by 1. Trivially, in the ring $(R, \circledast, \oplus)$, for any a and $b \in R$ it can be shown that

$$0 \circledast a = a \circledast 0 = 0,$$

$$a \circledast (-b) = (-a) \circledast b$$
$$= -(a \circledast b),$$

and

$$(-a) \circledast (-b) = a \circledast b.$$

Here $(-a)$ is the inverse of a wrt $\oplus$, $a \oplus (-b)$ is written as $a - b$ and if $1 \in R$, then 1 is unique.

Example 5.3.1

$(\mathbb{Z}_n, *, +)$ is a commutative ring with unity for any $n \in \mathbb{N}$. //

In a general modulo system $(\mathbb{Z}_n, *, +)$ we cannot guarantee to be able to perform 'division' and essentially this is the difference between a field and a commutative ring with unity. To highlight this difference and justify the introduction of an intermediate structure, we return briefly to consideration of the ring $(\mathbb{Z}_6, *, +)$ which, as will be shown subsequently, is not a field.

In $\mathbb{Z}_6$ there are 15 ways in which a product of two elements can give zero, namely

$$(0,0),$$
$$(0,1),(0,2),(0,3),(0,4),(0,5),$$
$$(1,0),(2,0),(3,0),(4,0),(5,0),$$
$$(2,3),(3,4),$$
$$(3,2),(4,3).$$

Obviously, there is something not quite right for 'normal' arithmetic calculations. In factorizing expressions we implicitly use the fact that

$$a*b=0 \quad \text{iff } a=0 \text{ or } b=0.$$

In the ring $(R, \circledast, \oplus)$ non-zero elements x and y whose product $(x \circledast y)$ is zero are called *zero divisors.* Since R may not be commutative we may need to specify that x is a *left zero divisor* and y is a *right zero divisor.* It is easily shown (see Exercises 5.3) that $\mathbb{Z}_6$ has zero divisors because 6 is not prime and that $\mathbb{Z}_p$ has no zero divisors iff p is prime.

Now it happens that in the group $(G, \circledast)$, if

$$a \circledast b = a \circledast c,$$

then

$$b = c;$$

however, in a ring this is not generally the case.

Theorem. Cancellation can take place in a ring R iff R contains no zero divisors.

Proof. ($\Leftarrow$) Assume R has no zero divisors. Then, if $x \circledast y = x \circledast z$ and $x \neq 0$,

$$\begin{aligned}(x \circledast y)-(x \circledast z) &= (x \circledast y)-(x \circledast y) \quad \text{(assumption)}\\ &= 0\end{aligned}$$

and

$$\begin{aligned}(x \circledast y)-(x \circledast z) &= (x \circledast y) \oplus (x \circledast (-z))\\ &= x \circledast (y \oplus (-z)) \quad \text{(distributivity)}\\ &= x \circledast (y-z),\end{aligned}$$

so

$$x \circledast (y-z)=0,$$

but since there are no zero divisors and $x \neq 0$ it follows that $y-z=0$ and hence $y=z$. Similarly, if $y \circledast x = z \circledast x$, then $y=z$. So cancellation is possible.

(⇒) Suppose cancellation is possible and

$$x \circledast y = 0,$$

then

$$x \circledast y = x \circledast 0 \quad \text{(see Exercises 5.3)}$$

and by cancellation, if $x \neq 0$ then $y = 0$. Similarly, if $y \neq 0$ then

$$x \circledast y = 0 = 0 \circledast y$$

and by cancellation $x = 0$. Thus

$$x \circledast y = 0 \text{ implies } x = 0 \text{ or } y = 0. \quad //$$

Now for the last structure before we reach fields.

Definition. An *integral domain* is a commutative ring with unity having no zero divisors; i.e., a set D with two binary operations $\circledast$ and $\oplus$ such that

(1) addition $\oplus$ is associative,
(2) addition is commutative,
(3) there is an additive identity 0,
(4) there are additive inverses, written $(-x)$,
(5) multiplication $\circledast$ is associative,
(6) multiplication is commutative,
(7) there is a multiplicative identity, written 1,
(8) multiplication distributes over addition:

$$x \circledast (y \oplus z) = (x \circledast y) \oplus (x \circledast z) \quad \text{for all } x, y, z \in D$$

(9) cancellation holds:

$$\text{if } x \neq 0 \quad \text{and} \quad x \circledast y = x \circledast z \quad \text{then} \quad y = z. \quad //$$

We are *almost* there. Every finite integral domain is in fact a field but there exist infinite examples which are not fields.

5.3.2 Fields

Having already *played* with the concepts of arithmetic we jump straight into the axiomatic definition of a field.

Definition. A *field* is a set F, together with two binary operations over F (called multiplication, $\circledast$, and addition, $\oplus$) usually written $(F, \circledast, \oplus)$ or simply F, such that the following nine properties hold.

(1) Addition is commutative:

$$x \oplus y = y \oplus x \quad \text{for all } x, y \in F.$$

(2) Addition is associative:

$$x \oplus (y \oplus z) = (x \oplus y) \oplus z \quad \text{for all } x, y, z \in F.$$

(3) There is an element in F, usually written 0, such that

$$x \oplus 0 = x \quad \text{for all } x \in F;$$

0 is the *additive identity*, called zero.

(4) To each $x \in F$ there corresponds an element $y \in F$ such that

$$x \oplus y = 0;$$

y, usually written $-x$, is the *additive inverse* of x.

(5) Multiplication is commutative:

$$x \circledast y = y \circledast x \quad \text{for all } x, y \in F.$$

(6) Multiplication is associative:

$$x \circledast (y \circledast z) = (x \circledast y) \circledast z \quad \text{for all } x, y, z \in F.$$

(7) There is an element of F, usually written 1, such that $1 \neq 0$ and

$$x \circledast 1 = x \quad \text{for all } x \in F;$$

1 is called the *multiplicative identity* or unity.

(8) For each $x \in F\backslash\{0\}$, there corresponds an element $y \in F$ such that

$$x \circledast y = 1;$$

y is the *multiplicative inverse* of x, usually written x^{-1}.

(9) Multiplication distributes over addition:

$$x \circledast (y \oplus z) = (x \circledast y) \oplus (x \circledast z) \quad \text{for all } x, y, z \in F. \quad /\!/$$

Example 5.3.2

$(\mathbb{R}, *, +)$ is a field and hence $(\mathbb{R}, +)$ and $(\mathbb{R}\backslash\{0\}, *)$ are commutative groups. $(\mathbb{N}, *, +)$ is not since there are no additive identities nor inverses. $(\mathcal{P}(A), \cap, \cup)$ for a given set A is not a field since there are no inverses. //

In the preceding definition we have purposely used the symbols $\circledast$ and $\oplus$ to emphasize that the operations need not be the common multiplication and addition. However, as we shall often be concerned with the fields $(\mathbb{R}, *, +)$ and $(\mathbb{Q}, *, +)$, we shall henceforth use the symbols $*$ and $+$. The major facts which we shall require in subsequent sections will be stated in the form of a theorem but before giving this we recall (together with proofs) some important consequences of the restrictions imposed on the operations in a field.

Proposition. The identity elements in a field are unique.

Proof. Suppose (i) $x*e=x$ and
(ii) $x*e'=x$
for all $x\in F$.

Then

$$\begin{aligned} e &= e*e' && \text{by (ii)} \\ &= e'*e && \text{commutativity} \\ &= e' && \text{by (i).} \end{aligned}$$

So $e=e'=1$.

Similarly for the additive identity. //

Proposition. In a field, inverses are unique.

Proof. Again we consider the multiplicative case. Take $x\in F\backslash\{0\}$ and assume that we have $y, z\in F$ such that

$$x*y=1, \quad x*z=1.$$

By commutativity we also have

$$y*x=1=z*x.$$

So

$$\begin{aligned} y &= y*1 && \text{property of 1} \\ &= y*(x*z) && \text{assumption} \\ &= (y*x)*z && \text{associativity} \\ &= 1*z && \text{assumption} \\ &= z && \text{property of 1.} \end{aligned}$$

Hence $y=z=x^{-1}$.

The uniqueness of additive inverses is proved analogously. //

Now for the main results. Care must be taken in following through the arguments – there *is* something to prove!

Theorem. In the field $(F, *, +)$ for any a, $b\in F$ the following hold

(a) $a*0=0$,
(b) $(-a)=a*(-1)$ and $-(a+b)=(-a)+(-b)$,
(c) $-(-a)=a$ and $(-1)*(-1)=1$,
(d) if $a\neq 0$ then $(a^{-1})^{-1}=a$,
(e) $a*b=0\Rightarrow a=0$ or $b=0$,
(f) $(-a)*(-b)=a*b$.

Proof.

(a) $a*1=a$ and $a+0=a$

so

$$a+(a*0)=(a*1)+(a*0)$$
$$=a*(1+0) \quad \text{distributivity}$$
$$=a*1$$
$$=a.$$

Therefore $a*0$ is an additive identity, but this is unique so $a*0=0$.

(*b*) Similarly,

$$a+(a*(-1))=(a*1)+(a*(-1))$$
$$=a*(1+(-1))$$
$$=a*0$$
$$=0 \quad \text{by } (a).$$

Thus $-a=a*(-1)$. Using this identity,

$$-(a+b)=(a+b)*(-1)$$
$$=(a*(-1))+(b*(-1))$$
$$=(-a)+(-b).$$

(*c*) By definition,

$$(-a)+a=0 \quad \text{and} \quad (-a)+(-(-a))=0.$$

But inverses are unique so

$$a=-(-a),$$

thus

$$1=-(-1).$$

Now let x, say, be -1, then

$$1=-(x)$$
$$=x*(-1) \quad \text{by } (b)$$
$$=(-1)*(-1).$$

(*d*) First notice that $a^{-1}\neq 0$, otherwise

$$1=a*a^{-1}=a*0=0 \quad \text{by } (a)$$

which contradicts the field properties. Therefore $a^{-1}\neq 0$ and the argument proceeds as in (*c*).

(*e*) Take $a\neq 0$, then a^{-1} is defined and

$$b=1*b$$
$$=(a^{-1}*a)*b$$
$$=a^{-1}*(a*b) \quad \text{associativity}$$
$$=a^{-1}*0 \quad \text{assumption}$$
$$=0 \quad \text{by } (a).$$

(f) From (b)

$$(-a)=a*(-1)$$

and

$$(-b)=b*(-1).$$

So

$$\begin{aligned}(-a)*(-b)&=(a*(-1))*(b*(-1))\\ &=a*((-1)*(-1))*b \quad \text{commutativity and associativity}\\ &=a*1*b \quad \text{by } (c)\\ &=a*b. \quad /\!/\end{aligned}$$

In order to simplify the writing of expressions in fields we shall henceforth adopt the usual convention that, except where brackets dictate otherwise, multiplication shall take operational precedence over addition. Hence $a+b*c$ means $a+(b*c)$.

One consequence of the field axioms (the field rules) is so important as to demand special attention; this is the solvability of linear equations. Indeed it can be sensibly argued that this is the main property of fields.

A *linear equation* in x over the field F is of the form $a*x+b=0$ where 0, a, $b \in F$.

Theorem. Any such equation with $a \neq 0$ is uniquely soluble in F; i.e. there is exactly one element of the field which can be substituted for x so that the identity is true.

Proof.

$$\begin{aligned}&a*x+b=0,\\ &a*x+b+(-b)=0+(-b),\\ &a*x+(b+(-b))=(-b) && \text{(associativity and property of 0)},\\ &a*x+0=(-b) && \text{(definition of 0)},\\ &a*x=(-b) && \text{(property of 0)},\\ &a^{-1}*(a*x)=a^{-1}*(-b) && (a\neq 0),\\ &(a^{-1}*a)*x=(-b)*a^{-1} && \text{(associativity and commutativity)},\\ &1*x=(-b)*a^{-1} && \text{(definition of 1)},\\ &x=(-b)*a^{-1} && \text{(property of 1)}.\end{aligned}$$

By the closure properties of F, $(-b)*a^{-1}$ is in F and hence gives a solution of the equation. Moreover, since inverses in F are unique $-b$ and a^{-1} are uniquely derived from the equation and hence the solution is unique. //

Equations derived from polynomials of higher degree, such as quadratics, viz.

$$a*x*x+b*x+c=0,$$

with coefficients (a, b and c) in $\mathbb{R}$ are not generally soluble in $\mathbb{R}$. Indeed we must move into a larger field, the complex numbers, in order to guarantee that we can find solutions. However, polynomial equations whose coefficients are complex numbers are fully soluble in the field of complex numbers – no larger field is required. Investigation of this interesting fact, that the complex numbers are in some way a more algebraically complete system, would take us far away from more relevant topics.

5.3.3 Finite fields

So far all the fields mentioned have been infinite (the underlying sets having cardinaltiy $\aleph_0$ or $\aleph_1$). We now discuss the possibility of the existence of finite fields based on finite sets. The main facts about the size of such fields are stated and some are proved (the proof of others lies outside the scope of the present text). First we need to say a few more things about fields in general.

Definitions. If we take $a \in F$ then a, $a+a$, $a+a+a, \ldots$ are all members of the field and we can denote them by a, $2a$, $3a, \ldots, na, \ldots$. (This does not necessarily mean that $n \in F$.) Similarly, a, $a*a$, $a*a*a, \ldots$ are also in the field and we denote these by a, a^2, $a^3, \ldots, a^n, \ldots$. Now assume $a \neq 0$; *if* there is an integer $n \in \mathbb{N}$ such that $na=0$ (and there is no smaller integer $r \in \mathbb{N}$ such that $ra=0$) then n is called the *additive order* of a, and if there is $m \in \mathbb{N}$ such that $a^m=1$ (and there is no smaller integer $r \in \mathbb{N}$ such that $a^r=1$) then m is called the *multiplicative order* of a. //

Theorem. The additive order of all non-zero elements of a field F is the same.

Proof. Take $a, b \in F\backslash\{0\}$ and suppose the additive orders of a and b to be n and m respectively. Then

$$\begin{aligned} nb &= n(a*a^{-1})*b \\ &= (na)*(a^{-1}*b) \quad \text{distributivity} \\ &= 0*a^{-1}*b \\ &= 0. \end{aligned}$$

So $m \leqslant n$.

Similarly,

$$\begin{aligned} ma &= m(b*b^{-1})*a \\ &= (mb)*(b^{-1}*a) \\ &= 0*b^{-1}*a \\ &= 0. \end{aligned}$$

Therefore $n \leqslant m$ and so $m = n$. //

Definition. If in a field F all non-zero elements have additive order n then F is said to have *characteristic* n; if no such additive order exists then the field is said to have characteristic 0. //

If $|F| = m \in \mathbb{N}$, i.e. F has m elements, then F is said to be *finite*; if F has characteristic 0 then it must be infinite. (See Exercises 5.3.)

Theorem. The characteristic of any finite field is prime.

Proof. Suppose the finite field F has characteristic n and $n = p*q$ with $p, q < n$ and $p, q \in \mathbb{N}$. Take

$$a \in F \backslash \{0\},$$

then

$$\begin{aligned} 0 = na &= (p*q)a \\ &= p(qa). \end{aligned}$$

Now $qa \in F$, so if $qa = 0$ then since the order of a is n we must have $n \leqslant q$; otherwise $qa \in F\backslash\{0\}$ and the order of qa is also n so $n \leqslant p$. Both these possibilities violate the assumptions about p and q, hence no such p and q exist and thus n must be prime. //

Thus we obtain the crucial result.

Theorem. A finite field F has characteristic p (prime) and $|F| = p^n$ for some $n \in \mathbb{N}$.

Proof. We already know that F has characteristic p and p is prime. Let $|F| = q$. If $p = q$ then the result is trivially true; otherwise take $a_1 \in F\backslash\{0\}$ and let

$$\mathscr{F}_1 = \{y \colon y = na_1 \colon n \in \mathbb{N} \text{ and } 1 \leqslant n \leqslant p\}$$

$$|\mathscr{F}_1| = p.$$

Now take $a_2 \in F \backslash \mathscr{F}_1$ and let

$$\mathscr{F}_2 = \{y \colon y = na_1 + ma_2 \colon m, n \in \mathbb{N} \text{ and } 1 \leqslant n \leqslant p, 1 \leqslant m \leqslant p\}.$$

If $\mathscr{F}_2 = F$ then we are done, otherwise take $a_3 \in F \setminus \mathscr{F}_2$ and so on. Since F is finite the process eventually stops and we have $\mathscr{F}_1, \mathscr{F}_2, \ldots, \mathscr{F}_n$ for some $n \in \mathbb{N}$.

Now each $f \in F$ can be written uniquely in the form

$$m_1 a_1 + m_2 a_2 + \ldots + m_n a_n$$

where each m_i is such that $1 \leqslant m_i \leqslant p$. (Prove this as an exercise.) Consequently, there are p^n such expressions and thus $|F| = p^n$. //

So any finite field must have p^n elements for some $p, n \in \mathbb{N}$ where p is prime. In fact for any such p and n there *is* a field of order p^n but this is not particularly easy to show. Of course so far we do not know of any finite field therefore let us look at one, $(\mathbb{Z}_3, *, +)$ where $*$ and $+$ are defined by tables in Figure 5.3. Requirements 1 to 8 for a field are

Fig. 5.3

*	0	1	2
0	0	0	0
1	0	1	2
2	0	2	1

+	0	1	2
0	0	1	2
1	1	2	0
2	2	0	1

obviously true. Proof of distributivity is somewhat long-winded but nevertheless possible. In contrast examine the corresponding tables for $\mathbb{Z}_4$ (Figure 5.4). It is apparent that there is no multiplicative inverse of

Fig. 5.4

*	0	1	2	3
0	0	0	0	0
1	0	1	2	3
2	0	2	0	2
3	0	3	2	1

+	0	1	2	3
0	0	1	2	3
1	1	2	3	0
2	2	3	0	1
3	3	0	1	2

2 and hence $\mathbb{Z}_4$ with the obvious multiplication is not a field. Even though there *is* a field of order 4 ($=2^2$) it is not $\mathbb{Z}_4$ (see Exercises 5.3). Indeed the construction of fields of a given permissible order is dependent on the theory of polynomials over other structures (not fields) and can be very involved. Such fields are invaluable in coding theory but we regard this application as being too specialized for general study and hence we shall not elaborate.

5.3.4 Ordered fields

We have seen already that the set $\mathbb{R}$ together with the usual operations of addition and multiplication constitute a field. However, the structure of a field does not itself give rise to any of the ordering properties which we commonly associate with $\mathbb{R}$. Not all fields can be ordered so we must examine what extra conditions must be satisfied before the notion of ordering can be sensibly considered. As is to be expected there are many ways in which the usual order properties can be derived; we shall start by defining the concept of positivity. This leads directly into the definition of an order relation on a field and then to the notion of length. The main results of this section will be given in two theorems and subsidiary ones included as exercises.

Definitions. A field F is said to be *ordered* if it contains a non-empty subset P which is closed wrt addition and multiplication, and such that if $x \in F$ then exactly one of the following holds:

$$x \in P \backslash \{0\},$$

$$x = 0,$$

or

$$-x \in P \backslash \{0\}.$$

P is the set of all *positive elements* of F. (At this stage note that 0 can be either included or excluded from P. We choose to include 0 so as to conform with our earlier work.)

If $x \in P$ then we say that x is *positive* and write $x \geqslant 0$; if $-x \in P\backslash\{0\}$ then x is *negative* and we write $x < 0$. Similarly, if x and y are elements of F we say that x *is less than or equal to* y, written $x \leqslant y$, iff $y - x \in P$, and x *is less than* y, $x < y$, iff $y - x \in P\backslash\{0\}$. The symbols can also be written in the reverse direction with the obvious meanings. //

From these, possibly deceptively simple, definitions we can now justify that '$\leqslant$' is an order relation and that all the expected properties follow.

Theorem. If F is an ordered field and

$$a, b, c, d \in F,$$

then

(i) $a \leqslant b$ and $b \leqslant c$ imply that $a \leqslant c$,
(ii) $a \leqslant a$,
(iii) $a \leqslant b$ and $b \leqslant a$ imply that $a = b$,
(iv) if $a \neq 0$ then $a^2 > 0$,
(v) $1 > 0$,
(vi) if $a \leqslant b$ then $a + c \leqslant b + c$,
(vii) if $a \leqslant b$ and $c \leqslant d$ then $a + c \leqslant b + d$,
(viii) if $a \leqslant b$, $0 < c$ and $d < 0$, then $a * c \leqslant b * c$ and $b * d \leqslant a * d$,
(ix) if $0 < a$ then $0 < a^{-1}$ and if $b < 0$ then $b^{-1} < 0$.

Proof.

(i) $a \leqslant b$ and $b \leqslant c \Rightarrow (b-a) \in P$ and $(c-b) \in P$ by definition.

Now,

$$c-a=(c-b)+(b-a) \in P,$$

since P is closed under addition. Hence

$$a \leqslant c.$$

(ii) Trivially, $a-a=0 \in P$.

(iii) $a \leqslant b$ and $b \leqslant a$, so let $b-a=x$, then $x \in P$ and $-x \in P$ which contradicts the definition of P unless $x=0$. Hence $x=0$ and $a=b$.

(iv) If $a \neq 0$ then $a \in P$ or $-a \in P$. Since P is closed under multiplication $a \in P \Rightarrow a^2 \in P$ and, as shown in Section 5.3.2, $(-a)*(-a)=a^2$ so $-a \in P \Rightarrow a^2 \in P$.

(v) $1^2=1$ so by (iv) $1>0$.

(vi) This follows immediately from
$b-a=(b+c)-(a+c)$.

(vii) Using the closure properties of P the result follows from
$(b-a)+(d-c)=(b+d)-(a+c)$.

(viii) Similarly, consider
$(b-a)*c=b*c-a*c$ and $(b-a)*(-d)=a*d-b*d$.

(ix) $0<a \Rightarrow a \in P\backslash\{0\}$. If $a^{-1}=0$ then $1=a*a^{-1}=a*0=0$. ※
If $a^{-1} \notin P$, then by (viii) we have $1=a^{-1}*a<0*a=0$ ※ (contradicts part (v)). So we must have $a^{-1} \in P\backslash\{0\}$ and hence $0<a^{-1}$. The last part follows analogously. //

Having disposed with inequalities rather abruptly we shall now proceed to do the same with the concept of size within an ordered field.

Definition. If F is an ordered field, the *absolute value* (or *size*, or *length* or *modulus*) function is defined to be

$$x \mapsto \begin{cases} x \text{ if } x \geqslant 0 \\ -x \text{ if } x<0. \end{cases}$$

Traditionally this function is denoted by $|x|$ (where x is the argument) and read 'mod x'. //

The main results concerning $|x|$ are now stated but the proofs are left as an exercise.

Theorem. If F is an ordered field and $a, b \in F$ then:

(i) $|a|=0$ iff $a=0$,

(ii) $|-a|=|a|$,

(iii) $|a*b|=|a|*|b|$,

(iv) If $0 \leqslant b$ then $|a| \leqslant b$ iff $-b \leqslant a \leqslant b$,

(v) $-|a| \leqslant a \leqslant |a|$,

(vi) (The triangle inequality)

$$||a|-|b|| \leqslant |a \pm b| \leqslant |a|+|b|. \quad //$$

Exercises 5.3

1. Prove that in the ring $(R, *, +)$
   (i) $0*a=a*0=0$,
   (ii) $a*(-b)=(-a)*(b)=-(a*b)$ and
   (iii) $(-a)*(-b)=(a*b)$.
2. Show that in the ring $(R, *, +)$ if for each $a \in R$

   $a*a=a$,

   then R is commutative.
3. In the ring $\mathbb{Z}_n$ show that the zero divisors are exactly those elements which have (non-trivial) factors in common with n. Hence prove $\mathbb{Z}_p$ (p prime) has no zero divisors.
4. Show that every finite integral domain is a field.
5. Show that $(\mathbb{Z}, *, +)$ is an integral domain but not a field.
6. If $p<q$ and $(\mathbb{Z}_p, *_p, +_p)$ and $(\mathbb{Z}_q, *_q, +_q)$ are the usual modulo-p and modulo-q systems then show that although the systems are commutative rings and $\mathbb{Z}_p \subset \mathbb{Z}_q$, $(\mathbb{Z}_p, *_p, +_p)$ is not a subring of $(\mathbb{Z}_q, *_q, +_q)$. Explicitly, demonstrate that $*_q$ and $+_q$ are not closed on $\mathbb{Z}_p$.
7. Without reference to the theorems show that $(\mathbb{Z}_6, *, +)$ is not a field ($*$ and $+$ are modulo-6 operations).
8. Show that a finite field has a non-zero characteristic and that a field with characteristic zero is infinite.
9. If $a_1, \ldots, a_n$ are derived as in the proof of the last theorem of Section 5.3.3 prove that every expression of the form

   $m_1a_1+m_2a_2+\ldots+m_na_n$

   defines an element of the field and that this representation is unique.
10. In the field $(F, *, +)$ with operations defined below, solve the simultaneous linear equations

    $x+d*y=c$,

    $x*d+y=b$.

$*$	a	b	c	d
a	a	a	a	a
b	a	b	c	d
c	a	c	d	b
d	a	d	b	c

$+$	a	b	c	d
a	a	b	c	d
b	b	a	d	c
c	c	d	a	b
d	d	c	b	a

11. Show that if $a*b>0$ with $a,\ b\in F$ an ordered field then either $a>0$ and $b>0$, or $a<0$ and $b<0$.
12. Show that in an ordered field $a^2+b^2=0$ iff $a=b=0$.
13. If F is an ordered field and $a,\ b\in F$ such that $0\leqslant a\leqslant b$ show that $a^2\leqslant b^2$.
14. Prove that every field is an integral domain.
15. In an ordered field show, by adding inequalities of the forms

    $-|a|\leqslant a\leqslant|a|$

    and

    $-|a|\leqslant -a\leqslant|a|,$

    that (i)

    $|a\pm b|\leqslant|a|+|b|$

    and (ii)

    $||a|-|b||\leqslant|a\pm b|.$

5.4 Linear algebra

In many elementary texts, vectors are defined to be entities with the attributes of 'magnitude' and 'direction'. This stems from applications in geometry and physics and these aspects are discussed formally in Section 5.4.2. We begin by giving a more general definition of vectors for which neither of the concepts of magnitude or direction needs be meaningful.

5.4.1 Vector spaces and linear transformations

Definition. Let F be a field and V a set with binary operation $+$. Suppose that for each $a\in F$ and $\mathbf{x}\in V$ there is defined an element $a\mathbf{x}\in V$, then if

(i) $(V,+)$ is a commutative group

and for all $\mathbf{x},\ \mathbf{y}\in V$ and $a,\ b\in F$

(ii) $(a+b)\mathbf{x}=a\mathbf{x}+b\mathbf{x}$

$a(\mathbf{x}+\mathbf{y})=a\mathbf{x}+a\mathbf{y}$

$(ab)\mathbf{x}=a(b\mathbf{x})$

$1_F\mathbf{x}=\mathbf{x}$, where 1_F is multiplicative identity in F,

V is said to be a *vector space* over F. Elements of V are called *vectors*, $+$ is called *vector addition* and the mapping

$$\Lambda: F \times V \to V$$

defined by $\Lambda(a, \mathbf{x}) = a\mathbf{x}$ is known as *scalar multiplication of vectors*. //

A vector space over F may be regarded as a triple $(V, +, \Lambda)$ satisfying the axioms above. The additive zero of a vector space is written $\mathbf{0}$. It follows from the axioms (see Exercises 5.4) that

$$0_F \mathbf{x} = \mathbf{0} \quad \text{for all } \mathbf{x} \in V$$

where 0_F is the additive identity in F, and

$$a\mathbf{0} = \mathbf{0} \quad \text{for all } a \in F.$$

A diverse class of sets has a natural vector space structure as illustrated by the following examples.

Examples 5.4.1

(i) $F^n (n \in \mathbb{N})$ is a vector space over F under the operations

$$(a_1, \ldots, a_n) + (b_1, \ldots, b_n) = (a_1 + b_1, \ldots, a_n + b_n),$$

$$a(a_1, \ldots, a_n) = (aa_1, \ldots, aa_n).$$

The zero of F^n is $(0_F, \ldots, 0_F)$. The entries $a_1, \ldots, a_n$ of the vector $\mathbf{a} = (a_1, \ldots, a_n)$ are called the *components* of $\mathbf{a}$.

(ii) If $\mathscr{F}$ denotes the set of all mappings $f: [a, b] \to \mathbb{R}$ then $\mathscr{F}$ is a vector space over $\mathbb{R}$ under the operations

$$(f+g)(x) = f(x) + g(x) \quad \text{for all } f, g \in \mathscr{F}$$

$$(af)(x) = af(x) \quad \text{for all } a \in \mathbb{R}.$$

(iii) If $\mathscr{C} \subset \mathscr{F}$ is the subset of continuous elements of $\mathscr{F}$ then $\mathscr{C}$ is a vector space under the operations defined in $\mathscr{F}$. //

A set $U \subseteq V$ is a vector *subspace* of V if it is a vector space in its own right under the operations it inherits from V. The subset

$$\{(a_1, \ldots, a_{n-1}, 0_F): a_i \in F\}$$

of F^n is a vector subspace; $\mathscr{C}$ is a vector subspace of $\mathscr{F}$. If $U \subseteq V$ is a vector subspace of V then $\mathbf{0} \in U$.

The vector spaces $\mathbb{R}^n$ $(1 \leq n \leq 4)$ arise naturally in Chapter 10. The operations in $\mathbb{R}^n$ have a geometrical interpretation as illustrated in Figure 5.5 for $\mathbb{R}^2$. If $\mathbf{r} = (x, y) \in \mathbb{R}^2$ it is conventional to measure the components x and y along orthogonal lines from their point of intersection O. The y component is represented along a line OY (the y-axis) at $90°$ to the line OX for x (the x-axis) where the angle between the lines is measured anticlockwise from OX as depicted in the figure. An axis system of this

Fig. 5.5

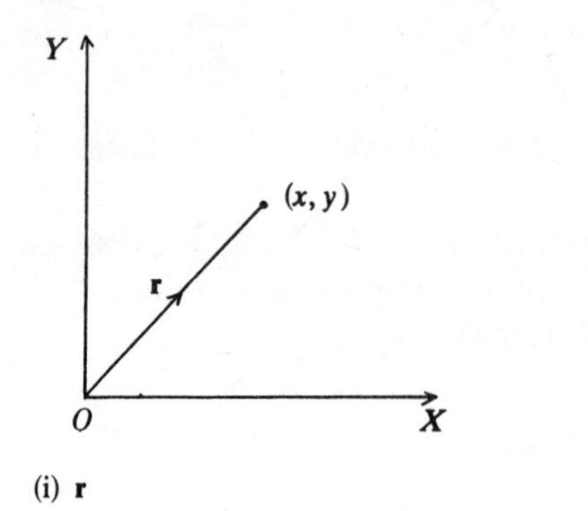

(i) $\mathbf{r}$

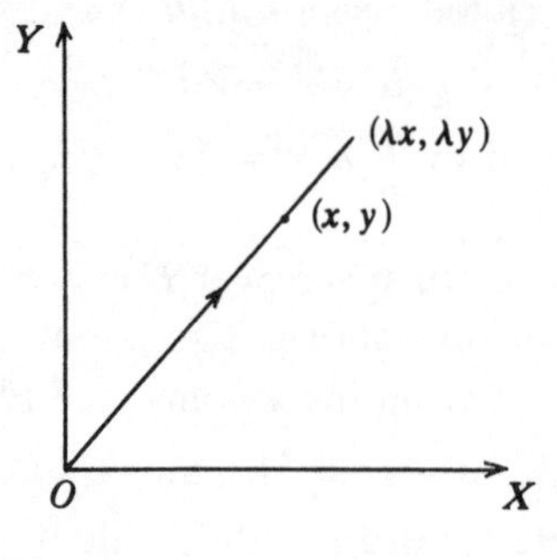

(ii) Scalar multiplication of $\mathbf{r}$ ($\lambda > 1$)

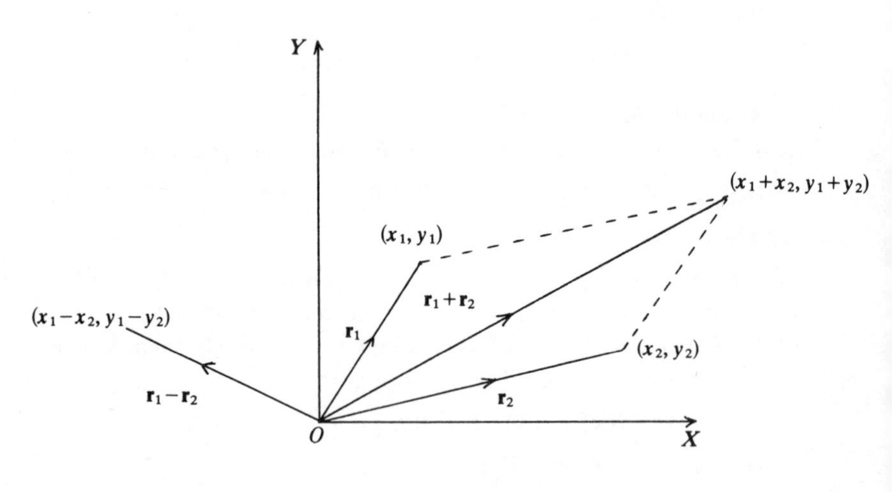

(iii) Vector addition

type is called a *right handed* system for $\mathbb{R}^2$. The vector addition in $\mathbb{R}^2$ corresponds geometrically to the 'parallelogram law' as shown in (iii) of Figure 5.5.

The geometry of the vector spaces $\mathbb{R}^n$ is considered further in Section 5.4.2 but first we consider the more general concepts of basis and dimension. If V is a vector space over F and $S \subseteq V$ then sums of the form

$$\sum_{i=1}^{k} a_i \mathbf{x}_i \quad \text{for } a_i \in F \text{ and } \mathbf{x}_i \in S$$

are called *linear combinations* of the vectors of S.

A finite set $\{\mathbf{x}_i\colon 1 \leqslant i \leqslant k\} \subseteq V$ is said to be *linearly independent* if

$$\sum_{i=1}^{k} a_i \mathbf{x}_i = \mathbf{0} \Rightarrow a_1 = a_2 = \ldots = a_k = 0_F,$$

otherwise the set is *linearly dependent.* A *spanning set* for V is a subset $S \subseteq V$ with the property that all elements of V may be represented as linear combinations of the elements of S. A *basis* for V is an ordered linearly independent spanning set.

Example 5.4.2

In $\mathbb{R}^3$ the vector $(5, 5, \sqrt{2})$ is a linear combination of the vectors $(1, 1, 0)$ and $(0, 0, 3)$ for

$$(5, 5, \sqrt{2}) = 5(1, 1, 0) + \frac{\sqrt{2}}{3}(0, 0, 3).$$

$L = \{(1, 1, 0), (0, 0, 3)\}$ is a linearly independent set in $\mathbb{R}^3$ for

$$a(1, 1, 0) + b(0, 0, 3) = (a, a, 3b) = \mathbf{0} \text{ iff } a = 0 \text{ and } b = 0.$$

L is not a basis for it determines only the vector subspace

$$\{(x, x, y): x, y \in \mathbb{R}\} \subset \mathbb{R}^3.$$

$B = L \cup \{(1, 0, 0)\}$ 'extends' L to a basis for $\mathbb{R}^3$. //

It is easy to show that each element of a vector space has a *unique* representation in a fixed basis, for if V has basis $B = \{\mathbf{e}_1, \ldots, \mathbf{e}_n\}$ and

$$\mathbf{x} = \sum_{i=1}^{n} a_i \mathbf{e}_i = \sum_{i=1}^{n} b_i \mathbf{e}_i,$$

then

$$\mathbf{0} = \mathbf{x} - \mathbf{x} = \sum_{i=1}^{n} (b_i - a_i)\mathbf{e}_i.$$

But B is linearly independent, hence $b_i = a_i$ for all $1 \leqslant i \leqslant n$ and the representation is unique. The next result leads to an important conclusion concerning bases.

Proposition. If $S = \{\mathbf{x}_1, \ldots, \mathbf{x}_m\}$ is a spanning set for V and $L = \{\mathbf{y}_1, \ldots, \mathbf{y}_l\}$ is a linearly independent set in V then $m \geqslant l$.

Proof. Assume $m < l$. Now S spans V hence there are elements $a_1, \ldots, a_m \in F$ with

$$\mathbf{y}_1 = a_1\mathbf{x}_1 + \ldots + a_m\mathbf{x}_m.$$

But $\mathbf{y}_1 \neq \mathbf{0}$ because L is linearly independent (see Exercises 5.4) hence not all the scalars $a_1, \ldots, a_m$ can be zero. Assume that $a_1 \neq 0_F$ (this can always be arranged by re-ordering) then

$$\mathbf{x}_1 = a_1^{-1}\mathbf{y}_1 - a_1^{-1}a_2\mathbf{x}_2 - \ldots - a_1^{-1}a_m\mathbf{x}_m$$

so that $\mathbf{x}_1$ is a linear combination of $\{\mathbf{y}_1, \mathbf{x}_2, \ldots, \mathbf{x}_m\}$. Now S spans V therefore by the above

$$\{\mathbf{y}_1, \mathbf{x}_2, \ldots, \mathbf{x}_m\}$$

spans V. A similar argument gives the fact that

$$\{\mathbf{y}_1, \mathbf{y}_2, \mathbf{x}_3, \ldots, \mathbf{x}_m\}$$

spans V and repeating m times we obtain

$$\{\mathbf{y}_1, \ldots, \mathbf{y}_m\}$$

as a spanning set. Therefore

$$\mathbf{y}_{m+1} = p_1\mathbf{y}_1 + p_2\mathbf{y}_2 + \ldots + p_m\mathbf{y}_m$$

for $p_1, \ldots, p_m \in F$ not all zero (because $\mathbf{y}_{m+1}$ cannot be zero), hence

$$\mathbf{y}_{m+1} - p_1\mathbf{y}_1 - p_2\mathbf{y}_2 - \ldots - p_m\mathbf{y}_m = \mathbf{0},$$

but this is impossible because $\{\mathbf{y}_1, \ldots, \mathbf{y}_l\}$ is linearly independent hence $m \geqslant l$. //

Proposition. If B and B' are bases for a vector space V over F then $|B| = |B'|$.

Proof. If $B = \{\mathbf{e}_1, \ldots, \mathbf{e}_n\}$ and $B' = \{\mathbf{e}'_1, \ldots, \mathbf{e}'_m\}$ then $n \geqslant m$ and $n \leqslant m$ by the previous proposition hence $m = n$. //

The cardinality of a basis of a vector space V (invariant by the above) is called the *dimension* of V and written $\dim(V)$.

Proposition. $\dim(F^n) = n$.

Proof. We define $B = \{\mathbf{e}_1, \ldots, \mathbf{e}_n\}$ where

$$\mathbf{e}_i = (0, \ldots, 0, 1_F, 0, \ldots, 0)$$

↑

*i*th position

and show that B is a basis for F^n. Clearly

$$(a_1, \ldots, a_n) = \sum_{i=1}^{n} a_i\mathbf{e}_i$$

so that B spans F^n and

$$\sum_{i=1}^{n} b_i\mathbf{e}_i = \mathbf{0} \Rightarrow (b_1, \ldots, b_n) = \mathbf{0}$$

$$\Rightarrow b_1 = 0_F, b_2 = 0_F, \ldots, b_n = 0_F,$$

hence B is a basis for F^n and $\dim(V) = |B| = n$. //

A basis, as defined above, is always finite and such a set does not exist for all vector spaces ($\mathscr{F}$ and $\mathscr{C}$ being examples). The concepts of basis and dimension can be extended to all vector spaces but we have no requirement for such generality here. If V has a basis of the type defined above it is said to be *finite dimensional.*

We now consider the homomorphic mappings between vector spaces.

Definition. If V_1 and V_2 are vector spaces over a field F, a mapping $T: V_1 \to V_2$ is said to be *linear* if for all $\mathbf{x}, \mathbf{y} \in V$ and $a \in F$

$$T(\mathbf{x}+\mathbf{y}) = T\mathbf{x} + T\mathbf{y}$$

$$T(a\mathbf{x}) = a(T\mathbf{x}).$$

If $V_2 = V_1$, then T is called a *linear transformation* of V_1. //

Our interest henceforth is in finite dimensional vector spaces over $\mathbb{R}$ and their linear transformations. Throughout the remainder of this section V denotes such a vector space and End(V) denotes the set of all linear transformations of V (the endomorphisms of V). We remark in passing that much of the discussion does hold more generally.

We switch our attention from the algebra of V to the algebra of End(V) and show that End(V) is closed under natural addition, multiplication and scalar multiplication operations. First we notice that the identity map I_V and the zero map 0_V on V are linear, for by definition

$$I_V\mathbf{x} = \mathbf{x} \quad \text{for all } \mathbf{x} \in V$$

and

$$0_V\mathbf{x} = \mathbf{0} \quad \text{for all } \mathbf{x} \in V,$$

hence for all $\mathbf{x}, \mathbf{y} \in V$ and $\lambda \in \mathbb{R}$ we have

$$I_V(\mathbf{x}+\mathbf{y}) = \mathbf{x}+\mathbf{y} = I_V\mathbf{x} + I_V\mathbf{y},$$

$$I_V(\lambda\mathbf{x}) = \lambda\mathbf{x} = \lambda(I_V\mathbf{x})$$

and

$$0_V(\mathbf{x}+\mathbf{y}) = \mathbf{0} = \mathbf{0}+\mathbf{0} = 0_V\mathbf{x} + 0_V\mathbf{y},$$

$$0_V(\lambda\mathbf{x}) = \mathbf{0} = \lambda\mathbf{0} = \lambda 0_V\mathbf{x}.$$

If $S, T \in \text{End}(V)$ then the *sum* $S+T$ and the *Product* $S \circ T$ are defined by

$$(S+T)\mathbf{x} = S\mathbf{x} + T\mathbf{x} \quad \text{for all } \mathbf{x} \in V$$

and

$$(S \circ T)\mathbf{x} = S(T\mathbf{x}) \quad \text{for all } \mathbf{x} \in V.$$

The properties of End(V) under these operations are summarized in the following result.

Proposition. (End $(V), \circ, +)$ is a ring with unity.

Sketch proof. We need to show that

(i) $S, T \in \text{End}(V) \Rightarrow S+T \in \text{End}(V)$ and $S \circ T \in \text{End}(V)$,

(ii) $(\text{End}(V), +)$ is a commutative group

and if $S, T, U \in \text{End}(V)$, then

(iii) $S \circ (T \circ U) = (S \circ T) \circ U$,

(iv) $S \circ (T+U) = S \circ T + S \circ U$,

(v) $I_V \circ T = T \circ I_V = T$.

(i)
$$\begin{aligned}(S+T)(\mathbf{x}+\mathbf{y}) &= S(\mathbf{x}+\mathbf{y}) + T(\mathbf{x}+\mathbf{y}) && \text{by definition of } S+T \\ &= S\mathbf{x}+S\mathbf{y}+T\mathbf{x}+T\mathbf{y} && S, T \text{ linear} \\ &= (S\mathbf{x}+T\mathbf{x})+(S\mathbf{y}+T\mathbf{y}) && (V,+) \text{ is a commutative group} \\ &= (S+T)\mathbf{x}+(S+T)\mathbf{y}.\end{aligned}$$

Similarly,

$$\begin{aligned}(S+T)(\lambda\mathbf{x}) &= S\lambda\mathbf{x}+T\lambda\mathbf{x} \\ &= \lambda S\mathbf{x}+\lambda T\mathbf{x} \\ &= \lambda(S\mathbf{x}+T\mathbf{x}) \\ &= \lambda(S+T)\mathbf{x}.\end{aligned}$$

The proof that $S \circ T \in \text{End}(V)$ is left as an exercise.

(ii)
$$\begin{aligned}(S+(T+U))\mathbf{x} &= S\mathbf{x}+(T+U)\mathbf{x} \\ &= S\mathbf{x}+(T\mathbf{x}+U\mathbf{x}) \\ &= (S\mathbf{x}+T\mathbf{x})+U\mathbf{x} \\ &= (S+T)\mathbf{x}+U\mathbf{x} \\ &= ((S+T)+U)\mathbf{x}\end{aligned}$$

hence + is associative. $0_V \in \text{End}(V)$ satisfies

$$T+0_V = 0_V+T = T \quad \text{for all } T \in \text{End}(V)$$

and is the additive identity of End (V). For $T \in \text{End}(V)$, we define $(-T)\colon V \to V$ to be the map

$$(-T)\mathbf{x} = -(T\mathbf{x}) \quad \text{for all } \mathbf{x} \in V.$$

It is easy to show that $-T \in \text{End}(V)$ and

$$(-T+T) = T+(-T) = 0_V;$$

$-T$ is the additive inverse of T. Commutativity of $(\text{End}(V), +)$ follows from the commutativity of $(V, +)$.

(iii) This follows from the work of Chapter 3.

(iv) For $\mathbf{x} \in V$ we have

$$\begin{aligned}(S \circ (T+U))\mathbf{x} &= S((T+U)\mathbf{x}) \\ &= S(T\mathbf{x}+U\mathbf{x}) \\ &= S(T\mathbf{x})+S(U\mathbf{x}) \\ &= (S \circ T)\mathbf{x}+(S \circ U)\mathbf{x} = (S \circ T + S \circ U)\mathbf{x}.\end{aligned}$$

(v) This is trivial. //

If $T \in \mathrm{End}\,(V)$ and $\lambda \in \mathbb{R}$ we define a mapping $\lambda T: V \to V$ by

$$(\lambda T)\mathbf{x} = \lambda(T\mathbf{x}) \quad \text{for all } \mathbf{x} \in V.$$

It is easy to show that $\lambda T \in \mathrm{End}\,(V)$. The mapping

$$\Lambda : \mathbb{R} \times \mathrm{End}\,(V) \to \mathrm{End}\,(V)$$

defined by $\Lambda(\lambda, T) = \lambda T$ is called *scalar multiplication* of transformations.

Proposition. $(\mathrm{End}\,(V), +, \Lambda)$ is a vector space over $\mathbb{R}$.

Proof. From the previous proposition we have that $(\mathrm{End}\,(V), +)$ is a commutative group, hence we just need to show that scalar multiplication satisfies

$$\begin{aligned}&(\lambda+\mu)T = \lambda T + \mu T, \\ &\lambda(S+T) = \lambda S + \lambda T, \\ &(\lambda\mu)T = \lambda(\mu T), \\ &1_{\mathbb{R}} T = T,\end{aligned}$$

where $\lambda, \mu \in \mathbb{R}$ and $S, T \in \mathrm{End}\,(V)$.

$$\begin{aligned}((\lambda+\mu)T)\mathbf{x} &= (\lambda+\mu)(T\mathbf{x}) && \text{by definition of } \Lambda \\ &= \lambda(T\mathbf{x})+\mu(T\mathbf{x}) && V \text{ is a vector space} \\ &= (\lambda T)\mathbf{x}+(\mu T)\mathbf{x} && \text{by definition of } \Lambda.\end{aligned}$$

The remainder follows similarly. //

Proposition. The ring multiplication $\circ$ and the scalar multiplication Λ on $\mathrm{End}\,(V)$ satisfy

$$\lambda(S \circ T) = (\lambda S) \circ T = S \circ (\lambda T)$$

for $\lambda \in \mathbb{R}$ and $S, T \in \mathrm{End}\,(V)$.

Proof.

$$\begin{aligned}(\lambda(S \circ T))\mathbf{x} &= \lambda((S \circ T)\mathbf{x}) \\ &= \lambda(S(T\mathbf{x})) \\ &= (\lambda S)(T\mathbf{x}) \\ &= ((\lambda S) \circ T)\mathbf{x}\end{aligned}$$

and

$$\begin{aligned}(\lambda(S\circ T))\mathbf{x} &= \lambda((S\circ T)\mathbf{x})\\ &= \lambda(S(T\mathbf{x}))\\ &= S(\lambda(T\mathbf{x}))\\ &= (S\circ(\lambda T))\mathbf{x}. \quad //\end{aligned}$$

Algebraic structures that satisfy the properties shown for End (V) are known as linear algebras; we define these formally as follows.

Definition. A quadruple $(X, +, \circ, \Lambda)$ is a *linear algebra* over $\mathbb{R}$ if $\Lambda: \mathbb{R}\times X \to X$ and

(i) $(X, +, \Lambda)$ is a vector space over $\mathbb{R}$,

(ii) $(X, \circ, +)$ is a ring,

(iii) Λ and $\circ$ satisfy

$$\lambda(\mathbf{x}_1\circ\mathbf{x}_2) = (\lambda\mathbf{x}_1)\circ\mathbf{x}_2 = \mathbf{x}_1\circ(\lambda\mathbf{x}_2) \quad \text{for all } \lambda\in\mathbb{R} \text{ and } \mathbf{x}_1, \mathbf{x}_2\in X. \quad //$$

Our results for End (V) may be summarized in the following.

Proposition. Under the operations defined End (V) is a linear algebra with multiplicative identity. //

If $T\in$ End (V) and there is a transformation $S: V\to V$ of V with

$$S\circ T = T\circ S = I_V,$$

then (see Exercises 5.4) $S\in$ End (V) and T is said to be *invertible* with inverse S, written T^{-1}. Aut (V) denotes the set of all invertible transformations of End (V), i.e. the set of automorphisms of V.

Proposition. (Aut (V), $\circ$) is a group.

Proof. $I_V\in$ Aut (V) for $I_V\circ I_V = I_V$, hence I_V^{-1} exists and equals I_V. If $S\in$ Aut (V), then

$$S^{-1}\circ S = S\circ S^{-1} = I_V,$$

so that $(S^{-1})^{-1}$ exists and equals S, hence $S^{-1}\in$ Aut (V). Now if S, $T\in$ Aut (V) then

$$(S\circ T)\circ(T^{-1}\circ S^{-1}) = S\circ(T\circ T^{-1})\circ S^{-1} = S\circ S^{-1} = I_V$$

and similarly

$$(T^{-1}\circ S^{-1})\circ(S\circ T) = I_V,$$

so that $(S\circ T)^{-1}$ exists; therefore S, $T\in$ Aut (V) implies $S\circ T\in$ Aut (V). Associativity of $\circ$ has already been shown. //

5.4.2 Structure maps for $\mathbb{R}^n$

By equipping $\mathbb{R}^n$ with a suitable map, 'magnitude' and 'direction' become meaningful terms for the vectors of $\mathbb{R}^n$. Returning to our geometrical picture for $\mathbb{R}^2$ we see that if $\mathbf{r}=(x, y)\in\mathbb{R}^2$ then the distance of the point (x, y) from $(0, 0)$ is

$$(x^2+y^2)^{1/2}.$$

We write this as $\|\mathbf{r}\|$; $\|\ \|$ may be regarded as a mapping

$$\|\ \|:\mathbb{R}^2\to\mathbb{R}$$

and is called the *length, modulus* or *norm* mapping. If $\mathbf{r}_1=(x_1, y_1)$ and $\mathbf{r}_2=(x_2, y_2)$ are as shown in Figure 5.6 then the distance between the

Fig. 5.6

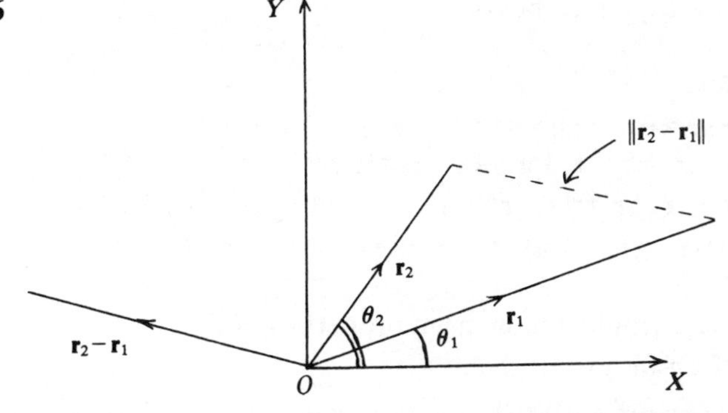

points $\mathbf{r}_1$ and $\mathbf{r}_2$ is $\|\mathbf{r}_2-\mathbf{r}_1\|$ and if θ_1 and θ_2 are the angles in the interval $[0, \pi]$ made by the vectors $\mathbf{r}_1$ and $\mathbf{r}_2$ with the positive x-axis, then the angle θ between $\mathbf{r}_1$ and $\mathbf{r}_2$ is clearly given by $\theta=\theta_2-\theta_1$. Now

$$\begin{aligned}\cos\theta&=\cos(\theta_2-\theta_1)\\&=\cos\theta_1\cos\theta_2+\sin\theta_1\sin\theta_2\\&=\frac{x_1}{\|\mathbf{r}_1\|}\frac{x_2}{\|\mathbf{r}_2\|}+\frac{y_1}{\|\mathbf{r}_1\|}\frac{y_2}{\|\mathbf{r}_2\|}\\&=\frac{x_1x_2+y_1y_2}{\|\mathbf{r}_1\|\|\mathbf{r}_2\|}\end{aligned}$$

so that the expression,

$$x_1x_2+y_1y_2$$

may be used to express the concepts of length and angle in $\mathbb{R}^2$. Formally we define the mapping

$$\Phi:\mathbb{R}^2\times\mathbb{R}^2\to\mathbb{R}$$

by

$$\Phi(\mathbf{r}_1, \mathbf{r}_2)=x_1x_2+y_1y_2$$

in terms of which

$$\|\mathbf{r}\| = (\Phi(\mathbf{r}, \mathbf{r}))^{1/2}$$

and

$$\cos\theta = \frac{\Phi(\mathbf{r}_1, \mathbf{r}_2)}{(\Phi(\mathbf{r}_1, \mathbf{r}_1)\Phi(\mathbf{r}_2, \mathbf{r}_2))^{1/2}}.$$

The angle θ between two vectors in $\mathbb{R}^2$ is uniquely determined if we impose the condition $0 \leqslant \theta \leqslant \pi$. When $\theta = 0$ or π, $\mathbf{r}_1$ and $\mathbf{r}_2$ are said to be *parallel.* It is conventional in applied mathematics to denote $\Phi(\mathbf{r}_1, \mathbf{r}_2)$ by $\mathbf{r}_1 \cdot \mathbf{r}_2$ and refer to it as the *scalar product* of $\mathbf{r}_1$ and $\mathbf{r}_2$. If $\mathbf{r}_1 \neq \mathbf{0}$ and $\mathbf{r}_2 \neq \mathbf{0}$ then $\mathbf{r}_1 \cdot \mathbf{r}_2 = 0$ iff $\theta = \pi/2$ and $\mathbf{r}_1$ and $\mathbf{r}_2$ are said to be *mutually orthogonal, perpendicular* or *normal.* Some further properties of the scalar product are expressed below.

Proposition.

$\mathbf{r} \cdot \mathbf{r} \geqslant 0$ for all $\mathbf{r} \in \mathbb{R}^2$ and $\mathbf{r} \cdot \mathbf{r} = 0$ iff $\mathbf{r} = (0, 0)$,

(ii) $\mathbf{r}_1 \cdot \mathbf{r}_2 = \mathbf{r}_2 \cdot \mathbf{r}_1$ for all $\mathbf{r}_1, \mathbf{r}_2 \in \mathbb{R}^2$,

(iii) $\mathbf{r}_1 \cdot (\mathbf{r}_2 + \mathbf{r}_3) = \mathbf{r}_1 \cdot \mathbf{r}_2 + \mathbf{r}_1 \cdot \mathbf{r}_3$ for all $\mathbf{r}_1, \mathbf{r}_2, \mathbf{r}_3 \in \mathbb{R}^2$,

(iv) $\lambda(\mathbf{r}_1 \cdot \mathbf{r}_2) = (\lambda \mathbf{r}_1) \cdot \mathbf{r}_2 = \mathbf{r}_1 \cdot (\lambda \mathbf{r}_2)$ for all $\mathbf{r}_1, \mathbf{r}_2 \in \mathbb{R}^2$ and $\lambda \in \mathbb{R}$.

Proof. These results follow from properties of $\mathbb{R}$.

(i) If $\mathbf{r} = (x, y) \in \mathbb{R}^2$ then

$$\begin{aligned}\mathbf{r} \cdot \mathbf{r} = (x^2 + y^2) &\geqslant 0 \quad \text{for all } x, y \in \mathbb{R} \\ &= 0 \quad \text{iff } x = 0 \text{ and } y = 0.\end{aligned}$$

(ii)
$$\begin{aligned}\mathbf{r}_1 \cdot \mathbf{r}_2 &= x_1x_2 + y_1y_2 \\ &= x_2x_1 + y_2y_1 \quad \text{because } \mathbb{R} \text{ is a field} \\ &= \mathbf{r}_2 \cdot \mathbf{r}_1 \quad \text{by definition.}\end{aligned}$$

(iii) and (iv) are proved similarly and are left as exercises. //

More generally, if V is a vector space over $\mathbb{R}$ and $\Phi: V \times V \to \mathbb{R}$ is a mapping satisfying properties (i) to (iv) of the proposition above then (V, Φ) becomes a space for which the concepts of length and angle may be discussed. The mapping Φ is called an *inner product* for V and (V, Φ) is called an *inner product space.* In particular if we define $\cdot : \mathbb{R}^n \times \mathbb{R}^n \to \mathbb{R}$ for all $n \in \mathbb{N}$ by

$$\mathbf{a} \cdot \mathbf{b} = \sum_{i=1}^{n} a_i b_i,$$

where $\mathbf{a} = (a_1, \ldots, a_n)$ and $\mathbf{b} = (b_1, \ldots, b_n)$ then $\cdot$ has the required properties. We define the length of a vector $\mathbf{a} \in \mathbb{R}^n$ by

$$\|\mathbf{a}\| = (\mathbf{a} \cdot \mathbf{a})^{1/2}$$

and the cosine of the angle between two vectors **a** and **b** by

$$\frac{\mathbf{a} \cdot \mathbf{b}}{\|\mathbf{a}\|\|\mathbf{b}\|}$$

and the angle is defined to be in the range $[0, \pi]$.

Other inner products may be defined on the spaces $\mathbb{R}^n$; the one above is called the *usual* or *Euclidean* inner product and gives rise to values for lengths and angles that one would expect intuitively.

When $n=1$ $\cdot$ is clearly just multiplication in $\mathbb{R}$ and the angle between two vectors $x, y \in \mathbb{R}$ is determined by

$$\cos^{-1} \frac{xy}{|x||y|},$$

which is either 0 or π depending on the sign of xy. $\| \ \|$ generalizes the modulus function $| \ |$ of $\mathbb{R}$ and has similar properties; for example it can be shown that

$$\begin{aligned} \|\mathbf{a}\| &\geqslant 0 \quad \text{for all } \mathbf{a} \in \mathbb{R}^n \\ &= 0 \quad \text{iff } \mathbf{a} = \mathbf{0}, \\ \|\lambda \mathbf{a}\| &= |\lambda| \|\mathbf{a}\| \text{ for all } \mathbf{a} \in \mathbb{R}^n \text{ and } \lambda \in \mathbb{R} \end{aligned}$$

and

$$\|\mathbf{a}+\mathbf{b}\| \leqslant \|\mathbf{a}\| + \|\mathbf{b}\| \text{ for all } \mathbf{a}, \mathbf{b} \in \mathbb{R}^n.$$

A vector $\mathbf{a} \in \mathbb{R}^n$ with $\|\mathbf{a}\|=1$, or equivalently $\mathbf{a} \cdot \mathbf{a}=1$, is called a *unit vector*; if $\mathbf{a} \in \mathbb{R}^n \backslash \{\mathbf{0}\}$ then $\mathbf{a}/\|\mathbf{a}\|$ is a unit vector parallel to **a**. A unit vector is usually 'decorated' thus: $\hat{\mathbf{a}}$. If $B = \{\hat{\mathbf{e}}_1, \ldots, \hat{\mathbf{e}}_n\}$ is a basis for $\mathbb{R}^n$ with

$$\begin{aligned} \hat{\mathbf{e}}_i \cdot \hat{\mathbf{e}}_j &= 0 \quad \text{if } i \neq j \\ &= 1 \quad \text{if } i = j, \end{aligned}$$

then B is said to be *orthonormal*. The orthonormal basis for $\mathbb{R}^n$ defined by

$$\hat{\mathbf{e}}_i = (0, \ldots, 0, \underset{\substack{\uparrow \\ i\text{th place}}}{1}, 0, \ldots, 0), \quad 1 \leqslant i \leqslant n$$

is called the *standard basis* for $\mathbb{R}^n$. It is conventional to write the standard bases of $\mathbb{R}^2$ and $\mathbb{R}^3$ as $\{\hat{\mathbf{i}}, \hat{\mathbf{j}}\}$ and $\{\hat{\mathbf{i}}, \hat{\mathbf{j}}, \hat{\mathbf{k}}\}$ and to interpret them geometrically as follows: $\hat{\mathbf{i}}$ and $\hat{\mathbf{j}}$ define a right handed system of axes for $\mathbb{R}^2$ and the third axis OZ has its positive direction defined by the direction of motion of a right handed screw placed at O (see Figure 5.7) perpendicular to the plane containing $\hat{\mathbf{i}}$ and $\hat{\mathbf{j}}$ and turned from OX to OY. This is known as the *right hand rule* and an axis system for $\mathbb{R}^3$ of this type is said to be *right handed*.

Fig. 5.7

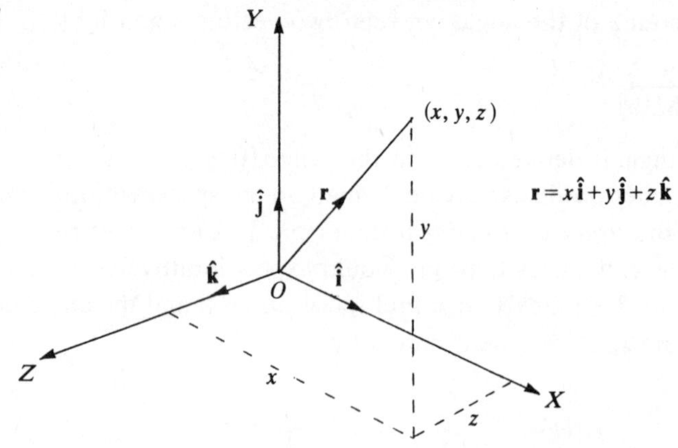

Definition. If $\mathbf{a}=(a_1, a_2, a_3)\in\mathbb{R}^3$ and $\mathbf{b}=(b_1, b_2, b_3)\in\mathbb{R}^3$, the *vector product* (cross product) of $\mathbf{a}$ and $\mathbf{b}$, written $\mathbf{a}\times\mathbf{b}$, is defined to be the vector

$$\mathbf{a}\times\mathbf{b}=(a_2b_3-a_3b_2,\, a_3b_1-a_1b_3,\, a_1b_2-a_2b_1); \quad /\!/$$

$\times$ may be regarded as a mapping from $\mathbb{R}^3\times\mathbb{R}^3$ onto $\mathbb{R}^3$.

Proposition. If $\mathbf{a}, \mathbf{b}\in\mathbb{R}^3$ then

(i) $\|\mathbf{a}\times\mathbf{b}\|=\|\mathbf{a}\|\|\mathbf{b}\|\sin\theta$, where θ is the angle between $\mathbf{a}$ and $\mathbf{b}$.

(ii) $\mathbf{a}\times\mathbf{b}$ is orthogonal to both $\mathbf{a}$ and $\mathbf{b}$.

Proof.

(i)
$$\begin{aligned}\|\mathbf{a}\times\mathbf{b}\|^2 &= (a_2b_3-a_3b_2)^2+(a_3b_1-a_1b_3)^2+(a_1b_2-a_2b_1)^2\\ &= a_2^2b_3^2+a_3^2b_2^2-2a_2b_3a_3b_2+a_3^2b_1^2+a_1^2b_3^2\\ &\quad -2a_3b_1a_1b_3+a_1^2b_2^2+a_2^2b_1^2-2a_1b_2a_2b_1\\ &= (a_1^2+a_2^2+a_3^2)(b_1^2+b_2^2+b_3^2)-(a_1b_1+a_2b_2+a_3b_3)^2\\ &= \|\mathbf{a}\|^2\|\mathbf{b}\|^2-(\mathbf{a}\cdot\mathbf{b})^2\\ &= \|\mathbf{a}\|^2\|\mathbf{b}\|^2\left(1-\frac{(\mathbf{a}\cdot\mathbf{b})^2}{\|\mathbf{a}\|^2\|\mathbf{b}\|^2}\right)\\ &= \|\mathbf{a}\|^2\|\mathbf{b}\|^2(1-\cos^2\theta)\\ &= \|\mathbf{a}\|^2\|\mathbf{b}\|^2\sin^2\theta.\end{aligned}$$

(ii) It is easy to show that $\mathbf{a}\cdot(\mathbf{a}\times\mathbf{b})=0$ and $\mathbf{b}\cdot(\mathbf{a}\times\mathbf{b})=0$, hence the result. //

To interpret $\mathbf{a}\times\mathbf{b}$ geometrically we note that if

$$\mathbf{a}=(a_1, 0, 0)$$

and

$$\mathbf{b}=(b_1, b_2, 0),$$

then

$$\mathbf{a}\times\mathbf{b}=(0, 0, a_1b_2),$$

so that if $a_1 > 0$ we have

$$\mathbf{a}\times\mathbf{b} = |a_1 b_2|\hat{\mathbf{k}} \quad \text{for } b_2 > 0,$$
$$= -|a_1 b_2|\hat{\mathbf{k}} \quad \text{for } b_2 < 0$$

and the direction of $\mathbf{a}\times\mathbf{b}$ is thus determined by the right hand rule turning from **a** to **b**. This argument is quite general, for given an arbitrary pair of vectors a right handed axis system can always be chosen in such a way that the vectors take the form given for **a** and **b**.

The result of this is that the vector product has the geometric form

$$\mathbf{a}\times\mathbf{b} = \|\mathbf{a}\|\|\mathbf{b}\| \sin\theta\hat{\mathbf{n}},$$

where $\hat{\mathbf{n}}$ is the unit vector orthogonal to both **a** and **b** with direction given by the right hand rule. If $\mathbf{a}\times\mathbf{b} = \mathbf{0}$ then the vectors **a** and **b** are linearly dependent and if $\|\mathbf{a}\| > 0$ and $\|\mathbf{b}\| > 0$ then $\mathbf{a}\times\mathbf{b} = \mathbf{0}$ implies that **a** and **b** are parallel. If **a** and **b** are as depicted in Figure 5.8, then $\|\mathbf{a}\times\mathbf{b}\|$ is the area of the parallelogram $OACB$, and $\mathbf{a}\times\mathbf{b}$ may be regarded as a *vector* area.

Fig. 5.8

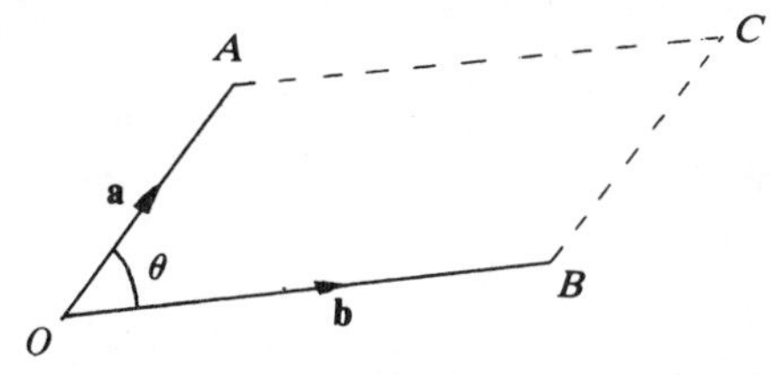

Some further properties of the vector product are summarized below, the proofs being left as an exercise for the reader.

Proposition.

(i) $\mathbf{a}\times\mathbf{b} = -\mathbf{b}\times\mathbf{a}$,

(ii) $\mathbf{a}\times(\mathbf{b}+\mathbf{c}) = \mathbf{a}\times\mathbf{b} + \mathbf{a}\times\mathbf{c}$,

(iii) $(\lambda\mathbf{a})\times\mathbf{b} = \mathbf{a}\times(\lambda\mathbf{b}) = \lambda(\mathbf{a}\times\mathbf{b})$,

(iv) $\hat{\mathbf{i}}\times\hat{\mathbf{j}} = \hat{\mathbf{k}}, \hat{\mathbf{j}}\times\hat{\mathbf{k}} = \hat{\mathbf{i}}, \hat{\mathbf{k}}\times\hat{\mathbf{i}} = \hat{\mathbf{j}}$,

(v) $\mathbf{a}\times(\mathbf{b}\times\mathbf{c}) = (\mathbf{a}\cdot\mathbf{c})\mathbf{b} - (\mathbf{a}\cdot\mathbf{b})\mathbf{c}$,

(vi) $\mathbf{a}\cdot(\mathbf{b}\times\mathbf{c}) = \mathbf{b}\cdot(\mathbf{c}\times\mathbf{a}) = \mathbf{c}\cdot(\mathbf{a}\times\mathbf{b})$
$= -\mathbf{a}\cdot(\mathbf{c}\times\mathbf{b}) = -\mathbf{b}\cdot(\mathbf{a}\times\mathbf{c}) = -\mathbf{c}\cdot(\mathbf{b}\times\mathbf{a})$. //

The expression $\mathbf{a}\times(\mathbf{b}\times\mathbf{c})$ is often referred to as the vector triple product of **a**, **b** and **c** and $\mathbf{a}\cdot(\mathbf{b}\times\mathbf{c})$ as the scalar triple product. Geometrically, $\mathbf{a}\cdot(\mathbf{b}\times\mathbf{c})$ is the volume of the parallelepiped defined by the vectors **a**, **b** and **c**.

Proposition. The set $\{\mathbf{a}, \mathbf{b}, \mathbf{c}\} \subset \mathbb{R}^3$ is linearly dependent iff $\mathbf{a}\cdot(\mathbf{b}\times\mathbf{c}) = 0$.

Proof. Assuming **a**, **b**, **c** to be linearly dependent implies that there exist $\lambda, \mu, \sigma \in \mathbb{R}$ not all zero with

$$\lambda \mathbf{a} + \mu \mathbf{b} + \sigma \mathbf{c} = \mathbf{0}.$$

Assume, without loss of generality, that $\lambda \neq 0$, then

$$\mathbf{a} = -\lambda^{-1}(\mu \mathbf{b} + \sigma \mathbf{c})$$

and

$$\begin{aligned} \mathbf{a} \cdot (\mathbf{b} \times \mathbf{c}) &= -\lambda^{-1}(\mu \mathbf{b} + \sigma \mathbf{c}) \cdot (\mathbf{b} \times \mathbf{c}) \\ &= -\lambda^{-1}[\mu \mathbf{b} \cdot (\mathbf{b} \times \mathbf{c}) + \sigma \mathbf{c} \cdot (\mathbf{b} \times \mathbf{c})] \\ &= 0. \end{aligned}$$

Conversely, if $\mathbf{a} \cdot (\mathbf{b} \times \mathbf{c}) = 0$, then either

(i) **a** or **b** or **c** is zero, in which case the result follows, or
(ii) **a** is orthogonal to $\mathbf{b} \times \mathbf{c}$.

But **b** and **c** are orthogonal to $\mathbf{b} \times \mathbf{c}$, thus

$$\mathbf{a} = \lambda' \mathbf{b} + \mu' \mathbf{c} \quad \text{for some } \lambda', \mu' \in \mathbb{R}$$

and $\{\mathbf{a}, \mathbf{b}, \mathbf{c}\}$ is a linearly dependent set. //

We close this section with a brief discussion on the differentiation of 'vector valued' functions. With the usual norm on $\mathbb{R}^n$ we may define the derivative of functions of the form

$$\mathbf{f}: \mathbb{R} \to \mathbb{R}^n.$$

Generalizing the case $n = 1$ we say that **f** is differentiable at t if there is a vector $\mathbf{F}(t) = (F_1(t), \ldots, F_n(t)) \in \mathbb{R}^n$ such that

$$\left\| \frac{\mathbf{f}(t+h) - \mathbf{f}(t)}{h} - \mathbf{F}(t) \right\| \to 0$$

as $h \to 0$, or equivalently if **f** has components $f_1, \ldots, f_n$

$$\left\| \left(\frac{f_1(t+h) - f_1(t)}{h} - F_1(t), \ldots, \frac{f_n(t+h) - f_n(t)}{h} - F_n(t) \right) \right\| \to 0$$

as $h \to 0$. Evidently, each component must tend to zero as $h \to 0$ so that $\mathrm{d}\mathbf{f}/\mathrm{d}t$ exists iff $\mathrm{d}f_1/\mathrm{d}t, \ldots, \mathrm{d}f_n/\mathrm{d}t$ exist and

$$\frac{\mathrm{d}\mathbf{f}}{\mathrm{d}t} = \left(\frac{\mathrm{d}f_1}{\mathrm{d}t}, \ldots, \frac{\mathrm{d}f_n}{\mathrm{d}t} \right).$$

In other words, to differentiate a vector valued function we just differentiate it componentwise. For example if $\mathbf{f}: \mathbb{R} \to \mathbb{R}^3$ is defined by

$$\mathbf{f}(t) = (2t^2, \log_e t, \sin^2 t),$$

then

$$\frac{\mathrm{d}\mathbf{f}(t)}{\mathrm{d}t} = \left(4t, \frac{1}{t}, 2 \sin(t) \cos(t) \right).$$

If $\mathbf{f}:\mathbb{R}\to\mathbb{R}^3$ and $\mathbf{g}:\mathbb{R}\to\mathbb{R}^3$ we define functions $\mathbf{f}\cdot\mathbf{g}:\mathbb{R}\to\mathbb{R}$ and $\mathbf{f}\times\mathbf{g}:\mathbb{R}\to\mathbb{R}^3$ by

$$(\mathbf{f}\cdot\mathbf{g})(t)=\mathbf{f}(t)\cdot\mathbf{g}(t)$$

and

$$(\mathbf{f}\times\mathbf{g})(t)=\mathbf{f}(t)\times\mathbf{g}(t).$$

The behaviour of these functions under differentiation is as follows:

$$\frac{\mathrm{d}}{\mathrm{d}t}(\mathbf{f}\cdot\mathbf{g})=\mathbf{f}\cdot\frac{\mathrm{d}\mathbf{g}}{\mathrm{d}t}+\frac{\mathrm{d}\mathbf{f}}{\mathrm{d}t}\cdot g$$

$$\frac{\mathrm{d}}{\mathrm{d}t}(\mathbf{f}\times\mathbf{g})=\mathbf{f}\times\frac{\mathrm{d}\mathbf{g}}{\mathrm{d}t}+\frac{\mathrm{d}\mathbf{f}}{\mathrm{d}t}\times\mathbf{g}$$

The verification of these formulae is left as an exercise.

Exercises 5.4

1. Show that if V is a vector space over a field F then

   $0_F\,\mathbf{x}=\mathbf{0}$ for all $\mathbf{x}\in V$

   and

   $a\mathbf{0}=\mathbf{0}$ for all $a\in F$.

2. Express the vector $(a,1,3)\in\mathbb{R}^3$, where $a\in\mathbb{R}$, as a linear combination of the vectors of the set

   $S=\{(1,1,0),(0,2,0),(0,0,4)\}$

   and show that S is a linearly independent set. Is S a basis for $\mathbb{R}^3$?

3. Show that if $\{\mathbf{x}_1,\ldots,\mathbf{x}_m\}$ is a linearly independent subset of a vector space V then $\mathbf{x}_i\neq\mathbf{0}$ for any $1\leqslant i\leqslant m$.

4. (i) Which of the following transformations of $\mathbb{R}^2$ are linear?

   $T_1(x,y)=(a,y)$ where $a\in\mathbb{R}\backslash\{0\}$,
   $T_2(x,y)=(\lambda x+y,\sigma y)$ for $\lambda,\sigma\in\mathbb{R}\backslash\{0\}$,
   $T_3(x,y)=(x^2,0)$,
   $T_4(x,y)=(x,0)$.

   (ii) Determine the products $T_2\circ T_4$ and $T_4\circ T_2$.
   (iii) Show that if $T\in\mathrm{End}\,(V)$ then $T\mathbf{0}=\mathbf{0}$.

5. (i) If V is a vector space, a *projection* of V is a transformation $P\colon V\to V$ with the property

   $(P\circ P)\mathbf{x}=P\mathbf{x}$ for all $\mathbf{x}\in V$.

Show that the transformation $P: \mathbb{R}^2 \to \mathbb{R}^2$ defined by

$$P(x, y) = \left(\frac{1}{a-b}(ax - y + c), \frac{b}{a-b}(ax - y + c) + c \right)$$

for $a, b, c \in \mathbb{R}$ and $a \neq b$ is a projection of $\mathbb{R}^2$. Under what conditions do we have $P \in \text{End}(\mathbb{R}^2)$?

(ii) Which of the transformations defined in Question 4 above are projections?

6. If $T \in \text{End}(V)$, the *null space* of T written $\mathcal{N}(T)$ is defined by

$$\mathcal{N}(T) = \{\mathbf{x} \in V: T\mathbf{x} = \mathbf{0}\}.$$

Show that $\mathcal{N}(T)$ is a vector subspace of V. Show also that the range of T is a vector subspace of V.

7. If V is a vector space over $\mathbb{R}$ and $T \in \text{End}(V)$ then T has a real *eigenvalue* $\lambda \in \mathbb{R}$ if there is a vector $\mathbf{x} \in V$ such that $\mathbf{x} \neq \mathbf{0}$ and

$$T\mathbf{x} = \lambda\mathbf{x},$$

$\mathbf{x}$ is called an *eigenvector* of T corresponding to the eigenvalue λ. Show that

(i) if $T \in \text{End}(\mathbb{R}^2)$ is defined by

$$T(x, y) = (x + ay, y)$$

then any vector of the form $(p, 0)$, for $p \neq 0$, is an eigenvector of T. What are the eigenvalues?

(ii) If $T \in \text{End}(V)$ then if V_λ denotes the set of eigenvectors of T corresponding to eigenvalue λ then $V_\lambda \cup \{\mathbf{0}\}$ is a vector subspace of V. Interpret $\mathcal{N}(T)$ as such.

8. Determine the eigenvalues and eigenvectors of the transformations $T_1, T_2 \in \text{End}(\mathbb{R}^2)$ defined by

$$T_1(x, y) = (-y, x),$$

$$T_2(x, y) = (x, -y)$$

and interpret T_1 and T_2 geometrically.

9. Show that

(i) if $S, T \in \text{End}(V)$ then $S \circ T \in \text{End}(V)$

(ii) if for $T \in \text{End}(V)$ there is a transformation $S: V \to V$ such that

$$S \circ T = T \circ S = I_V$$

then $S \in \text{End}(V)$.

(iii) $T \in \text{Aut}(V)$ implies $\mathcal{N}(T) = \{\mathbf{0}\}$.

10. Show that if $\mathbf{r}_1, \mathbf{r}_2, \mathbf{r}_3 \in \mathbb{R}^2$ and $\lambda \in \mathbb{R}$ then

(i) $\mathbf{r}_1 \cdot (\mathbf{r}_2 + \mathbf{r}_3) = \mathbf{r}_1 \cdot \mathbf{r}_2 + \mathbf{r}_1 \cdot \mathbf{r}_3$,

(ii) $\lambda(\mathbf{r}_1 \cdot \mathbf{r}_2) = (\lambda \mathbf{r}_1) \cdot \mathbf{r}_2 = \mathbf{r}_1 \cdot (\lambda \mathbf{r}_2)$,
(iii) $|\mathbf{r}_1 \cdot \mathbf{r}_2| \leqslant \|\mathbf{r}_1\| \, \|\mathbf{r}_2\|$,
(iv) $\|\mathbf{r}_1 - \mathbf{r}_2\|^2 = \|\mathbf{r}_1\|^2 + \|\mathbf{r}_2\|^2 - 2\|\mathbf{r}_1\| \, \|\mathbf{r}_2\| \cos \theta$,
where θ is the angle between $\mathbf{r}_1$ and $\mathbf{r}_2$
(v) $|\|\mathbf{r}_1\| - \|\mathbf{r}_2\|| \leqslant \|\mathbf{r}_1 + \mathbf{r}_2\| \leqslant \|\mathbf{r}_1\| + \|\mathbf{r}_2\|$
(this is known as the triangle inequality)

and interpret these results geometrically. In fact the above hold in any inner product space, in particular they are valid in $\mathbb{R}^n$ ($n \in \mathbb{N}$) with the usual inner products.

11. Compute unit vectors parallel to
(i) $\mathbf{a} = (1, 1, 1)$,
(ii) $\mathbf{b} = (1, p, 0)$ for $p \in \mathbb{R}$
and determine a unit vector orthogonal to both $\mathbf{a}$ and $\mathbf{b}$.

12. If $\mathbf{a}, \mathbf{b}, \mathbf{c} \in \mathbb{R}^3$ and $\lambda \in \mathbb{R}$ show that
(i) $\mathbf{a} \times \mathbf{b} = -\mathbf{b} \times \mathbf{a}$,
(ii) $\mathbf{a} \times (\mathbf{b} + \mathbf{c}) = \mathbf{a} \times \mathbf{b} + \mathbf{a} \times \mathbf{c}$,
(iii) $(\lambda \mathbf{a}) \times \mathbf{b} = \mathbf{a} \times (\lambda \mathbf{b}) = \lambda(\mathbf{a} \times \mathbf{b})$,
(iv) $\hat{\mathbf{i}} \times \hat{\mathbf{j}} = \hat{\mathbf{k}}, \quad \hat{\mathbf{j}} \times \hat{\mathbf{k}} = \hat{\mathbf{i}}, \quad \hat{\mathbf{k}} \times \hat{\mathbf{i}} = \hat{\mathbf{j}}$,
(v) $\mathbf{a} \times (\mathbf{b} \times \mathbf{c}) = (\mathbf{a} \cdot \mathbf{c})\mathbf{b} - (\mathbf{a} \cdot \mathbf{b})\mathbf{c}$,
(vi) $\mathbf{a} \cdot (\mathbf{b} \times \mathbf{c}) = \mathbf{b} \cdot (\mathbf{c} \times \mathbf{a}) = \mathbf{c} \cdot (\mathbf{a} \times \mathbf{b})$
$= -\mathbf{a} \cdot (\mathbf{c} \times \mathbf{b}) = -\mathbf{b} \cdot (\mathbf{a} \times \mathbf{c}) = -\mathbf{c} \cdot (\mathbf{b} \times \mathbf{a})$.

13. Using the results of Question 12 above, show that $\times : \mathbb{R}^3 \times \mathbb{R}^3 \to \mathbb{R}^3$ is *not* an associative operation, i.e. $\mathbf{a} \times (\mathbf{b} \times \mathbf{c}) \neq (\mathbf{a} \times \mathbf{b}) \times \mathbf{c}$ in general.

14. If $\mathbf{f}, \mathbf{g}: \mathbb{R} \to \mathbb{R}^3$ show that
(i) $\dfrac{\mathrm{d}}{\mathrm{d}t}(\mathbf{f} \cdot \mathbf{g}) = \mathbf{f} \cdot \dfrac{\mathrm{d}\mathbf{g}}{\mathrm{d}t} + \dfrac{\mathrm{d}\mathbf{f}}{\mathrm{d}t} \cdot \mathbf{g}$,

(ii) $\dfrac{\mathrm{d}}{\mathrm{d}t}(\mathbf{f} \times \mathbf{g}) = \mathbf{f} \times \dfrac{\mathrm{d}\mathbf{g}}{\mathrm{d}t} + \dfrac{\mathrm{d}\mathbf{f}}{\mathrm{d}t} \times \mathbf{g}$

$$= -\frac{\mathrm{d}\mathbf{g}}{\mathrm{d}t} \times \mathbf{f} - \mathbf{g} \times \frac{\mathrm{d}\mathbf{f}}{\mathrm{d}t},$$

(iii) if $\|\mathbf{f}(t)\| = a$ for all t, where $a \in \mathbb{R}$ is constant, then $\mathrm{d}\mathbf{f}/\mathrm{d}t$ is orthogonal to $\mathbf{f}$ for all t.

Perform the calculations with
$\mathbf{f}(t) = (a \cos t, a \sin t, 0)$
$\mathbf{g}(t) = (0, 1, t)$.

5.5 Lattices and Boolean algebras

Boolean algebra is one mathematical subject that may well have been encountered already by the reader in an introductory computing course. As we shall see there is not just *one* Boolean algebra, there are many, and therefore any earlier exposure to the subject may be a little confusing; nevertheless, the familiar examples will appear towards the end of the section. However, we begin by introducing a more general structure called a lattice.

Certain lattices are important in abstract theories of computation developed from the notion of approximation; one program may approximate another by virtue of performing the same computation on a subset of the data acceptable to the other or, regarding program output as a list, for a specific input one program may produce a sublist of the output list produced by another. These are simple, and probably obvious, possible ways of approximating programs. A moment's thought about the requirements of approximations will confirm that they are nothing more than elements linked by some *order* relation.

Informally, a computation is the outcome of running a program on certain data and a program is written in a certain language (see Chapter 8). To be of general interest, properties of computations need to be derived not in terms of the program in question but the language used to write the program. Turning this process upside down we should, in principle, be able to determine the result of executing a 'Language X' program by reference to its formal definition. Unfortunately, the amount of detail required to describe even simple examples is considerable and hence none are included. However, we do give, in Section 5.5.1, the theoretical underpinning of various key results. This is followed (5.5.2) by a formal treatment of Boolean algebras and subsequently (5.5.3) by a brief look at some common applications.

5.5.1 Lattices

Recall that the binary relation ρ on the set S is a (partial) order relation if it is reflexive, transitive and antisymmetric. Consequently (S, ρ) is a partially ordered set (po-set) and, if no ambiguity occurs ρ will be written $\leqslant$ and $(S, \leqslant)$ denoted simply by S. Certain po-sets are totally ordered in the sense that given any $x, y \in S$ then either $x \leqslant y$ or $y \leqslant x$, or both. In fact any partial order can be equated with a union of total orders and this gives rise to a natural and useful way of depicting order relations.

First, notice that a (finite) total order relation $(A, \leqslant)$, also called a *chain* for reasons that will soon become apparent, can be used to label the elements of the set A so that $a_i \leqslant a_i$ and $a_i \leqslant a_j$ for all $i < j$ and

$a_i, a_j \in A$. This gives rise to the abbreviated form

$$a_1 \leqslant a_2 \leqslant a_3 \leqslant \ldots \leqslant a_n,$$

where reflexivity and transitivity are taken for granted. Written in this form, it is easy to see how Figure 5.9 is an 'obvious' representation of $(A, \leqslant)$. Alternatively we write A as $(a_1, a_2, \ldots, a_n)$.

Fig. 5.9

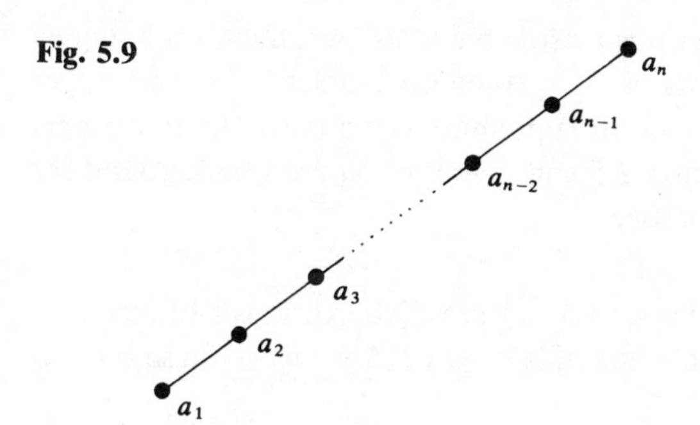

Proposition. A partial ordering on a finite set can be represented as a union of total orderings on suitable subsets.

The proof of this is left as an exercise to the reader. //

By this result we can represent any partial order (finite *or* infinite, but infinite relations are somewhat difficult to draw on finite pieces of paper in a finite time!) by *overdrawing* a set of suitable chain representations; the resulting picture is called a *Hasse diagram.*

Example 5.5.1

$\rho = \{(x, y)$: x is a factor of $y\}$ defined on the set $\{1, 2, 3, 4, 6, 10, 12, 20\}$ gives the Hasse diagram in Figure 5.10 and can be decomposed into the totally ordered sets $\{(1, 2, 4, 12), (1, 3, 6, 12), (2, 6), (4, 20), (2, 10, 20)\}$. Equivalently the following set of total orders will also suffice;

Fig. 5.10

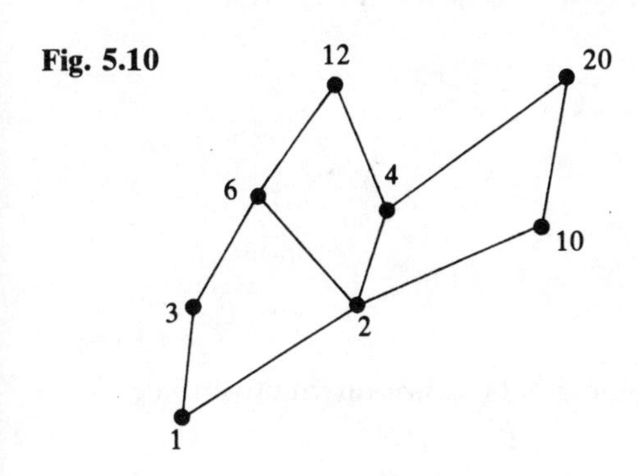

$\{(2, 6, 12), (1, 3, 6), (1, 2, 10, 20), (2, 4, 20), (4, 12)\}$. Notice, in contrast with the general relational diagrams in Section 2.2, that although $1\rho x$ for all $x \in \{1, 2, 3, 4, 6, 10, 12, 20\}$ we do not have an overcluttered picture with 8 arrows emanating from 1. This is achieved by *implicit* representation of the reflexive and transitive properties. //

Given $(A, \leqslant)$ and $B \subseteq A$ it is reasonable to ask whether B is bounded above (below) by elements of A; further we can seek the least upper bound (greatest lower bound) or supremum (infimum). These concepts were fully defined in Section 2.5 and give the required background for the characterization of a lattice.

Definition. A *lattice* is a po-set $(A, \leqslant)$ in which every pair of elements has a supremum and an infimum. Given $x, y \in A$ we write these bounds as

$$x \wedge y = \inf(\{x, y\}),$$
$$x \vee y = \sup(\{x, y\}). \quad //$$

Not all po-sets are lattices. The po-set in Example 5.5.1 is *not* a lattice since $12 \vee 20$ is not defined.

Having defined $\wedge$ and $\vee$ between pairs of elements in a po-set we can extend this notation in a natural way to give $\bigwedge X = \bigwedge_{x \in X} x = \inf X$ and $\bigvee X = \bigvee_{x \in X} x = \sup X$ to indicate $\inf X$ and $\sup X$ where X is a finite non-empty set.

As is to be expected there are many special kinds of lattices in which different calculations can be performed. We limit consideration to three such types.

Definition. The lattice L, now denoted by $(L, \wedge, \vee)$, is *distributive* if it obeys the distributive laws

$$x \wedge (y \vee z) = (x \wedge y) \vee (x \wedge z)$$

and

$$x \vee (y \wedge z) = (x \vee y) \wedge (x \vee z)$$

for all $x, y, z \in L$. //

Not all lattices are distributive.

Example 5.5.2

The lattice shown in Figure 5.11 is not distributive since

$$b \wedge (d \vee c) = b \wedge e = b,$$

Fig. 5.11

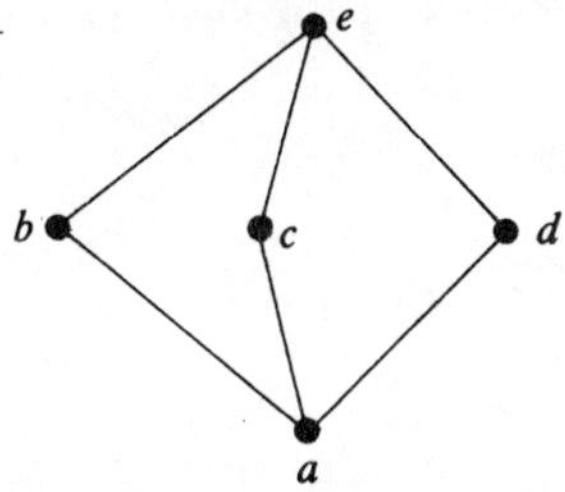

whereas

$$(b \wedge d) \vee (b \wedge c) = a \vee a = a. \quad //$$

Proposition. In a distributive lattice $(L, \wedge, \vee)$

$$x \vee y = x \vee z$$

and

$$x \wedge y = x \wedge z$$

together imply that $y = z$.

Proof. First notice that by definition of inf and sup

(i) $a \wedge b = b \wedge a$,
$a \vee b = b \vee a$,

(ii) $a \wedge b \leqslant a \leqslant a \vee b$,

(iii) $(a \wedge b) \vee a = a$,
$a \wedge (a \vee b) = a$.

So

$$\begin{aligned}
y &= y \vee (y \wedge x) && \text{by (iii) and (i)} \\
&= y \vee (z \wedge x) && \text{by assumption} \\
&= (y \vee z) \wedge (y \vee x) && \text{by distributivity} \\
&= (z \vee y) \wedge (z \vee x) && \text{by (i) and assumption} \\
&= z \vee (y \wedge x) && \text{by distributivity} \\
&= z \vee (z \wedge x) && \text{by assumption} \\
&= z && \text{by (iii) and (i).} \quad //
\end{aligned}$$

Definition. Suppose that $(L, \wedge, \vee)$ is a lattice and $0, 1 \in L$ such that $0 \leqslant x \leqslant 1$ for all $x \in L$ (we shall say more about 0 and 1 later), then

$$x \vee 1 = 1,$$

$$x \wedge 1 = x,$$

$$x \wedge 0 = 0$$

and

$$x \vee 0 = x,$$

for any $x \in L$. Such a lattice is *complemented* if for each $x \in L$, there is $\bar{x} \in L$ such that

$$x \wedge \bar{x} = 0,$$
$$x \vee \bar{x} = 1$$

($\bar{x}$ is called the complement of x). //

Proposition. If $(L, \wedge, \vee)$ is a complemented distributive lattice then complements are unique.

Proof. Suppose $x, y, z \in L$ and

$$x \vee y = x \vee z \quad (=1),$$
$$x \wedge y = x \wedge z \quad (=0).$$

Then by the previous proposition $y = z$. //

The third and final special type of lattice that we shall define is peculiar in that for finite lattices it tells us nothing new. To emphasize the key point at issue we note an alternative definition of a lattice; in the form of a proposition we have:

Proposition. L is a lattice iff $\bigvee X$ and $\bigwedge X$ exist for every non-empty finite subset X of L.

(This can be proved by induction on the size of X and is left as an exercise.) //

If L is a lattice and $\bigwedge L$ is defined then it can be denoted by 0 and called the *bottom* of L. Similarly $1 = \bigvee L$ is the *top* of L if it exists and by convention $\bigvee \varnothing = 0$.

Definition. A lattice L is *complete* if $\bigvee X$ and $\bigwedge X$ exist for *all* subsets X of L. //

It follows that all finite lattices are complete; however, consider $\mathbb{Q}$ ordered by the usual inequality $\leqslant$ and the (infinite) set of approximations to π each having one more significant digit. The sup of this sequence is obviously π but $\pi \in \mathbb{R} \backslash \mathbb{Q}$ and hence $(\mathbb{Q}, \leqslant)$ is not a complete lattice. Now just as $(\mathbb{R}, \leqslant)$ is complete as a lattice and $\mathbb{Q} \subseteq \mathbb{R}$ it can also be shown that *any* lattice can be extended to a complete lattice, but the construction lies outside the scope of this book.

5.5.2 Boolean algebras

We have already dealt extensively with the algebra of sets and made mention of a related 'logical' arithmetic. Now we give a formal treatment of a general Boolean algebra, named after the 19th Century mathematician George Boole. Set algebra is a particular Boolean algebra and although different Boolean algebras are structurally very similar it should be noted that they do not all involve sets in the usual way.

Definition. A *Boolean algebra* is a set $\mathscr{B}$ together with three operations $\vee$, $\wedge$ and $^{-}$ ($\vee$ is called the *or* operation; $\wedge$ the *and* operation and $^{-}$ is the complementation or *not* operator). The binary operators $\vee$ and $\wedge$ and the unary operator $^{-}$ (usually written over its operand thus $\bar{a}$) together with two distinct elements of $\mathscr{B}$, written 0 and 1, satisfy the following axioms.

For arbitrary elements a, b and c in $\mathscr{B}$

(1) $a \vee b = b \vee a$,
(2) $a \vee (b \vee c) = (a \vee b) \vee c$,
(3) $a \vee 0 = a$,
(4) $a \vee \bar{a} = 1$,
(5) $a \wedge b = b \wedge a$,
(6) $a \wedge (b \wedge c) = (a \wedge b) \wedge c$,
(7) $a \wedge 1 = a$,
(8) $a \wedge \bar{a} = 0$,
(9) $a \wedge (b \vee c) = (a \wedge b) \vee (a \wedge c)$,
(10) $a \vee (b \wedge c) = (a \vee b) \wedge (a \vee c)$.

Thus *and* and *or* are commutative, associative and distributive one over the other; each has an identity element and when an element is combined with its complement using *or*/*and* it gives the identity wrt *and*/*or* respectively. (The complement is *not* an inverse.)

The operator names used above are those most immediately associated with computer hardware logic. In other Boolean algebras it may be more appropriate to read $\vee$ as the join, least upper bound or infimum, and $\wedge$ as the meet, greatest lower bound or supremum. Similarly $\bar{a}$ may be written as a' or $\neg a$. //

Equivalently, a Boolean algebra can be defined as a complemented, distributive lattice hence by the results in the previous section and by analogy with the Boolean algebra $(\mathscr{P}(X), \cup, \cap, ')$ for a given non-empty set X we can derive many *consequences* some of which should be proved as exercises. The most important of these are:

The involution law. The complement of the complement of x ($\in \mathscr{B}$ the Boolean algebra in question) is x.

The absorption laws. For $a, b \in \mathscr{B}$

$$a \wedge (a \vee b) = a$$

and

$$a \vee (a \wedge b) = a.$$

The idempotent laws. Any element of $\mathscr{B}$ is idempotent wrt both $\wedge$ and $\vee$.

De Morgan's laws. For $a, b \in \mathscr{B}$

$$\overline{a \wedge b} = \bar{a} \vee \bar{b}$$
$$\overline{a \vee b} = \bar{a} \wedge \bar{b}.$$

As a consequence of the involution law and De Morgan's laws it follows that if we take a Boolean algebra $\mathscr{B}$, more properly $(\mathscr{B}, \vee, \wedge, \bar{\ })$, and create a new system either by mapping each x in $\mathscr{B}$ to $\bar{x}$ or by interchanging $\wedge$ and $\vee$ then $(\mathscr{B}, \wedge, \vee, \bar{\ })$ is also a Boolean algebra. This is known as the *principle of duality*; indeed each of De Morgan's laws can be obtained directly from the other by mapping each element to its complement.

We now derive a theorem that shows how Boolean expressions can be rewritten in certain uniform ways, and an immediate consequence of this is the affirmation that two specific combinatory operators are (separately) sufficient to describe all binary functions. As usual we need some terminology.

Two commonly used logical connectives which we have not yet defined are the following. Boolean formulae A and B are said to be *equivalent* (written $A \equiv B$ or $A \leftrightarrow B$) if they generate the same function. Trivially this is an equivalence relation. Similarly the formula A is said to *imply* formula B, written $A \rightarrow B$, if the table in Figure 5.12 is satisfied. (A table of this sort is often referred to as a *truth table*.)

Fig. 5.12

A	B	$A \rightarrow B$
0	0	1
0	1	1
1	0	0
1	1	1

The arrow symbols are used in many different contexts and so extreme care is needed in deciphering their meaning. So

$$A \rightarrow B$$

is the same as

$$(\neg A) \vee B$$

and

$$A \equiv B$$

is the same as

$$(A \to B) \wedge (B \to A).$$

Logically $A \to B$ means 'if A then B' (i.e. if A is true then so is B) and is normally read as such. Alternative names for the symbols $\to$ and $\leftrightarrow$ are the *conditional* and *biconditional* operators.

It is also usual to allow negation of binary connectives thus

$$A \not\equiv B = \neg(A \equiv B) \quad \text{(non-equivalence)},$$
$$A \not\to B = \neg(A \to B) \quad \text{(non-implication)},$$
$$A \uparrow B = \neg(A \wedge B) \quad \text{(alternative denial)},$$
$$A \downarrow B = \neg(A \vee B) \quad \text{(joint denial)}.$$

It is now possible to describe all binary operations $\{0, 1\}^2 \to \{0, 1\}$ (here the arrow is between two sets and hence indicates the sets from which the domain and range are to be taken; do not confuse with the logical operator '$\to$') in a short form as shown in Figure 5.13. Moreover, we

Fig. 5.13

A	0	1	0	1	
B	0	0	1	1	Function
f_0	0	0	0	0	0
f_1	0	0	0	1	$A \wedge B$
f_2	0	0	1	0	$B \not\to A$
f_3	0	0	1	1	B
f_4	0	1	0	0	$A \not\to B$
f_5	0	1	0	1	A
f_6	0	1	1	0	$A \not\equiv B$
f_7	0	1	1	1	$A \vee B$
f_8	1	0	0	0	$A \downarrow B$
f_9	1	0	0	1	$A \equiv B$
f_{10}	1	0	1	0	$\neg A$
f_{11}	1	0	1	1	$A \to B$
f_{12}	1	1	0	0	$\neg B$
f_{13}	1	1	0	1	$B \to A$
f_{14}	1	1	1	0	$A \uparrow B$
f_{15}	1	1	1	1	1

can describe all mappings from $\{0, 1\}^n$ to $\{0, 1\}$ by using only $\uparrow$ or only $\downarrow$ (the former of these is variously called the Pierce function, or the Sheffer stroke, also written '|'. In computing contexts $\uparrow$ is called **nand**, not-and, and $\downarrow$ is called **nor**, not-or). Notice that $\uparrow$ and $\downarrow$ are effectively shorthand and that, for instance, they are not associative. Hence $A \uparrow B \uparrow C$ is defined to be $(A \wedge B \wedge C)'$ and not, for instance, $((A \wedge B)' \wedge C)'$. The sets of operators $\{\uparrow\}$ and $\{\downarrow\}$ are said to be *adequate*. We now give the theorem and a constructive proof. (This not only shows the theorem to be true but also gives a method for deriving the result.)

Theorem. Any mapping $\{0, 1\}^n \to \{0, 1\}$ can be written as a formula containing (a) only the operator $\uparrow$, or (b) only the operator $\downarrow$.

Proof. Consider the mapping $f: \{0, 1\}^n \to \{0, 1\}$ for some $n \in \mathbb{N}$ expressed in terms of the 'variables' $p_1, \ldots, p_n$. The function can be fully defined by a truth table having 2^n lines. (These are the columns in Figure 5.13.)

Trivially, if all lines give the result 1 then $f = 1$ (a constant mapping); similarly if all lines give 0 then $f = 0$. Otherwise there are m rows of the table having the result 1; let these be $B_1, \ldots, B_m$ each B_i corresponding to an ordered n-tuple $(B_{i1}, \ldots, B_{in})$ where each B_{ij} is either p_j or $\neg p_j$ depending on whether p_j is 1 or 0 for that line of the table. Now

$$\begin{aligned} f &= B_1 \vee B_2 \vee \ldots \vee B_m \\ &= (B_1 \vee B_2 \vee \ldots \vee B_m)'' \\ &= (B_1' \wedge B_2' \wedge \ldots \wedge B_m')' \quad \text{De Morgan} \\ &= (B_1' \uparrow B_2' \uparrow \ldots \uparrow B_m') \end{aligned}$$

and

$$\begin{aligned} B_i' &= (B_{i1} \wedge B_{i2} \wedge \ldots \wedge B_{in})' \\ &= B_{i1} \uparrow B_{i2} \uparrow \ldots \uparrow B_{in}. \end{aligned}$$

Moreover if any of the B_{ij} correspond to a zero in the table then we need $\neg p_j$ and this can be obtained by using the identity

$$(\neg a) \equiv (a \uparrow a).$$

Hence the result follows.

It may be written in a concise way, using prefix notation, as

(a)
$$f = \bigvee_{i=1}^{m} \left(\bigwedge_{j=1}^{n} B_{ij} \right) \qquad (*)$$

and

$$f = \mathop{\uparrow}_{i=1}^{m} \left(\mathop{\uparrow}_{j=1}^{n} B_{ij} \right).$$

(b) Similarly, by applying De Morgan's laws to (a) or, by taking rows yielding 0,

$$f = \mathop{\downarrow}_{i=1}^{2^n - m} \left(\mathop{\downarrow}_{j=1}^{n} \neg B_{ij} \right).$$

A consequence of this result is that the number of different expressions derivable, and hence the number of elements in a Boolean algebra *generated* by $p_1, \ldots, p_n$ is 2^n. //

Example 5.5.3

Derive an expression, using only the $\uparrow$ connective, for the function f whose truth table is given in Figure 5.14.

Fig. 5.14

x	y	z	f
0	0	0	0
0	0	1	1
0	1	0	1
0	1	1	0
1	0	0	1
1	0	1	1
1	1	0	0
1	1	1	1

Working through the method

$$\begin{aligned} f = & (x' \wedge y' \wedge z) \\ & \vee (x' \wedge y \wedge z') \\ & \vee (x \wedge y' \wedge z') \\ & \vee (x \wedge y' \wedge z) \\ & \vee (x \wedge y \wedge z). \end{aligned}$$

Thus

$$\begin{aligned} f' = & (x' \wedge y' \wedge z)' \\ & \wedge (x' \wedge y \wedge z')' \\ & \wedge (x \wedge y' \wedge z')' \\ & \wedge (x \wedge y' \wedge z)' \\ & \wedge (x \wedge y \wedge z)' \end{aligned}$$

and, so

$$\begin{aligned} f = & (x' \uparrow y' \uparrow z) \\ & \uparrow (x' \uparrow y \uparrow z') \\ & \uparrow (x \uparrow y' \uparrow z') \\ & \uparrow (x \uparrow y' \uparrow z) \\ & \uparrow (x \uparrow y \uparrow z), \end{aligned}$$

with

$$\begin{aligned} x' &= x \uparrow x, \\ y' &= y \uparrow y, \\ z' &= z \uparrow z. \quad // \end{aligned}$$

The form generated by part (a) of the theorem (the line marked (*)) is called the *disjunctive normal form* (DNF), it being the *disjunction* of *conjunctions* (*ors* of *ands*) and the form derived in part (b) is the *conjunctive normal form* (CNF).

Finally we note that a statement in a Boolean algebra (a logical formula) that always gives the value 1 is called a *tautology* and one which always gives zero is a *contradiction.*

5.5.3 Some applications of Boolean algebra

There are numerous problems of a combinatory or 'logical' nature that can best be solved by construction of a suitable Boolean algebra. Another way to describe this process is to say that we *model* the real situation or, alternatively, we *interpret* a Boolean algebra – of the right size – in terms that relate to the problem.

This method can be applied in areas which are so widely spread as to defy classification. Hence we single out only one application for special consideration, namely combinatorial and switching circuits. Other areas will be represented by a selection of illustrative examples and exercises.

At this point it must be stressed that we are dealing only with the manipulative aspect of these problems; formal treatment of the deductive aspects lies outside the scope of the current text. First, let us see how expressions in a Boolean algebra can be used to describe and define (segments of) circuits consisting of current-carrying wires and on–off switches. The circuit drawn in Figure 5.15 consists of five switches, a

Fig. 5.15

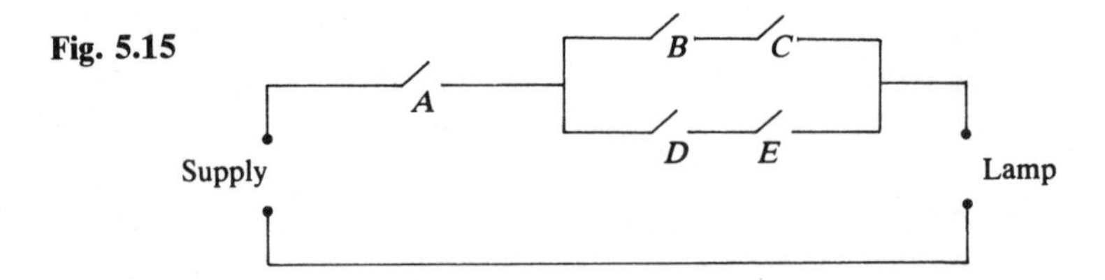

power supply, a lamp and interconnecting wires. As will be readily observed from the diagram current will flow round the top portion of the circuit only if

switch A is closed

and either

switches B and C are both closed,

or

switches D and E are both closed.

Algebraically this can be written as

$$A \wedge ((B \wedge C) \vee (D \wedge E)).$$

Before going on we make some remarks about the diagrammatic representation of these switching circuits. The only segment of the circuit that will differ between examples is the upper section that contains the switches, hence we need not depict the rest of the circuit. Within a circuit several switches may be interconnected (Figure 5.16*a*) and possibly be such that when one particular switch is on another must be off and vice versa (Figure 5.16*b*). The basic methods of combining switches are in *series* (Figure 5.16*c*) which corresponds to the **and** connective, $A \wedge B$, and in *parallel* (Figure 5.16*d*) which corresponds to the **or** connective,

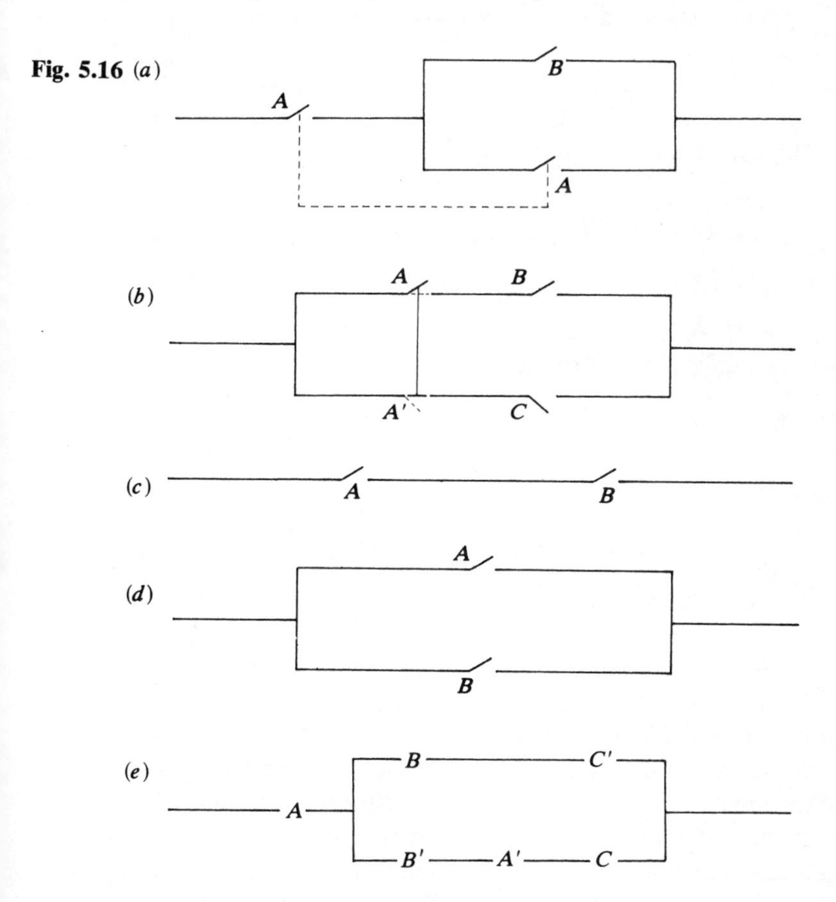

$A \vee B$. Finally, we note that diagrammatically the switches themselves are completely superfluous and can be denoted by their 'names' to give a convenient abbreviated notation (Figure 5.16*e*). Thus we can represent any operation in a Boolean algebra by a switching configuration and vice versa. Hence, to obtain equivalent circuits, all we need to do is to pass from the original circuit, to a representative expression in the algebra, to a 'simpler' (or in some way more desirable) expression and then to the circuit corresponding to this expression.

Example 5.5.4

Simplify the switching network given in Figure 5.17.

Fig. 5.17

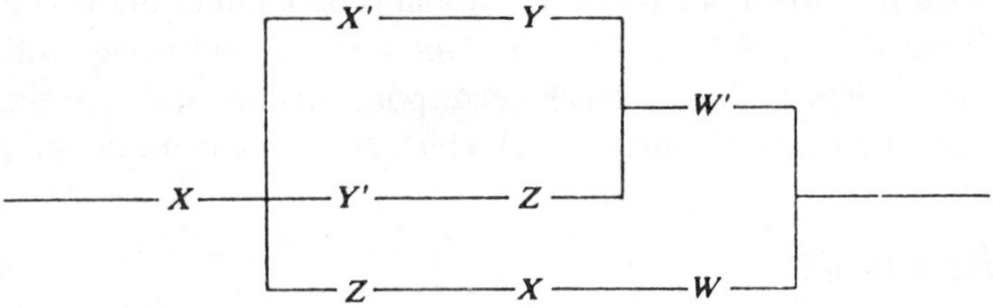

The circuit is representable by

$$\begin{aligned}
&X \wedge ((((X' \wedge Y) \vee (Y' \wedge Z)) \wedge W') \vee (Z \wedge X \wedge W)) \\
&= X \wedge ((X' \wedge Y \wedge W') \vee (Y' \wedge Z \wedge W') \vee (Z \wedge X \wedge W)) \\
&= (X \wedge X' \wedge Y \wedge W') \vee (X \wedge Y' \wedge Z \wedge W') \vee (X \wedge Z \wedge X \wedge W) \\
&= (X \wedge Y' \wedge Z \wedge W') \vee (X \wedge Z \wedge W) \\
&= (X \wedge Z) \wedge ((Y' \wedge W') \vee W).
\end{aligned}$$

This gives the circuit shown in Figure 5.18. //

Fig. 5.18

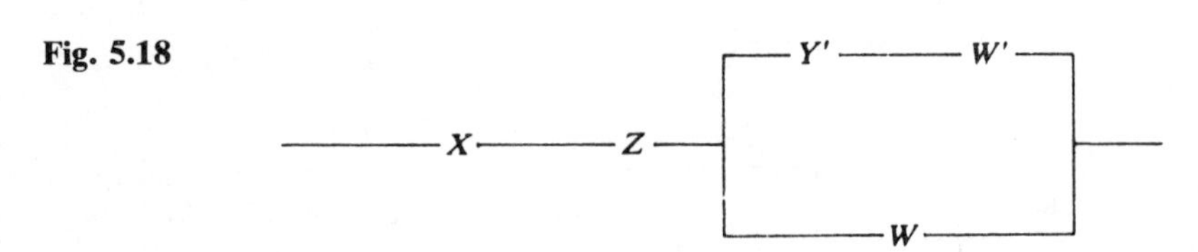

So much for switching circuits, now for combinatorial circuits. The principal components of such circuits are called gates. In its most general form a *gate* is a device having n inputs and m outputs ($m, n \in \mathbb{N}$). Each of the inputs receives one of two binary signals; we shall represent these as 0 and 1 but any system of two distinct values will suffice. Typically, in a real situation, we may have signals in a band around 0 volts (for example, from 0 volts to 0.8 volts) to represent 0 and another band around 4 volts (between 3 volts and 5 volts, say) to represent 1. From these the output values (0s and 1s again) are defined. You might expect that we will need very many different gates to cope with all the possible functions $\{0, 1\}^n \to \{0, 1\}$ but, as we shall see, this is not so. For circuits not involving time delay, and these are the only ones which we shall consider here, a single type of gate is adequate. However, this is not obvious and the structure of circuits can be more clearly seen by first considering the three gates depicted in Figure 5.19. The simplest gate

Fig. 5.19

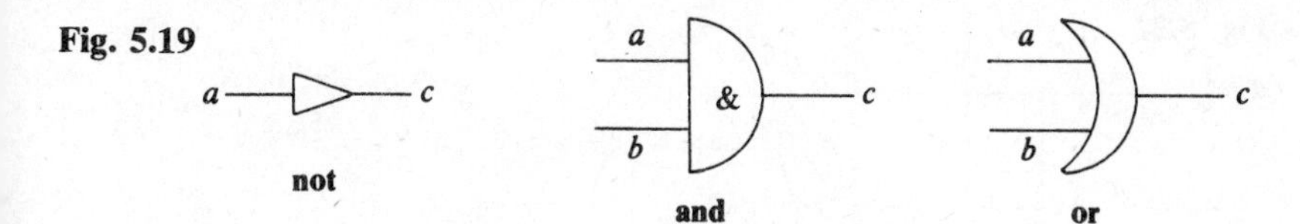

of all is the **not** gate; its effect is defined by the identity

$$c = \neg a \quad (c = \textbf{not } a)$$

which is given in Figure 5.20.

Fig. 5.20

INPUT (a)	OUTPUT (c)
0	1
1	0

Similarly the (two input) **and** and **or** gates are defined by the tables in Figure 5.21.

Fig. 5.21

INPUTS		OUTPUT (c)	
a	b	**and**	**or**
0	0	0	0
0	1	0	1
1	0	0	1
1	1	1	1

The **and** and **or** gates also have multiple-input versions defined analogously, i.e. output of an **and** gate is 1 iff *all* inputs are 1, the output of an **or** gate is 1 if *any* input is 1. Algebraically the output of a **not** gate is $\neg$ (the input), the output of an **and** gate is (first input) $\wedge$ (second input) $\wedge \ldots$, and the output of an **or** gate is (first input) $\vee$ (second input) $\vee \ldots$. To represent more complex logical functions the gates can be 'plugged together' in numerous ways.

Example 5.5.5

Consider the circuit (more properly a partial circuit, but it is the only really interesting part) in Figure 5.22. Tracing 'backwards' from the outputs we obtain

$$x = ((a \vee (a \wedge b')) \wedge (y))'$$

and

$$y = (b \wedge c)'.$$

So

$$\begin{aligned} x &= (((a \vee a) \wedge (a \vee b')) \wedge (b \wedge c)')' \\ &= ((a \wedge (a \vee b')) \wedge (b \wedge c)')' \\ &= (a \wedge (b \wedge c)')' \quad \text{by absorption.} \end{aligned}$$

Fig. 5.22

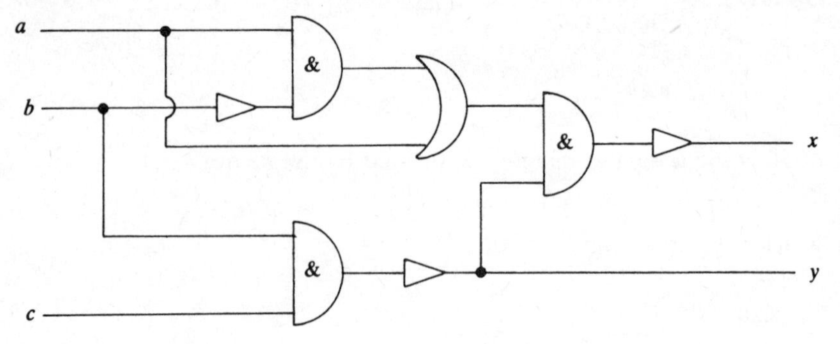

Thus one feasible simplification is the circuit in Figure 5.23. //

Fig. 5.23

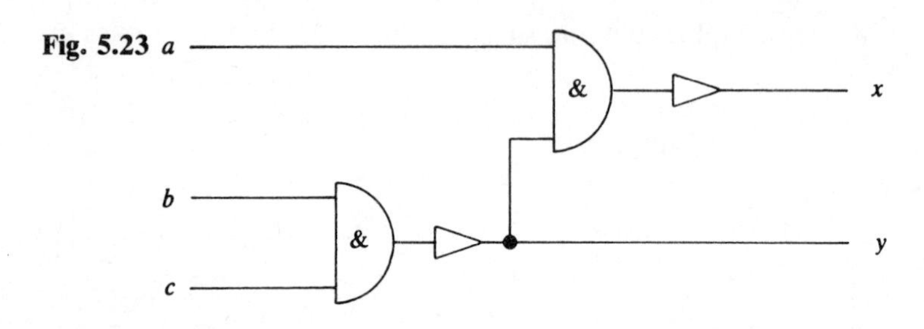

Hence we can use Boolean expressions to represent, specify and simplify combinatory circuits. The example given here merely shows how such manipulations can be carried out.

Hardware devices that reflect the operation of the other logical connectives also exist. Of particular importance are **nand** and **nor** gates as depicted in Figure 5.24. With these multiple input gates we can use the DNF (CNF) of a Boolean expression to obtain a suitable circuit directly from the truth table.

Fig. 5.24

Example 5.5.3 *(continued)*

Without formality, each bracketed expression in the final form of f (see earlier) represents an input to the last gate in Figure 5.25 and is itself computed by a gate in the preceding row. (The portions enclosed in broken lines are in practice not required since with many electrical components the complement of the output is also available from a

Fig. 5.25

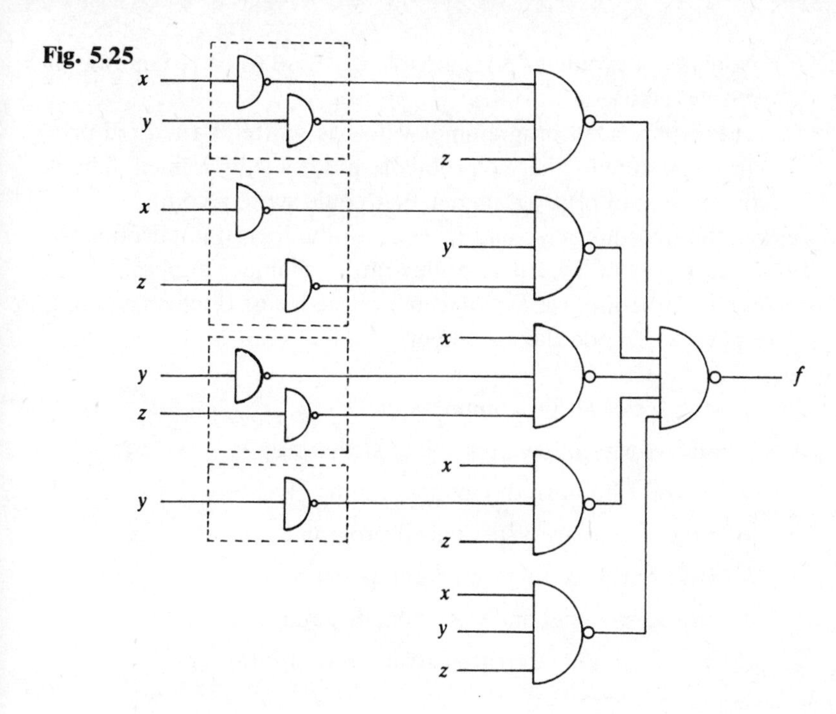

subsidiary output.) Notice that the circuit produced will not in general be minimal in terms of the number of components used. //

Finally, as an indication of other kinds of logical problem that can be tackled by using Boolean algebra we give two illustrative examples.

Example 5.5.6

Let us assume that the following statements are true.

(a) A knowledge of data structures is necessary in developing a disciplined mind.

(b) Only experience in programming can produce a disciplined mind.

(c) In order to write a compiler one must be able to analyse problems.

(d) No undisciplined mind can analyse problems.

(e) No one who has not written structured programs can be regarded as an experienced programmer.

From these assumptions is it possible to determine the validity of the statements below?

(A) Experience in writing structured programs is necessary in order to be able to write a compiler.

(B) A knowledge of data structures is part of programming experience.

(*C*) Analysis of problems is impossible by those who are ignorant of data structures.

(*D*) The experienced programmer who has written structured programs, is able to analyse problems and has a disciplined mind is the only kind of programmer that could write a compiler.

To answer the question we could investigate the logical implications in an *ad-hoc* fashion but, in order to illustrate techniques applicable in a more complex situation, we exploit our knowledge of Boolean algebra. First, we need to encode the statements.

Let

$\mathscr{E}$ = the set of all programmers,

U = those who know about data structures,

V = those who have disciplined minds,

W = those who are experienced programmers,

X = those who could write a compiler,

Y = those who can analyse problems, and

Z = those who have written structured programs.

Then we have

(*a*) $U \supseteq V$,
(*b*) $W \supseteq V$,
(*c*) $X \subseteq Y$,
(*d*) $V \supseteq Y$,
(*e*) $W \subseteq Z$.
(*A*) $Z \supseteq X$,
(*B*) $U \supseteq W$,
(*C*) $Y \subseteq U$,
(*D*) $W \cap Z \cap Y \cap V \supseteq X$.

Now (*A*) follows directly from (*c*), (*d*), (*b*) and (*e*) by transitivity; similarly, (*C*) follows from (*d*) and (*a*). Also

$$\begin{aligned} W \cap Z \cap Y \cap V &= W \cap Y \cap V && \text{absorption } (e) \\ &= Y \cap V && \text{absorption } (b) \\ &= Y && \text{absorption } (d) \\ &\supseteq X && \text{by } (c). \end{aligned}$$

So (*D*) is also true. Statement (*B*) cannot be derived from (*a*)–(*e*) since we only know that $W \subseteq Z$ and the chain ends there. //

Example 5.5.7

Alice said that what Barbara and Claire say is always right, and Claire said that Elspeth and Fiona were either both liars or both truthful.

On the other hand, Debbie reckoned that at least one of Alice and Barbara was truthful whereas Barbara stated that only one of Elspeth and Fiona was truthful. Elspeth then commented that Alice and Barbara always tell the truth but Fiona was sure that Barbara and Claire could not both be relied upon to tell the truth. Who can we believe?

Consider the set $\mathscr{E}$ of all statements of the form

"Alice does (not) tell the truth

and Barbara does (not) tell the truth

and ...

..."

and let A contain exactly the statements in which Alice tells the truth, and let B, C, D, E and F be similarly defined. So if S is a non-empty set of statements such that

$$S \subseteq A \cap B' \cap C,$$

then S includes the assertions in which

"Alice is truthful",
"Barbara is a liar"

and

"Claire is truthful".

From the first statement in the passage we can deduce that if Alice tells the truth then so do Barbara and Claire. This can be encoded as

if $x \in A$ **then** $x \in B \cap C$,

consequently,

$$A \subseteq B \cap C. \qquad \text{(i)}$$

Similarly, the other statements from the passage can be written, in order, as

$$C \subseteq (E \cap F) \cup (E' \cap F'), \qquad \text{(ii)}$$

$$D \subseteq A \cup B, \qquad \text{(iii)}$$

$$B \subseteq (E \cup F) \backslash (E \cap F) = (E \cap F') \cup (E' \cap F), \qquad \text{(iv)}$$

$$E \subseteq A \cap B, \qquad \text{(v)}$$

$$F \subseteq (B \cap C)'. \qquad \text{(vi)}$$

From (ii), (iv) and (i) it follows that $B \cap C = \varnothing$ so $A = \varnothing$ and then from (v) $E = \varnothing$. This is about as far as we can go from the bare facts. The interpretation so far is effectively that what a truthful person says is true but we know nothing about the validity of what a liar says. If, however, we restrict the meaning associated with these words so that *everything* that a liar says is false then we can replace the set inclusions of (i)–(vi) by equality – so that, for example, if Barbara and Claire are truthful then, since Alice says that this is so, Alice is truthful – therefore

$$A = B \cap C, \tag{ia}$$

$$C = (E \cap F) \cup (E' \cap F'), \tag{iia}$$

$$D = A \cup B, \tag{iiia}$$

$$B = (E \cap F') \cup (E' \cap F), \tag{iva}$$

$$E = A \cap B, \tag{va}$$

$$F = (B \cap C)'. \tag{via}$$

As before $A = \varnothing$ and $E = \varnothing$, so Alice and Elspeth never tell the truth. From (ii*a*), (iv*a*) and (vi*a*) $F = \mathscr{E}$ and thus by (iii*a*) and (iv*a*) $D = B = F = \mathscr{E}$, so by substitution in (ii*a*) $C = \varnothing$.

It therefore follows that the only possible 'statement point' x must be such that

$$x \in A' \cap B \cap C' \cap D \cap E' \cap F$$

which means that Alice, Claire and Elspeth are liars and Barbara, Debbie and Fiona tell the truth. //

Exercises 5.5

1. In the Boolean algebra $(\mathscr{B}, *, +, ')$ show that for $x, y, z \in \mathscr{B}$

(*a*) $(x+y)*(x'+y) = y$,

(*b*) $(z+x)*(z'+y) = (z*y)+(z'*x)$,

(*c*) $((x*z)+(y*z'))' = (x'*z)+(y'*z')$.

2. Show that if, in $(\mathscr{B}, *, +, ')$

$$x*y = x*z$$

and

$$x+y = x+z$$

then

$$y = z.$$

3. Derive 'simple' equivalent circuits for those depicted in figures (*a*) and (*b*).

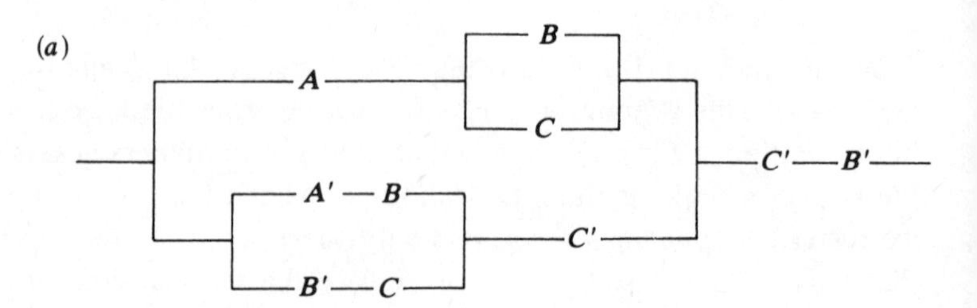

(b)

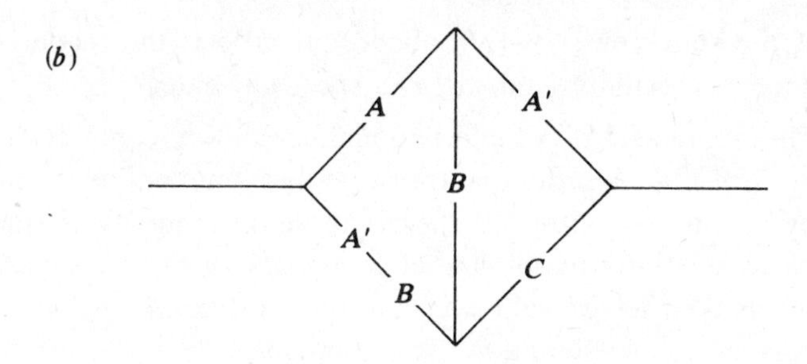

4. Derive an algebraic representation of the combinatory circuit in figure (*c*) and give a tabular representation of the outputs *x* and *y* in terms of inputs *a* and *b*. Call this circuit *C*. Use two *C* circuits as the basis for a circuit that corresponds to the functions tabulated in (*d*). Relate these to the arithmetic of Section 4.3.

(c)

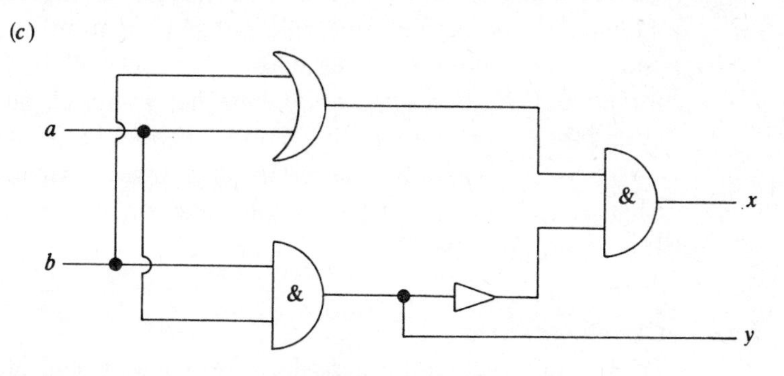

(*d*)

INPUTS			OUTPUTS	
P	*Q*	*R*	*S*	*T*
0	0	0	0	0
0	0	1	0	1
0	1	0	0	1
0	1	1	1	0
1	0	0	0	1
1	0	1	1	0
1	1	0	1	0
1	1	1	1	1

5. Construct a **nand** circuit that accepts a four bit binary number $b_3 b_2 b_1 b_0$ and gives an output of 1 iff the input number is prime.

6. (i) All speedway riders are impulsive.
   (ii) All good programmers are contemplative.
   (iii) Nobody can be both impulsive and contemplative.
   (iv) The reader is contemplative.

If the statements (i)–(iv), above, are true is the reader (a) a good programmer and/or (b) a speedway rider?

7. The organizers of an international computing conference have decided that, in order that the occasion should not be dominated by commercial interests, there will be only one trade stand to be used jointly by all computer manufacturers. The manufacturers themselves will be responsible for determining who takes part in the presentation. Ten companies, five European, five American, make known their desire to participate. (We denote the companies by A, B, C, D, E and F, G, H, I, J respectively.) However, because of contractual obligations and sales policy, various restrictions have to be observed. European regulations dictate that G and I cannot both take part and similarly neither can all three of F, G and J. Restrictions by the American manufacturers' organization rule out participation by both A and D and, unless G takes part, C and D will similarly be excluded. If E attends the conference then so must J but if both attend then B cannot. Finally, H and C cannot both participate.

   Can an agreement be reached in which three companies from each group can join together without violation of the conditions? If so, who takes part?

5.6 Closed semirings

To end this chapter we describe a very special and important structure which allows us to perform manipulations on relational information on infinite sets. Until we have some knowledge of matrices and graphs we cannot embark on discussions of applications and hence we shall give only an axiomatic definition and highlight the crucial difference between a (closed) semiring and superficially similar fields.

Definition. A (*closed*) *semiring* is a set S together with two binary operations $\circledast$ and $\oplus$ such that

(1) $\oplus$ is associative,

(2) there is an identity element wrt $\oplus$ which we write as 0,

(3) $\circledast$ is associative,

(4) there is an identity element wrt $\circledast$ which we write as 1,

(5) for all $x \in S$,

$$x \circledast 0 = 0 = 0 \circledast x,$$

(6) $\oplus$ is commutative:

$$x \oplus y = y \oplus x \quad \text{for all } x, y \in S,$$

(7) $\oplus$ is idempotent:

$$x \oplus x = x \quad \text{for all } x \in S,$$

(8) $\circledast$ distributes over $\oplus$:

$$x \circledast (y \oplus z) = (x \circledast y) \oplus (x \circledast z) \quad \text{for all } x, y, z \in S$$

and

$$(x \oplus y) \circledast z = (x \circledast z) \oplus (y \circledast z) \quad \text{for all } x, y, z \in S,$$

(9) the sum of a countable number of elements in S exists and is unique; i.e. it is independent of the order in which the addition takes place,

(10) $\circledast$ distributes over countably infinite sums, thus

$$\left(\sum_i a_i\right) \circledast \left(\sum_j b_j\right) = \sum_{i,j} (a_i \circledast b_j)$$

where

$$\sum_i a_i = a_1 \oplus a_2 \oplus \ldots,$$

$$\sum_j b_j = b_1 \oplus b_2 \oplus \ldots$$

and

$$\begin{aligned}\sum_{i,j} (a_i \circledast b_j) = (\sigma \ \circledast b_1) \oplus (a_1 \circledast b_2) \oplus \ldots \\ \oplus (a_2 \circledast b_1) \oplus (a_2 \circledast b_2) \oplus \ldots \\ \oplus (a_3 \circledast b_1) \oplus (a_3 \circledast b_2) \oplus \ldots \\ \oplus \ldots . \quad //\end{aligned}$$

This rather weird structure is invaluable in *closure* algorithms for graphs (Chapters 7 and 8). To illustrate the results derivable in closed semirings we compare the systems $(\mathbb{Z}_2, *, +)$ and $(\mathbb{Z}_2, \wedge, \vee)$ where the operations are those defined in the tables in Figure 5.26. Consider now

Fig. 5.26

*	0	1
0	0	0
1	0	1

+	0	1
0	0	1
1	1	0

∧	0	1
0	0	0
1	0	1

∨	0	1
0	0	1
1	1	1

the definition of a *closure operation*, written a^* when applied to the element a and defined by

$$a^* = \sum_{i=0}^{\infty} a^i,$$

where $a^0 = 1$, $a^1 = a$, $a^2 = a * a$, and $a^i = a * a^{i-1}$ (in $(\mathbb{Z}_2, \wedge, \vee)$ the addition is $\vee$ and the multiplication $\wedge$). In the closed semiring $(\mathbb{Z}_2, \wedge, \vee)$ such a definition is meaningful and possible, in particular

$$\begin{aligned} 1^* &= 1^0 \vee 1^1 \vee 1^2 \vee 1^3 \vee \ldots \\ &= 1 \vee 1 \vee 1 \vee 1 \vee \ldots \\ &= 1 \end{aligned}$$

(by virtue of axioms 7 and 9).

However, if we attempt the same computation in the field $(\mathbb{Z}_2, *, +)$ then

$$\begin{aligned} 1^* &= 1^0 + 1^1 + 1^2 + 1^3 + \ldots \\ &= 1 + 1 + 1 + 1 + \ldots. \end{aligned}$$

But this can be summed either as

$$\begin{aligned} &\underbrace{1+1}+\underbrace{1+1}+\underbrace{1+1}+\ldots \\ &= \quad 0 \quad + \quad 0 \quad + \quad 0 \quad + \ldots \\ &= 0 \end{aligned}$$

or as

$$\begin{aligned} &1+\underbrace{1+1}+\underbrace{1+1}+\underbrace{1+1}+\ldots \\ &= 1 + \quad 0 \quad + \quad 0 \quad + \quad 0 \quad + \ldots \\ &= 1 \end{aligned}$$

and so we have ambiguity. Hence the notion of this kind of closure is meaningless when applied to the field $\mathbb{Z}_2$ (which violates semiring axioms 7 and 9).

Exercises 5.6

1. Verify that $(\{\varnothing, \mathscr{E}\}, \cap, \cup)$ is a closed semiring.

6 MATRICES

Our exposition of *finite* matrices (the plural of matrix) is more general than that adopted in most texts, the usual attack being via linear transformations on vector spaces using familiarity with coordinate geometry as an anchor. Although we shall return (in Section 6.3) to this interpretation, we take the view that matrices are essentially realizations of abstract algebraic structures for *computational* purposes. The algebra of the abstract structure then determines the ways in which the matrices are to be combined.

To get the ball rolling we define (in Section 6.1) a matrix representation of binary relations over finite sets. This is followed by a wider investigation into the properties required of the abstract system in order that a matrix realization should be *sensible*. Finally, we consider, in depth, the important case of vector spaces over $\mathbb{R}$, utilizing and extending the ground work begun in the previous chapter.

6.1 Matrices and binary relations on finite sets

Formally a (two-dimensional) *matrix* over a set S is a mapping

$$M : \mathbb{N}_p \times \mathbb{N}_q \to S \quad \text{for some } p, q \in \mathbb{N}.$$

It is usual to denote the image of (i, j) under M as M_{ij} and depict the entire function by the array of elements from S thus

$$M = \begin{pmatrix} M_{11} & M_{12} & \dots & M_{1q} \\ M_{21} & M_{22} & \dots & M_{2q} \\ \vdots & & & \\ M_{p1} & M_{p2} & \dots & M_{pq} \end{pmatrix}.$$

This matrix is said to have p *rows* and q *columns* and to be of *size* $p \times q$ (p by q). A $p \times q$ matrix has $p * q$ elements. (When $p = q$ the matrix is *square*.) The set of all $p \times q$ matrices over the set S is denoted by $\mathcal{M}(p, q, S)$. When $p = q$ this can be reduced to $\mathcal{M}(p, S)$.

Now consider a (binary) relation between sets A and B where

$$A=\{a_1, a_2, \dots, a_p\}$$

and

$$B=\{b_1, b_2, \dots, b_q\},$$

so

$$|A|=p \quad \text{and} \quad |B|=q.$$

(The ordering of elements specified in these sets is arbitrary but once chosen it must be kept fixed.) Let this relation, ρ, be specified by selecting pairs (a, b) with $a\in A$ and $b\in B$.

Now consider a matrix M over $\{0, 1\}$, so $M:\mathbb{N}_p\times\mathbb{N}_q\to\{0, 1\}$, and associate the elements of M with the relation ρ by the bijection

$$\phi:\mathcal{P}(A\times B)\to\mathcal{M}(p, q, \mathbb{Z}_2)$$

(ϕ maps *any* relation between A and B to a $p\times q$ matrix over $\{0, 1\}$),

$$\phi:\rho\mapsto M$$

such that

$$(\phi(\rho))_{ij}=M_{ij}=1 \quad \text{if } (a_i, b_j)\in\rho$$

and

$$(\phi(\rho))_{ij}=M_{ij}=0 \quad \text{if } (a_i, b_j)\notin\rho.$$

Often it is useful to emphasize that the matrix M has been derived from the relation ρ, in which case we can write $M(\rho)$.

Example 6.1.1

Take a specific case where $|A|=4$, and $|B|=3$ and

$$\rho=\{ \quad (a_1, b_2), (a_1, b_3), \quad (a_2, b_1), \quad (a_3, b_1), \quad (a_4, b_2) \quad \}.$$

The corresponding matrix M is then

$$\begin{pmatrix} 0 & 1 & 1 \\ 1 & 0 & 0 \\ 1 & 0 & 0 \\ 0 & 1 & 0 \end{pmatrix}. \quad //$$

Thus we have a way of tabulating or coding the relation and we can encode by ϕ or decode by ϕ^{-1} whenever necessary, the process involved being the i, j 'look up' in $A\times B$ or M respectively. This representation is more convenient than the set-theoretic way of denoting relations since it can be handled in a mechanical way. It becomes even more amenable to relational computations if we impose some structure on the set from which the matrix elements are drawn. Again take $\{0, 1\}$ and on this set define the logical addition (**or**) and multiplication (**and**) operations; then, if M and N are $p \times q$ matrices corresponding to relations ρ and σ, the matrix Q representing the relation τ, where

$$\tau = \{(a, b): (a, b) \in \rho \text{ or } (a, b) \in \sigma\},$$

is such that its (i, j)th element

$$Q_{ij} = M_{ij} \textbf{ or } N_{ij} \quad \text{(see p. 128)}$$
$$= M_{ij} + N_{ij} \quad \text{(logical addition).}$$

Hence it is customary to call Q the *sum* of M and N, and we write

$$Q = M + N,$$

inferring that Q, M and N are the same size and Q is computed componentwise by the rule

$$Q_{ij} = M_{ij} + N_{ij}.$$

This is an instance of using a commutative diagram, Figure 6.1, to derive an operation on one set using an operation on another set via a suitable mapping, ϕ. The usual identity associated with this diagram is

$$\phi(\rho \cup \sigma) = \phi(\rho) + \phi(\sigma).$$

Fig. 6.1

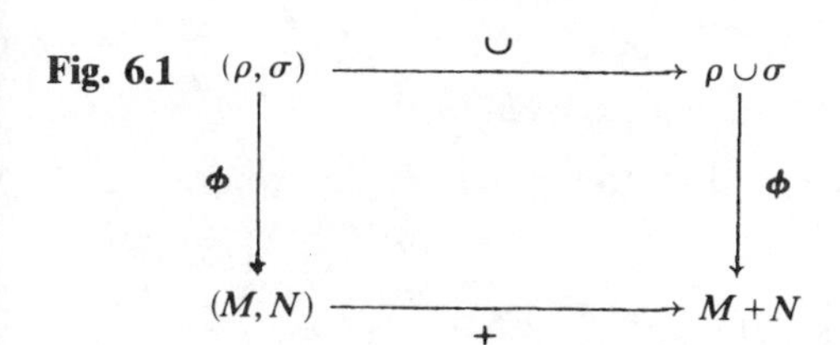

But this can be manipulated to give an explicit definition of matrix addition.

$$M + N = \phi(\rho) + \phi(\sigma)$$
$$= \phi(\rho \cup \sigma)$$
$$= \phi(\phi^{-1}(M) \cup \phi^{-1}(N)).$$

Example 6.1.1 *(continued)*

With A and B as before, let

$$\sigma = \{(a_1, b_1), (a_1, b_2),$$
$$(a_2, b_1),$$
$$(a_3, b_2)$$
$$\}.$$

So

$$N=\begin{pmatrix}1 & 1 & 0\\ 1 & 0 & 0\\ 0 & 1 & 0\\ 0 & 0 & 0\end{pmatrix}.$$

Further we see that

$$\begin{aligned}\rho\cup\sigma=\{&(a_1, b_1), (a_1, b_2), (a_1, b_3),\\ &(a_2, b_1),\\ &(a_3, b_1), (a_3, b_2),\\ &\qquad\qquad (a_4, b_2) \qquad\qquad \}\end{aligned}$$

and, equivalently,

$$M+N=\begin{pmatrix}1 & 1 & 1\\ 1 & 0 & 0\\ 1 & 1 & 0\\ 0 & 1 & 0\end{pmatrix}. \quad //$$

Moreover, if now we take a set $C=\{c_1, c_2, c_3, c_4, c_5\}$ and have a relation π between B and C defined by

$$\begin{aligned}\pi=\{&(b_1, c_1), \qquad\qquad\qquad\qquad (b_1, c_5),\\ &\qquad (b_2, c_2),\\ &\qquad\qquad\qquad\qquad (b_3, c_4), (b_3, c_5)\},\end{aligned}$$

then trivially this can be represented by the matrix P, say, where

$$P=\begin{pmatrix}1 & 0 & 0 & 0 & 1\\ 0 & 1 & 0 & 0 & 0\\ 0 & 0 & 0 & 1 & 1\end{pmatrix}.$$

Obviously the composite relation $\pi\circ\rho$ between A and C is well-defined and therefore should correspond to a 4×5 matrix. Call this matrix S. How do we calculate S? All we need is to be able to compute S_{ij} for all i, j where $1\leqslant i\leqslant 4$ and $1\leqslant j\leqslant 5$. By the bijection we know that $S_{ij}=1$ iff $(a_i, c_j)\in\pi\circ\rho$. But this is so only if there is some $b\in B$ such that $(a_i, b)\in\rho$ and $(b, c_j)\in\pi$. That is

$$\begin{aligned}(a_i, c_j)\in\pi\circ\rho\equiv\ &(a_i, b_1)\in\rho \ \textbf{and}\ (b_1, c_j)\in\pi\\ &\textbf{or}\\ &(a_i, b_2)\in\rho \ \textbf{and}\ (b_2, c_j)\in\pi\\ &\textbf{or}\\ &(a_i, b_3)\in\rho \ \textbf{and}\ (b_3, c_j)\in\pi.\end{aligned}$$

Equivalently

$$S_{ij} = M_{i1} * P_{1j} \quad \text{(the \textbf{and} operation of p. 128)}$$
$$+$$
$$M_{i2} * P_{2j}$$
$$+$$
$$M_{i3} * P_{3j}$$
$$= \sum_{k=1}^{3} M_{ik} * P_{kj}.$$

The matrix S, computed by this rule, is called the *product* of M and P, written $M * P$, or simply MP.

Again we have a natural (commutative) relationship between the derived operators as in Figure 6.2. Hence

$$M * P = \phi(\phi^{-1}(P) \circ \phi^{-1}(M)).$$

Fig. 6.2

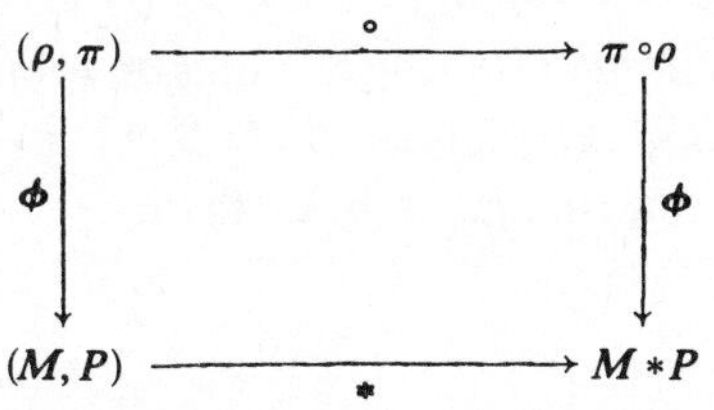

$$\begin{array}{ccc} (\rho, \pi) & \xrightarrow{\circ} & \pi \circ \rho \\ \downarrow \phi & & \downarrow \phi \\ (M, P) & \xrightarrow[*]{} & M * P \end{array}$$

Note. This order reversal under ϕ is a consequence of the way the matrix of a relation is defined; if instead we defined the matrix of a relation as

$$M_{ij} = 1 \quad \text{iff } (a_j, b_i) \in \rho,$$

then the order reversal would not take place. Although this would be mathematically more desirable it would break with the conventional practice for relations. The corresponding diagrams of Section 6.3 do *not* reverse order but follow conventions natural to the topics studied in *that* section.

***Example* 6.1.1** (continued)

Carrying out the calculations on the relations specified gives

$$MP = \begin{pmatrix} 0 & 1 & 1 \\ 1 & 0 & 0 \\ 1 & 0 & 0 \\ 0 & 1 & 0 \end{pmatrix} \begin{pmatrix} 1 & 0 & 0 & 0 & 1 \\ 0 & 1 & 0 & 0 & 0 \\ 0 & 0 & 0 & 1 & 1 \end{pmatrix} = \begin{pmatrix} 0 & 1 & 0 & 1 & 1 \\ 1 & 0 & 0 & 0 & 1 \\ 1 & 0 & 0 & 0 & 1 \\ 0 & 1 & 0 & 0 & 0 \end{pmatrix}. \quad //$$

To recap, if matrices M and N are the same size then their sum exists and is defined by

$$(M + N)_{ij} = M_{ij} + N_{ij}$$

and if M and P are compatible (M is p by q and P is q by r say) then multiplication is possible and is defined thus

$$(MP)_{ij} = \sum_{k=1}^{q} M_{ik} * P_{kj}.$$

Although these matrices are over $(\mathbb{Z}_2, \wedge, \vee)$, the symbols $*$ and $+$ have been used to suggest the possible application to matrices over different structures (see the next section). Henceforth $\wedge$ and $\vee$ will be used in cases where general operations are inadequate.

Throughout the remainder of the current section we restrict consideration to matrices representing relations *on* a finite set, A say, where $|A| = n$, then all matrices are compatible and so matrix sums and products are always defined.

As a direct consequence of the elementwise definition of addition it trivially follows that matrix addition is commutative and there is a *zero* $(n \times n)$ *matrix* 0: $0_{ij} = 0$ for all i, j such that $1 \leq i \leq n$ and $1 \leq j \leq n$. On the other hand, multiplication is in general not commutative but there is an identity, called the $n \times n$ *unit matrix* defined by I: $I_{ij} = 1$ if $i = j$ and 0 otherwise; for if X is an $n \times n$ matrix and $Y = XI$ then

$$Y_{ij} = \sum_{p=1}^{n} X_{ip} * I_{pj}$$

$$= \sum_{\substack{p=1 \\ (p \neq j)}}^{n} X_{ip} * I_{pj} + X_{ij} * I_{jj}.$$

Now all $I_{pj} = 0$ except when $p = j$ so the summation excluding the case $p = j$ results in zero (being the sum of products each of which has a zero factor) and $I_{jj} = 1$, the identity, thus

$$Y_{ij} = X_{ij}$$

and so

$$Y = X.$$

Therefore

$$X = XI,$$

similarly for IX and hence

$$IX = X = XI. \quad //$$

Unfortunately, multiplicative inverses may not exist but if they do they are unique. A square matrix having an inverse is said to be *invertible.*

Example 6.1.2

There is no matrix X such that

$$X\begin{pmatrix}0 & 0\\0 & 1\end{pmatrix}=I.$$

Proof. Evaluation of this product gives

$$\begin{pmatrix}0 & X_{12}\\0 & X_{22}\end{pmatrix}.$$

Hence, regardless of what values are present in X, the $(1, 1)$ element of the product can never be 1 and the result follows. //

Thus the set of square matrices of a given size, with the addition and multiplication as defined above, together constitute a ring.

Exploiting further the association between binary relations on a set and matrices over $(\mathbb{Z}_2, \wedge, \vee)$ we make the following definitions: the *transpose* of the matrix M is the matrix M^T such that

$$M_{ij}^T = M_{ji}$$

(so, if M is derived from the relation σ, M^T can be derived from σ^{-1}), and the *transitive closure* M^+ and the *reflexive closure* M^* (corresponding isomorphically to σ^+ and σ^*) are defined in the predictable way:

$$M^+ = \sum_{n=1}^{\infty} M^n$$

and

$$M^* = \sum_{n=0}^{\infty} M^n$$

where

$$M^0 = I,$$

$$M^1 = M$$

and

$$M^{n+1} = MM^n \quad (n \in \mathbb{N}).$$

(These closures may not be well-defined (convergent) in some cases but all goes well over $(\mathbb{Z}_2, \wedge, \vee)$ since this is a closed semiring.)

Finally in this section we note that matrices can be partially ordered, elementwise, by

$$M \leqslant N \quad \text{iff } M_{ij} \leqslant N_{ij} \quad \text{for all } i, j,$$

the ordering being given by the definition

$$M \leqslant N \quad \text{iff } M + N = N$$

whenever $+$ is a 'maximum' operation such as **or**.

Exercises 6.1

1. If A is a set such that $|A| = n$ and M is a matrix over $(\mathbb{Z}_2, \wedge, \vee)$ corresponding to some binary relation on A show that

$$M^+ = \sum_{p=1}^{n} M^p.$$

(A consequence of this exercise is that in most practical cases instead of being concerned with the semiring $(\mathbb{Z}_2, \wedge, \vee)$ we can use the Boolean algebra $(\mathbb{B}, \wedge, \vee, \neg)$ where $\mathbb{B} = \{0, 1\}$. Hence we shall often refer to the set of $n \times n$ relational matrices as $\mathcal{M}(n, \mathbb{B})$ and call them *Boolean matrices.*)

2. Prove that if M is a finite square matrix over $(\mathbb{Z}_2, \wedge, \vee)$ then

$$M^* = (I + M)^+.$$

Hint: see Question 1.

3. Show that if the matrix M over $(\mathbb{Z}_2, \wedge, \vee)$ is such that

$$I \leqslant M$$

then

$$M^n \leqslant M^{n+1} \quad \text{for any } n \in \mathbb{N}.$$

Hence prove that if M is $m \times m$ and $p \geqslant m$ then

$$M^* = (I + M)^p$$

and, that $M^* = M^{2^q}$ for some q such that $2^q \geqslant m$.

4. Show that if the inverse M^{-1} of the matrix M exists (so $MM^{-1} = I = M^{-1}M$) then it must be unique. Furthermore, if N is invertible and compatible with M, show that

$$(NM)^{-1} = M^{-1}N^{-1}.$$

5. Show that if A, B and C are compatible matrices such that

$$A * B = 0 = A * C$$

then this does not imply that $B = C$.

6.2 Matrices over other algebraic structures

So much for matrices over the semiring $(\mathbb{Z}_2, \wedge, \vee)$ and $(\mathbb{B}, \wedge, \vee, \neg)$, but what about matrices whose elements are drawn from other structures? First we consider what exactly is required of the structure $(S, *, +)$ in order that matrices over S will behave in the ways we have come to expect. This is followed by discussion of a computational technique applicable in situations when the closure operation is well-defined.

6.2.1 General matrices

Let $(\mathcal{M}, \circledast, \oplus)$ denote the set $\mathcal{M}(n, (S, *, +))$ of $n \times n$ matrices over $(S, *, +)$ with $\circledast$ and $\oplus$ defined in terms of $*$ and $+$. Trivially matrices can be defined over any non-empty set, however the induced matrix operations may not be well-defined nor behave reasonably. First we consider addition.

If M and N are compatible matrices over S, then, by definition,

$$(M \oplus N)_{ij} = M_{ij} + N_{ij}.$$

Only one application of $+$ is used to obtain $\oplus$ so a matrix addition is easily obtained, but in order that $\oplus$ should be commutative and associative the same properties must be possessed by $(S, +)$. It also follows that zero matrices in $\mathcal{M}$ exist iff there is a two-sided identity, 0, wrt $+$ in S; likewise additive inverses in $\mathcal{M}$ depend on additive inverses in S. So addition $(\oplus)$ on $\mathcal{M}$ can always be defined and although many properties are required of $(S, +)$ in order to make $(\mathcal{M}, \oplus)$ a commutative group, much can be done in situations where $+$ is less well-behaved.

In contrast multiplication in $\mathcal{M}$ requires much more of S. By analogy with matrices over $(\mathbb{Z}_2, \wedge, \vee)$ we take

$$(M \circledast N)_{ij} = \sum_k M_{ik} * N_{kj}.$$

Here the summation is in S and therefore, for multiplication in $\mathcal{M}$ to be well-defined, we require that the order in which the summation is performed does not affect the resulting sum. This can be guaranteed if *addition* in S is associative.

We do not expect multiplication in $\mathcal{M}$ to be commutative, but – because of the dependence of $(\mathcal{M}, \circledast)$ on both $(S, *)$ and $(S, +)$ – even to ensure associativity in $(\mathcal{M}, \circledast)$ requires more than associativity in $(S, *)$. As exemplified by the following calculation, left and right distributivity of $*$ over $+$, associativity in $(S, *)$ and associativity and commutativity in $(S, +)$ will suffice.

Example 6.2.1

Take M, N and P as 1×2, 2×3 and 3×1 matrices respectively, then

$$((M \circledast N) \circledast P)_{11} = \sum_{j=1}^{3} (M \circledast N)_{1j} * P_{j1}$$

$$= \sum_{j=1}^{3} \left(\sum_{i=1}^{2} M_{1i} * N_{ij} \right) * P_{j1}$$

$$
\begin{aligned}
&= ((M_{11} * N_{11}) + (M_{12} * N_{21})) * P_{11} \\
&\quad + ((M_{11} * N_{12}) + (M_{12} * N_{22})) * P_{21} \\
&\quad + ((M_{11} * N_{13}) + (M_{12} * N_{23})) * P_{31} \\
&= ((M_{11} * N_{11}) * P_{11}) + ((M_{12} * N_{21}) * P_{11}) \\
&\quad + ((M_{11} * N_{12}) * P_{21}) + ((M_{12} * N_{22}) * P_{21}) \\
&\quad + ((M_{11} * N_{13}) * P_{31}) + ((M_{12} * N_{23}) * P_{31}) \\
&\qquad\qquad\qquad\qquad\qquad\qquad\qquad \text{(left distributivity)} \\
&= (M_{11} * (N_{11} * P_{11})) + (M_{12} * (N_{21} * P_{11})) \\
&\quad + (M_{11} * (N_{12} * P_{21})) + (M_{12} * (N_{22} * P_{21})) \\
&\quad + (M_{11} * (N_{13} * P_{31})) + (M_{12} * (N_{23} * P_{31})) \\
&\qquad\qquad\qquad\qquad\qquad\qquad\qquad (* \text{ associative}) \\
&= (M_{11} * (N_{11} * P_{11})) + (M_{11} * (N_{12} * P_{21})) \\
&\qquad\qquad + (M_{11} * (N_{13} * P_{31})) \\
&\quad + (M_{12} * (N_{21} * P_{11})) + (M_{12} * (N_{22} * P_{21})) \\
&\qquad\qquad + (M_{12} * (N_{23} * P_{31})) \\
&\qquad\qquad\qquad\qquad\qquad\qquad\qquad (+ \text{ commutative}) \\
&= M_{11} * ((N_{11} * P_{11}) + (N_{12} * P_{21}) + (N_{13} * P_{31})) \\
&\quad + M_{12} * ((N_{21} * P_{11}) + (N_{22} * P_{21}) + (N_{23} * P_{31})) \\
&\qquad\qquad\qquad\qquad\qquad\qquad\qquad \text{(right distributivity)} \\
&= \sum_{i=1}^{2} M_{1i} * \left(\sum_{j=1}^{3} N_{ij} * P_{j1} \right) \\
&= (M \circledast (N \circledast P))_{11}.
\end{aligned}
$$

Throughout this derivation + is assumed to be associative and hence some brackets have been omitted. //

All this looks rather depressing. Surely not many structures can satisfy the conditions laid down for $(S, *, +)$! As a final blow, notice also that in order for unit matrices to exist we require that

(a) $0 * x = 0 = x * 0$ for all $x \in S$
where 0 is the additive identity of S (recall that this is *not* a field axiom) and

(b) S should have a two-sided multiplicative identity, 1.

However, as the reader should verify, if $(S, *, +)$ is a *ring* or a *field* then all the above conditions are met and $(\mathcal{M}, \circledast, \oplus)$ is a *ring*.

The concepts of ordering and closure of matrices over more specialized structures, such as ordered fields, etc., become very involved and will not be discussed here.

We have already noted, in Exercise 6.1, that (infinite sums apart) we can regard $\mathcal{M}(n, (\mathbb{Z}_2, \wedge, \vee))$ as $\mathcal{M}(n, \mathbb{B})$. Similarly, given other related algebraic structures, we can generalize the calculations but care needs to be taken to stipulate which system is in use. For instance $\mathbb{Z}_2 \subseteq \mathbb{Z}_3$ and by the usual inclusion $\mathbb{Z}_2 \hookrightarrow \mathbb{Z}_3$ ($A \hookrightarrow B$ denotes the identity mapping on A where $A \subseteq B$) it follows that

$$A \in \mathcal{M}(n, \mathbb{Z}_2) \Rightarrow A \in \mathcal{M}(n, \mathbb{Z}_3).$$

Put another way, given a matrix of 0s and 1s is it defined over $\{0, 1\}$ or $\{0, 1, 2\}$? When it comes to doing arithmetic we need to know. Of course it all depends on what we are trying to calculate as we shall see in the next section. Similar identification problems arise with any inclusion especially $\mathcal{M}(n, \mathbb{B})$, $\mathcal{M}(n, \mathbb{Z}_2)$, $\mathcal{M}(n, \mathbb{Z}_m)$, $\mathcal{M}(n, \mathbb{Z})$, $\mathcal{M}(n, \mathbb{Q})$ and $\mathcal{M}(n, \mathbb{R})$.

6.2.2 Warshall's algorithm

In this section we describe a quick way to calculate the (transitive or reflexive) closure of square matrices over $(\mathbb{Z}_2, \wedge, \vee)$. This is a variation of a method due to Warshall and is presented in the form of a program. If M is an $n \times n$ matrix over $(\mathbb{Z}_2, \wedge, \vee)$ then it can be changed into M^+ by the following

for j **from** 1 **to** n **do**
  for i **from** 1 **to** n **do**
    if $i \neq j$ **and** $M_{ij} = 1$ **then**
      for k **from** 1 **to** n **do**
        $M_{ik} \leftarrow M_{ik} \vee M_{jk}$.

To compute M^* we can add I either at the beginning or at the end, using the result

$$M^* = (M + I)^+ = M^+ + I.$$

Example 6.2.2

$$\text{If } M = \begin{pmatrix} 0 & 1 & 1 & 0 & 0 \\ 0 & 0 & 0 & 1 & 1 \\ 0 & 1 & 0 & 0 & 0 \\ 0 & 0 & 0 & 0 & 0 \\ 0 & 0 & 1 & 1 & 0 \end{pmatrix}$$

then, proceeding through the program, the following elements are in turn changed from 0 to 1: (1, 4), (1, 5), (3, 4), (3, 5), (5, 2), (5, 5), (2, 2),

(2, 3) and(3, 3) to obtain

$$M^+ = \begin{pmatrix} 0 & 1 & 1 & 1 & 1 \\ 0 & 1 & 1 & 1 & 1 \\ 0 & 1 & 1 & 1 & 1 \\ 0 & 0 & 0 & 0 & 0 \\ 0 & 1 & 1 & 1 & 1 \end{pmatrix}. \quad //$$

We shall not offer a proof that the program does carry out the desired process but make the comment that the inner section merely says that

> if $i \neq j$ and $M_{ij} = 1$ then $(i, j) \in \sigma^p$ for some $p \leqslant n$ where σ is the relation associated with M, so adjust row i to row $i \vee$ row j to indicate that anything related to j is also related to i.

Obviously if $i = j$ there is nothing to be gained by executing this step. The difficult part of the validation of the program is to ensure that nothing is missed – after reading Chapter 7 you should be able to prove this for yourself.

Warshall's method is a great improvement over direct use of the definition of closure; it can be further improved but this requires use of more sophisticated data structures and hence we shall not pursue the topic further.

The method can be adapted in many ways to compute various costs associated with matrices/relations, the simplest of these being the 'distance' between σ-related elements of a set. Informally the distance between points x and y, $d(x, y)$, is the smallest n such that $n \in \mathbb{N}$ and $y \in \sigma^n(x)$. If M is the matrix corresponding to σ then replacing the body of the i-loop by the following gives the desired result.

> **if** $M_{ij} \neq 0$ **then**
>
>   **for** k **from** 1 **to** n **do**
>
>   **if** $M_{jk} \neq 0$ **and**
>
>   ($M_{ik} = 0$ **or** $M_{ik} > M_{ij} + M_{jk}$)
>
>   **then** $M_{ik} \leftarrow M_{ij} + M_{jk}$.

Here the arithmetic is over $\mathbb{Z}$ (not $\mathbb{Z}_2$).

Example 6.2.3

Applying this procedure to the matrix in the previous example gives

$$\begin{pmatrix} 0 & 1 & 1 & 2 & 2 \\ 0 & 3 & 2 & 1 & 1 \\ 0 & 1 & 3 & 2 & 2 \\ 0 & 0 & 0 & 0 & 0 \\ 0 & 2 & 1 & 1 & 3 \end{pmatrix}. \quad //$$

Exercises 6.2

1. ***Definition.*** Two $n \times n$ matrices X and Y, over a field F are said to be *similar* over F if there is an invertible matrix P over F such that

   $X = P^{-1}YP. \quad //$

   Show that the relation induced by similarity on $\mathcal{M}(n, F)$ is an equivalence relation and that I is similar only to itself.

2. If A, $B \in \mathcal{M}(n, F)$ for some given $n \in \mathbb{N}$ and some field F, then show that

   $(A+B)^T = A^T + B^T$

   and

   $(AB)^T = B^T A^T.$

3. ***Definition.*** A *stochastic matrix* is a real matrix over $(\{x: 0 \leqslant x \leqslant 1\}, *, +)$ and is subject to the restriction that the sum of elements in each row is 1. $//$

   Show that the set of 3×3 stochastic matrices is *not* closed wrt addition but *is* closed wrt multiplication. State explicitly what field properties (from $\mathbb{R}$) are used.

6.3 Matrices and vector spaces

The work of the previous section shows how matrices may be used to implement relations between finite sets of appropriate size. Our aim in this section is to show how matrices over the field $(\mathbb{R}, *, +)$, henceforth written $\mathbb{R}$, are used to implement certain relations, specifically linear transformations, on a vector space, V, over $\mathbb{R}$. In the notation established in Section 5.4, if $T \in \mathrm{End}(V)$, the set of linear transformations of V, then

$$\{(\mathbf{x}, T\mathbf{x}): \mathbf{x} \in V\} \subseteq V \times V$$

is a binary relation on V; we seek to implement linear transformations on V in $\mathcal{M}(n, \mathbb{R})$, where $n = \dim(V)$. Two generalizations are immediate; firstly, everything works in an arbitrary field and secondly, linear mappings between vector spaces of different dimensions may be implemented in a similar way in $\mathcal{M}(n, m, \mathbb{R})$. As we have no requirement for either of these in the sequel they will not be considered further.

6.3.1 Matrix representation of linear transformations

The operations of matrix addition and multiplication within $\mathcal{M}(n, \mathbb{R})$ have been defined implicitly in Section 6.2. If $A \in \mathcal{M}(n, \mathbb{R})$ has elements A_{ij} and $\lambda \in \mathbb{R}$ we define $\lambda A \in \mathcal{M}(n, \mathbb{R})$ to be the matrix with elements λA_{ij}; this is known as *scalar multiplication of matrices.* $\mathcal{M}(n, \mathbb{R})$ plays a central role in the remainder of this chapter and it is helpful to summarize its properties under the operations defined. The unit matrix of $\mathcal{M}(n, \mathbb{R})$ is written I for all $n \in \mathbb{N}$.

Proposition. $\mathcal{M}(n, \mathbb{R})$ is a linear algebra. //

No proof of this is given; we recommend that the reader verifies the axioms of a linear algebra for $\mathcal{M}(2, \mathbb{R})$ for he will then see how a general proof may be constructed. The steps are to show that

(i) $\mathcal{M}(n, \mathbb{R})$ is a vector space,
(ii) $\mathcal{M}(n, \mathbb{R})$ is a ring,
(iii) scalar multiplication is related to matrix multiplication by the following property

$$\lambda(AB) = (\lambda A)B = A(\lambda B) \quad \text{for all } \lambda \in \mathbb{R} \text{ and } A, B \in \mathcal{M}(n, \mathbb{R}).$$

It is useful to have a notation for the subset of $\mathcal{M}(n, \mathbb{R})$ of invertible matrices; we write

$$GL(n, \mathbb{R}) = \{A \in \mathcal{M}(n, \mathbb{R}) \colon A^{-1} \text{ exists}\}.$$

Each $GL(n, \mathbb{R})$, for $n \in \mathbb{N}$, determines a group under matrix multiplication. These groups are known as the *general linear* groups.

Throughout the remainder of this section V is a vector space of dimension n over $\mathbb{R}$ and $T \in \operatorname{End}(V)$. If $B = \{\mathbf{e}_1, \ldots, \mathbf{e}_n\}$ is a basis for V then clearly $T\mathbf{e}_i \in V$ for all $1 \leq i \leq n$, hence there must exist $t_{ij} \in \mathbb{R}$, $1 \leq i, j \leq n$ such that

$$\begin{aligned} T\mathbf{e}_1 &= t_{11}\mathbf{e}_1 + t_{21}\mathbf{e}_2 + \ldots + t_{n1}\mathbf{e}_n \\ &\vdots \\ T\mathbf{e}_n &= t_{1n}\mathbf{e}_1 + t_{2n}\mathbf{e}_2 + \ldots + t_{nn}\mathbf{e}_n. \end{aligned}$$

If A_T is the matrix

$$A_T = \begin{pmatrix} t_{11} & t_{12} & \dots & t_{1n} \\ t_{21} & t_{22} & \dots & t_{2n} \\ \vdots & & & \\ t_{n1} & t_{n2} & \dots & t_{nn} \end{pmatrix}$$

then A_T is called the *matrix of T with respect to the basis B*. Given B, A_T is unique and we may thus define a mapping

$$\phi^B: \mathrm{End}\,(V) \to \mathcal{M}(n, \mathbb{R})$$

by

$$\phi^B(T) = A_T.$$

Once A_T has been computed we may use it to implement T according to the following proposition.

Proposition. If $T \in \mathrm{End}\,(V)$ has matrix A_T with respect to the basis $B = \{\mathbf{e}_1, \dots, \mathbf{e}_n\}$ then if $\mathbf{x} \in V$ is the vector

$$\mathbf{x} = \sum_{i=1}^{n} \lambda_i \mathbf{e}_i$$

$T\mathbf{x} \in V$ is given by

$$T\mathbf{x} = \sum_{i=1}^{n} \mu_i \mathbf{e}_i,$$

where

$$\begin{pmatrix} \mu_1 \\ \vdots \\ \mu_n \end{pmatrix} = A_T \begin{pmatrix} \lambda_1 \\ \vdots \\ \lambda_n \end{pmatrix}.$$

Proof. If $\mathbf{x} = \sum_{i=1}^{n} \lambda_i \mathbf{e}_i$

then

$$\begin{aligned} T\mathbf{x} &= \sum_{i=1}^{n} \lambda_i T\mathbf{e}_i \\ &= \sum_{i=1}^{n} \lambda_i \left(\sum_{j=1}^{n} t_{ji} \mathbf{e}_j \right) \\ &= \sum_{i=1}^{n} \left(\sum_{j=1}^{n} t_{ij} \lambda_j \right) \mathbf{e}_i. \quad // \end{aligned}$$

Hence in a given basis for V a linear transformation T may be implemented by forming the matrix product $A_T \Lambda$ of the $n \times n$ matrix

A_T and the $n \times 1$ matrix (or coordinate vector)

$$\Lambda = \begin{pmatrix} \lambda_1 \\ \vdots \\ \lambda_n \end{pmatrix}$$

corresponding to each $\mathbf{x} \in V$. In fact much more is true and the important properties of ϕ^B are stated below.

Proposition. The mapping ϕ^B: End $(V) \to \mathcal{M}(n, \mathbb{R})$ is a linear algebra isomorphism under the usual operations on End (V) and $\mathcal{M}(n, \mathbb{R})$ over $\mathbb{R}$; further the restriction of ϕ^B to the group Aut $(V) \subseteq$ End (V) is a group isomorphism onto $GL(n, \mathbb{R})$.

Sketch proof. To avoid a 'sea' of indices we outline a proof for $n = 2$. The general case is conceptually identical. The linear algebra isomorphism follows quickly from the following,

(i) ϕ^B is bijective,

(ii) $\phi^B(I_V) = I$,

(iii) $\phi^B(ST) = \phi^B(S)\phi^B(T)$ for all $S, T \in$ End (V)

(iv) $\phi^B(\lambda S + \mu T) = \lambda\phi^B(S) + \mu\phi^B(T)$ for all $S, T \in$ End (V) and $\lambda, \mu \in \mathbb{R}$.

For these we provide partial proofs; the rest is left as an exercise. Throughout the proof $\{\mathbf{e}_1, \mathbf{e}_2\}$ is a basis for V.

(i) To show that ϕ^B is injective we need to demonstrate that

$$\phi^B(S) = \phi^B(T) \Rightarrow S = T.$$

If

$$S\mathbf{e}_1 = s_{11}\mathbf{e}_1 + s_{21}\mathbf{e}_2 \quad \text{and} \quad T\mathbf{e}_1 = t_{11}\mathbf{e}_1 + t_{21}\mathbf{e}_2$$

then

$$\phi^B(S) = \phi^B(T) \Rightarrow s_{11} = t_{11} \quad \text{and} \quad s_{21} = t_{21}$$

so that $S\mathbf{e}_1 = T\mathbf{e}_1$; similarly, $S\mathbf{e}_2 = T\mathbf{e}_2$, hence S and T are identical on the basis elements $\mathbf{e}_1$ and $\mathbf{e}_2$. But for all $\mathbf{x} \in V$ we have

$$\mathbf{x} = \lambda\mathbf{e}_1 + \mu\mathbf{e}_2$$

thus

$$\begin{aligned} S\mathbf{x} &= \lambda S\mathbf{e}_1 + \mu S\mathbf{e}_2 \\ &= \lambda T\mathbf{e}_1 + \mu T\mathbf{e}_2 = T(\lambda\mathbf{e}_1 + \mu\mathbf{e}_2) \\ &= T\mathbf{x}, \end{aligned}$$

i.e.

$$S = T.$$

We leave the proof that ϕ^B is surjective as an exercise.

(ii) $I_V\mathbf{x}=\mathbf{x}$ for all $\mathbf{x}\in V$, hence

$$I_V\mathbf{e}_1=\mathbf{e}_1+0\mathbf{e}_2$$
$$I_V\mathbf{e}_2=0\mathbf{e}_1+\mathbf{e}_2$$

and

$$\phi^B(I_V)=\begin{pmatrix}1 & 0\\ 0 & 1\end{pmatrix}$$
$$=I.$$

(iii) Let

$$\phi^B(S)=\begin{pmatrix}s_{11} & s_{12}\\ s_{21} & s_{22}\end{pmatrix} \quad\text{and}\quad \phi^B(T)=\begin{pmatrix}t_{11} & t_{12}\\ t_{21} & t_{22}\end{pmatrix},$$

then

$$T\mathbf{e}_1=t_{11}\mathbf{e}_1+t_{21}\mathbf{e}_2$$

and

$$\begin{aligned}ST\mathbf{e}_1&=t_{11}S\mathbf{e}_1+t_{21}S\mathbf{e}_2\\ &=t_{11}(s_{11}\mathbf{e}_1+s_{21}\mathbf{e}_2)+t_{21}(s_{12}\mathbf{e}_1+s_{22}\mathbf{e}_2)\\ &=(t_{11}s_{11}+t_{21}s_{12})\mathbf{e}_1+(t_{11}s_{21}+t_{21}s_{22})\mathbf{e}_2.\end{aligned}$$

Similarly,

$$ST\mathbf{e}_2=(t_{12}s_{11}+t_{22}s_{12})\mathbf{e}_1+(t_{12}s_{21}+t_{22}s_{22})\mathbf{e}_2.$$

Therefore

$$\phi^B(ST)=\phi^B(S)\phi^B(T)$$

as required.

(iv) Follows by an argument similar in spirit to that of (iii).

To show that the restriction of ϕ^B to Aut (V) is a group isomorphism we use properties (ii) and (iii).

If $T\in$ Aut (V) then

$$\begin{aligned}I=\phi^B(I_V)&=\phi^B(TT^{-1})\\ &=\phi^B(T)\phi^B(T^{-1})\end{aligned}$$

therefore

$$T\in \text{Aut}\,(V)\Rightarrow\phi^B(T)\in GL\,(n,\mathbb{R})$$

and

$$\phi^B(T)^{-1}=\phi^B(T^{-1}). \quad /\!/$$

This result should be pondered in depth for its consequences are far-reaching. To assist the reader we spell out some of the more important implications in detail. In a given basis the mapping $T\rightarrow A_T$ from End (V)

to $\mathcal{M}(n, \mathbb{R})$ is bijective (a unique matrix for each transformation and vice versa); further

$$A_{S\circ T} = A_S A_T$$

or, equivalently, the diagram of Figure 6.3 is commutative. In practice this simply means that the matrix of a product transformation $S\circ T$ in End (V) may be computed by multiplication of the matrices A_S and A_T in order; it is *not* necessary to determine $S\circ T$ explicitly and then compute $A_{S\circ T}$ from the definition. Similarly for $S+T$. In addition on restricting ϕ^B to the invertible transformations Aut (V) we have

$$A_{T^{-1}} = A_T^{-1};$$

this means that to compute $A_{T^{-1}}$ we just need to invert the matrix A_T.

Fig. 6.3

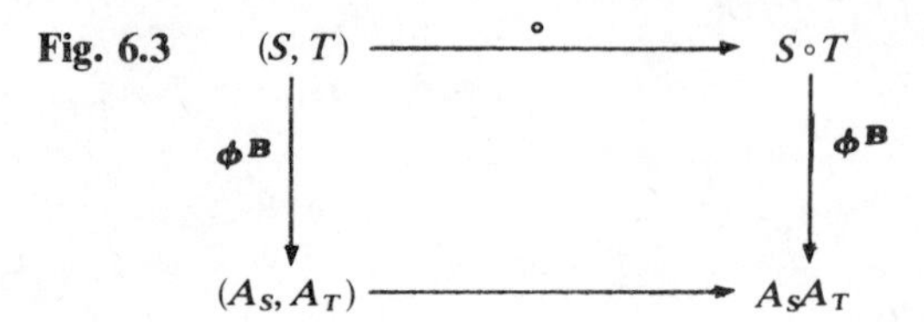

We stress again that the isomorphism ϕ^B is basis dependent, an alternative basis for V providing a second isomorphism between End (V) and $\mathcal{M}(n, \mathbb{R})$. An element of $\mathcal{M}(n, \mathbb{R})$ may thus be regarded as representative of an element of End (V) in a fixed basis, or as representative of distinct elements of End (V) in different bases. In Chapter 10, where the work of this section is applied extensively to $\mathbb{R}^n$ ($n=2, 3, 4$), the mapping ϕ^B is computed with respect to the standard basis

$$\{\hat{\mathbf{e}}_1, \ldots, \hat{\mathbf{e}}_n\}$$

as defined in Section 5.4, for which, if $\mathbf{x} = (x_1, \ldots, x_n) \in \mathbb{R}^n$, we have

$$\mathbf{x} = \sum_{i=1}^{n} x_i \hat{\mathbf{e}}_i$$

and

$$T\mathbf{x} = A_T \begin{pmatrix} x_1 \\ \vdots \\ x_n \end{pmatrix}.$$

If $T \in$ End (V) has eigenvector $\mathbf{x} \in V$ with eigenvalue $\sigma \in \mathbb{R}$ then if, in the basis

$$B = \{\mathbf{e}_1, \ldots, \mathbf{e}_n\}$$

of V, $\mathbf{x}$ has the form

$$\mathbf{x} = \sum_{i=1}^{n} \lambda_i \mathbf{e}_i,$$

we have

$$A_T\begin{pmatrix}\lambda_1\\ \vdots\\ \lambda_n\end{pmatrix}=\sigma\begin{pmatrix}\lambda_1\\ \vdots\\ \lambda_n\end{pmatrix},$$

so that the coordinate vector of $\mathbf{x}$ is an eigenvector of A_T with the same eigenvalue.

6.3.2 Some other matrix concepts

It is useful at this point to define the determinant mapping

$$\det: \mathcal{M}(n, \mathbb{R}) \to \mathbb{R}$$

on $\mathcal{M}(n, \mathbb{R})$. The 'correct' place to discuss det and its properties is really in Section 5.4 because det is basis independent. By this we mean that if $A_T \in \mathcal{M}(n, \mathbb{R})$ and $A'_T \in \mathcal{M}(n, \mathbb{R})$ represent $T \in \mathrm{End}\,(V)$ with respect to bases B and B' respectively then $\det A_T = \det A'_T$. However, to define det on End (V) would involve an excursion into tensor algebra; instead we give a familiar form for det on $\mathcal{M}(n, \mathbb{R})$ and state some of its important properties.

Our definition of det is recursive, we begin by defining it for $\mathcal{M}(2, \mathbb{R})$. If

$$A=\begin{pmatrix}a_{11} & a_{12}\\ a_{21} & a_{22}\end{pmatrix}$$

then

$$\det A = a_{11}a_{22} - a_{21}a_{12}.$$

If $A \in \mathcal{M}(n, \mathbb{R})$ has elements a_{ij}, $1 \leq i,j \leq n$ then the *minor* of an element a_{kl} is the matrix $A^{(k,l)} \in \mathcal{M}(n-1, \mathbb{R})$ obtained from A by omitting the kth row and lth column. Now $\det: \mathcal{M}(n, \mathbb{R}) \to \mathbb{R}$ may be defined recursively, for all $n \in \mathbb{N}$, as

$$\det A = \sum_{l=1}^{n} (-1)^{l+1} a_{1l} \det A^{(1,l)}.$$

This is sometimes called an expansion by the first row. It can be shown that if any other row or column of A is used to form a corresponding expression then the sum will still be $\det A$. In other words

$$\begin{aligned}\det A &= \sum_{l=1}^{n} (-1)^{l+r} a_{rl} \det A^{(r,l)} \quad \text{for all } 1 \leq r \leq n\\ &= \sum_{r=1}^{n} (-1)^{l+r} a_{rl} \det A^{(r,l)} \quad \text{for all } 1 \leq l \leq n.\end{aligned}$$

For small values of n one of these formulae may be appropriate for hand computation of $\det A$. To minimize the number of operations the expansion should proceed by the row or column containing the largest number of zeros. We stress, however, that these expansions are generally unsuitable for the evaluation of det and more efficient computational procedures are available.

Some important properties of det are summarized in the following result.

Proposition. The mapping $\det: \mathcal{M}(n, \mathbb{R}) \to \mathbb{R}$ satisfies

(i) $\det I = 1$,

(ii) $\det A = \det A^T$ for all $A \in \mathcal{M}(n, \mathbb{R})$,

(iii) $\det \lambda A = \lambda^n \det A$ for all $\lambda \in \mathbb{R}$ and for all $A \in \mathcal{M}(n, \mathbb{R})$,

(iv) $\det AB = \det A \det B$,

(v) $A \in GL(n, \mathbb{R})$ iff $\det A \neq 0$. //

No general proof of these results is given, the reader is invited to verify them for $\mathcal{M}(2, \mathbb{R})$. Notice also that (v) characterizes $GL(n, \mathbb{R})$ as $\det^{-1}(\mathbb{R}\backslash\{0\})$.

It is useful to give names to matrices with 'special' properties and the ones we use in the sequel are as follows. If $A \in \mathcal{M}(n, \mathbb{R})$ then if

$$A = A^T$$

A is said to be *symmetric*; and if

$$A = -A^T$$

A is called *antisymmetric.* If

$$AA^T = A^TA = I \quad (\text{i.e. } A^{-1} = A^T)$$

then A is said to be *orthogonal.* The set of all orthogonal matrices of $\mathcal{M}(n, \mathbb{R})$ is written $O(n)$. $SO(n)$ is the notation for the subset of $O(n)$ consisting of matrices with unit determinant. Elements of $SO(n)$ are referred to as *special orthogonal* matrices.

Proposition.

(i) $O(n)$ is a subgroup of $GL(n, \mathbb{R})$

(ii) $SO(n)$ is a subgroup of $O(n)$.

Proof.

(i) We just need to show that $O(n)$ is closed under multiplication and the inverse operation.

If $A, B \in O(n)$ then $AA^T = A^TA = I$ and $BB^T = B^TB = I$, hence

$$\begin{aligned}(AB)(AB)^T &= AB(B^TA^T)\\ &= A(BB^T)A^T\\ &= AA^T\\ &= I,\end{aligned}$$

similarly $(AB)^TAB = I$ and hence $AB \in O(n)$.

If $A \in O(n)$ then

$$\begin{aligned}(A^{-1})(A^{-1})^T &= A^T(A^T)^T\\ &= A^TA\\ &= I, \quad \text{etc.}\end{aligned}$$

hence $A^{-1} \in O(n)$.

(ii) The proof of this is similar and left as an exercise. //

We close this section with a brief discussion of functions of matrices; like det this really belongs in Section 5.4 but we did not develop the machinery for a discussion of infinite sums in End (V). The informal discussion in $\mathcal{M}(n, \mathbb{R})$ that follows is sufficient for our purposes.

If $A \in \mathcal{M}(n, \mathbb{R})$ then $A^2 \in \mathcal{M}(n, \mathbb{R})$ is the matrix AA. Similarly A^k can be defined for all $k \in \mathbb{N}$, $k > 2$ and A^0 is taken to be I; hence if

$$p : \mathbb{R} \to \mathbb{R}$$

is the polynomial function

$$p(x) = \sum_{i=0}^{N} \alpha_i x^i \quad \text{where } \alpha_i \in \mathbb{R} \quad \text{and } n \in \mathbb{N}$$

we define $p(A) \in \mathcal{M}(n, \mathbb{R})$ by

$$p(A) = \sum_{i=0}^{N} \alpha_i A^i$$

and this is well-defined because $\mathcal{M}(n, \mathbb{R})$ is a vector space. This idea can be extended: if

$$f : \mathbb{R} \to \mathbb{R}$$

has a convergent series expansion

$$f(x) = \sum_{i=0}^{\infty} a_i x^i$$

it is possible to make sense of the expression $\sum_{i=0}^{\infty} a_i A^i$. The set $\mathcal{M}(n, \mathbb{R})$ may be identified with $\mathbb{R}^{n^2}$ and we may define a norm on $\mathbb{R}^{n^2}$ as shown in Section 5.4; this induces a norm on $\mathcal{M}(n, \mathbb{R})$ and convergence of infinite sums can be discussed in a meaningful way. In particular, if $A \in \mathcal{M}(n, \mathbb{R})$, it can be shown that

$$\lim_{N\to\infty} \sum_{k=0}^{N} \frac{A^k}{k!}$$

always exists in $\mathcal{M}(n, \mathbb{R})$ and is written $\exp A$ or e^A. When $n=1$, exp is the familiar exponential function though when $n>1$ the behaviour of exp is quite different (see Exercises 6.3).

Exercises 6.3

1. Show that $\mathcal{M}(2, \mathbb{R})$ is a linear algebra under the usual operations.
2. (i) Determine the matrices of the linear transformations

   $T_1(x, y) = (2x+2y, -x-y)$,

   $T_2(x, y) = (2x+y, -x)$

   of $\mathbb{R}^2$ with respect to

   (*a*) the standard basis of $\mathbb{R}^2$,

   (*b*) the basis $B' = \{(1, 0), (1, 1)\}$ of $\mathbb{R}^2$.

   (ii) Compute the coordinate vector of $\mathbf{a} = (-1, 7)$ with respect to the basis B' and hence determine the vector $T_1 T_2 \mathbf{a}$ using the matrix representation.

   (iii) Use the determinant mapping to decide which of the transformations of (i) are invertible. Is the product $T_1 T_2$ invertible?
3. If

   $$A = \begin{pmatrix} 1 & 2 \\ 2 & 4 \end{pmatrix} \quad B = \begin{pmatrix} \cos\theta & -\sin\theta \\ \sin\theta & \cos\theta \end{pmatrix} \quad C = \begin{pmatrix} 0 & 2 \\ -2 & 0 \end{pmatrix}$$

   determine whether each is

   (*a*) symmetric,

   (*b*) antisymmetric,

   (*c*) orthogonal,

   (*d*) invertible.
4. Show that the diagonal entries of an antisymmetric matrix are all zero.
5. (i) Prove that the eigenvalues of a 2×2 symmetric matrix are always real.

   (ii) Prove that the eigenvalues of a non-zero 2×2 antisymmetric matrix are not real numbers.
6. If $A \in \mathcal{M}(n, \mathbb{R})$ show that

   $(A\mathbf{a}) \cdot \mathbf{b} = \mathbf{a} \cdot (A^T\mathbf{b})$ for all $\mathbf{a}, \mathbf{b} \in \mathbb{R}^n$

   and use this to show that if A is symmetric then eigenvectors of A corresponding to *distinct* eigenvalues are orthogonal.

7. (i) If

$$A = \begin{pmatrix} 1 & 2 \\ -1 & 3 \end{pmatrix}$$

and $p(x) = x^2 - 4x + 5$ show that $p(A) = 0$.

(ii) Use (i) to calculate A^{-1}.

8. (i) If

$$A = \begin{pmatrix} 1 & a \\ 0 & 1 \end{pmatrix}$$

use induction to show that

$$\begin{pmatrix} 1 & a \\ 0 & 1 \end{pmatrix}^n = \begin{pmatrix} 1 & na \\ 0 & 1 \end{pmatrix} \quad \text{for all } n \in \mathbb{N}$$

hence show that

$$e^A = eA$$

and write down an expression for $\det e^A$.

(ii) If

$$A = \begin{pmatrix} 0 & -1 \\ 1 & 0 \end{pmatrix}$$

and $\lambda \in \mathbb{R}$ determine the matrix $e^{\lambda A}$.

9. (i) If $A, B \in \mathcal{M}(n, \mathbb{R})$, under what conditions do we have $e^{(A+B)} = e^A e^B$.

(ii) Use (i) to demonstrate that e^A always has an inverse.

10. If $A \in \mathcal{M}(n, \mathbb{R})$ the *trace* of A, written $\operatorname{tr} A$, is defined by

$$\operatorname{tr} A = \sum_{i=1}^{n} A_{ii}.$$

If $A \in \mathcal{M}(n, \mathbb{R})$ is defined by

$$\begin{aligned} A_{ij} &= \lambda_i \quad \text{for } i = j, \\ &= 0 \quad \text{otherwise,} \end{aligned}$$

use induction to show that

$$\begin{aligned} \det A &= \prod_{i=1}^{n} \lambda_i \\ &= \lambda_1 \lambda_2 \ldots \lambda_n \end{aligned}$$

and hence prove that

$$\det e^A = e^{\operatorname{tr} A}.$$

11. If $A \in \mathcal{M}(n, \mathbb{R})$ has eigenvector $\mathbf{x} \in \mathbb{R}^n$ with eigenvalue $\lambda \in \mathbb{R}$, show that $\mathbf{x}$ is an eigenvector of e^A with eigenvalue e^λ.
12. (i) If $A \in GL(n, \mathbb{R})$ show that $\det A^{-1} = 1/\det A$.
    (ii) If $A \in O(n)$ show that $\det A = \pm 1$.
13. Show that $SO(2)$ is a subgroup of $O(2)$.

7 GRAPH THEORY

7.1 Introductory concepts

Many relations on finite sets may be usefully viewed as pictures (see Section 2.3) and manipulated by means of suitable matrices. Before defining the construction of these pictures we need to make sure that there is no ambiguity in what we are attempting. To help with this we define some notation.

Let V be a finite set and recall that

$$I_V = \{(v, v): v \in V\}$$

and define

$$V^2_{-} = V^2 \backslash I_V = \{(v_1, v_2): v_1 \neq v_2\}.$$

We further define a relation $\sim$ on V^2_{-} by

$$(v_1, v_2) \sim (w_1, w_2) \quad \text{if } (v_1, v_2) = (w_1, w_2) \text{ or } (v_1, v_2) = (w_2, w_1).$$

The important properties of the relation $\sim$ are summarized in the following proposition.

Proposition. $\sim$ is an equivalence relation on V^2_{-}. //

The proof is left as an exercise for the reader.

The set of equivalence classes thus determined is written $V^2_{-}/\sim$. Each equivalence class comprises exactly two elements, for if $(v_1, v_2) \in V^2_{-}$ then

$$[(v_1, v_2)] = \{(v_1, v_2), (v_2, v_1)\}.$$

We use the notation $[v_1, v_2]$ for the equivalence class $[(v_1, v_2)] \in V^2_{-}/\sim$. We are now in a position to define the concept of a graph precisely.

Definition. A *graph* G is a pair $G = (V, E)$ where V is a non-empty finite set of *vertices* and E is a subset of $V^2_{-}/\sim$. //

Alternatively, we can state that a *graph* G is a pair $G=(V, E)$ where V is a non-empty finite set of *vertices* and E is a set of unordered pairs of distinct vertices.

E is the set of *edges* of the graph,

$|V|$ denotes the number of vertices of G,

$|E|$ denotes the number of edges of G.

Our next result expresses the connection between graphs and a class of relations on finite sets.

Proposition.

(i) A graph $G=(V, E)$ determines an irreflexive and symmetric relation on V.

(ii) An irreflexive and symmetric relation on a finite set V determines a graph.

Proof.

(i) Let $G=(V, E)$ be a graph and define a relation $R(E)$ on V by

$v_1 R(E) v_2 \quad \text{iff } [v_1, v_2] \in E.$

$R(E)$ is irreflexive for

$vR(E)v \quad \text{iff } [v, v] \in E$

but $[v, v] \notin E$ because $(v, v) \notin V^2_-$. $R(E)$ is symmetric for

$v_1 R(E) v_2 \quad \text{iff } [v_1, v_2] \in E$

but $[v_1, v_2] = \{(v_1, v_2), (v_2, v_1)\} = [v_2, v_1]$. Therefore

$v_1 R(E) v_2 \quad \text{iff } v_2 R(E) v_1.$

(ii) If R is an irreflexive and symmetric relation on V then $R \subset V^2$.

R irreflexive means $(v, v) \notin R$ for any $v \in V$, so $R \subset V^2_-$.

R symmetric means that $(v_1, v_2) \in R$ iff $(v_2, v_1) \in R$.

Now define $E = R/\sim$ then $G=(V, E)$ is a graph. //

Graphs may be represented by matrices with Boolean entries. Many properties of graphs may be determined from their matrix representations by algebraic manipulation. This will become clear in the sequel.

Definition. The *adjacency matrix* $A \in \mathcal{M}(n, \mathbb{B})$ of a graph $G=(V, E)$ with $|V| = n$ has elements defined by

$$A_{ij} = 1 \quad \text{if } [v_i, v_j] \in E$$
$$= 0 \quad \text{otherwise.}$$

Vertices v_i and v_j are said to be *adjacent* if $A_{ij} = 1$.

Clearly, $A_{ii}=0$: $i \in \{1, \ldots, n\}$ and $A = A^T$; thus A is symmetric and, in the notation of Section 6.1, $A = A(R(E))$. //

A *picture* of a graph $G=(V, E)$ is obtained by drawing a distinct point in $\mathbb{R}^2$ for each $v \in V$ and if $[v, w] \in E$ we draw a line joining the vertices v and w.

Examples 7.1.1

(i) $V=\{v_1, v_2, v_3\}$,

$E=\{[v_1, v_2], [v_2, v_3], [v_1, v_3]\}$,

$|V|=3, \quad |E|=3.$

This graph can be represented as in Figure 7.1.

Fig. 7.1

$$A=\begin{pmatrix} 0 & 1 & 1 \\ 1 & 0 & 1 \\ 1 & 1 & 0 \end{pmatrix}$$

Adjacency matrix

v_2 v_1 v_3

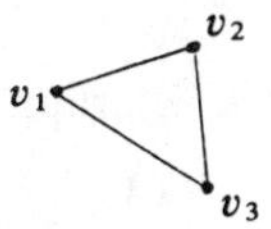

Picture

(ii) $V=\{v_1, v_2, v_3, v_4, v_5\}$,

$E=\{[v_1, v_2], [v_1, v_5], [v_2, v_3], [v_2, v_4], [v_3, v_4], [v_3, v_5], [v_4, v_5]\}$,

$|V|=5, \quad |E|=7.$

See Figure 7.2.

Fig. 7.2

$$A=\begin{pmatrix} 0 & 1 & 0 & 0 & 1 \\ 1 & 0 & 1 & 1 & 0 \\ 0 & 1 & 0 & 1 & 1 \\ 0 & 1 & 1 & 0 & 1 \\ 1 & 0 & 1 & 1 & 0 \end{pmatrix}$$

Adjacency matrix

v_1 v_5 v_2 v_4 v_3

or

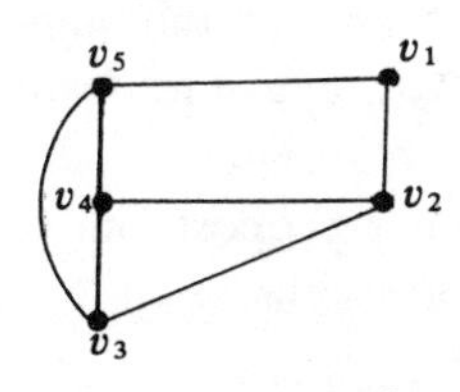

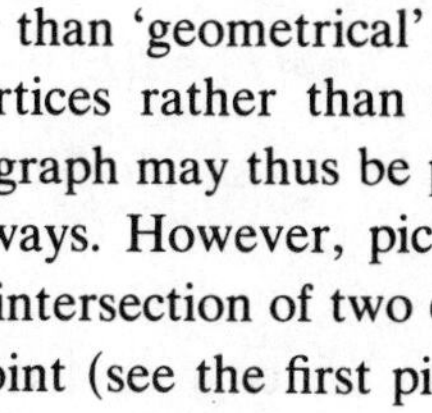

Picture //

Graphs are essentially 'topological' rather than 'geometrical' objects, i.e. they express relationships between vertices rather than defining positions for vertices and edges in space. A graph may thus be pictured in an infinity of different but 'equivalent' ways. However, pictures of graphs may be misleading; for example, the intersection of two edges in a picture does *not* imply a vertex at that point (see the first picture of

Figure 7.2). Clearly, the lower (or upper) triangular part of the adjacency matrix is sufficient to determine the graph, i.e. if A is the adjacency matrix of G then

$$A = \begin{bmatrix} 0 & & & \\ & 0 & & \\ & & 0 & \\ & & & \ddots \\ & & & & 0 \end{bmatrix}.$$

This triangular part determines G

The reader should by now be familiar with the concepts of substructure and isomorphism or equivalence within an algebraic system. In graph theory we make the following definitions.

Definition. A graph $H = (V_1, E_1)$ is said to be a *subgraph* of

$$G = (V, E) \quad \text{if } V_1 \subseteq V \quad \text{and} \quad E_1 \subseteq E.$$

If $V_1 = V$ then H is said to be a *spanning subgraph* of G. If V_1 is a non-empty subset of vertices of the graph (V, E) then the subgraph (V_1, E_1) *generated* by V_1 is defined by

$$[v, w] \in E_1 \Leftrightarrow v, w \in V_1 \quad \text{and} \quad [v, w] \in E. \quad /\!/$$

Definition.

(i) If $G_1 = (V_1, E_1)$ and $G_2 = (V_2, E_2)$ are graphs we say that G_1 and G_2 are *equivalent* if there is a bijection $f: V_1 \to V_2$ such that $vR(E_1)w \Rightarrow f(v)R(E_2)f(w)$.

(ii) If $G = (V, E)$ is any graph then we define a map

$\delta: V \to \mathbb{N} \cup \{0\}$

by $\delta(v) =$ the number of edges containing the vertex $v \in V$. We call $\delta(v)$ the *degree* of the vertex v. //

Our next proposition expresses two simple but general facts on the nature of graphs.

Proposition.

(i) $\sum_{v \in V} \delta(v) = 2|E|$.

(ii) In any graph the number of vertices of odd degree is even.

Proof.

(i) Each edge contributes twice to the sum.

(ii) Let $V_e \subseteq V$ be the vertices of even degree and $V_o \subseteq V$ be the vertices of odd degree. Then notice that

$$V = V_e \cup V_o \quad \text{and} \quad V_e \cap V_o = \varnothing,$$

hence

$$\sum_{v \in V} \delta(v) = \sum_{v \in V_e} \delta(v) + \sum_{v \in V_o} \delta(v), \text{ and}$$

$$2|E| = 2k + \sum_{v \in V_o} \delta(v)$$

$$\left(\sum_{v \in V_e} \delta(v) \text{ is clearly even, } 2k \text{ say} \right).$$

Thus

$$\sum_{v \in V_o} \delta(v) = 2(|E| - k)$$

which is even, but each $\delta(v)$ on the left hand side is odd so $|V_o|$ is even. //

In many applications of graph theory in computer science, the topology (V, E) of a graph is supplemented by additional information relating to either V or E or both. To make this more precise we define the concept of a labelled graph and give some examples.

Definitions.

(i) If S_V and S_E are sets, a *labelling* of the graph $G = (V, E)$ is a pair of functions

$f: V \to S_V$, vertex labelling,

$g: E \to S_E$, edge labelling.

(ii) If $G = (V, E)$ is labelled by f and g and $G_1 = (V_1, E_1)$ is labelled by f_1 and g_1 then G and G_1 are *equivalent labelled graphs* if there is a bijection $h: V \to V_1$ such that

(1) G and G_1 are equivalent as unlabelled graphs,

(2) $f(v) = f_1(h(v))$ for all $v \in V$ so corresponding vertices have the same label and

(3) $g[(v, w]) = g_1([h(v), h(w)])$ for all $v, w \in V$, i.e. corresponding edges have the same label.

Often only the edges or only the vertices will be labelled; the above still applies but with

$$\left.\begin{array}{l} f: V \to S_V \\ g = \text{constant} \end{array}\right\} \quad \text{vertex labelling only,}$$

$$\left.\begin{array}{l} f = \text{constant} \\ g: E \to S_E \end{array}\right\} \quad \text{edge labelling only.} \quad //$$

The edges or vertices (or both) of a labelled graph carry information that supplements or replaces the usual identifying names.

Examples 7.1.2

(i)

$V=\{v_1, v_2, v_3, v_4\}$,

$E=\{[v_1, v_3], [v_2, v_3], [v_3, v_4]\}$,

$f: V\to\{\text{UK cities}\}$,

$g: E\to\mathbb{N}$,

$f(v_1)=\text{London}$, $g([v_1, v_3])=105$,

$f(v_2)=\text{Cardiff}$, $g([v_2, v_3])=196$,

$f(v_3)=\text{Birmingham}$, $g([v_3, v_4])=292$,

$f(v_4)=\text{Edinburgh}$.

This can be drawn as in Figure 7.3.

Fig. 7.3

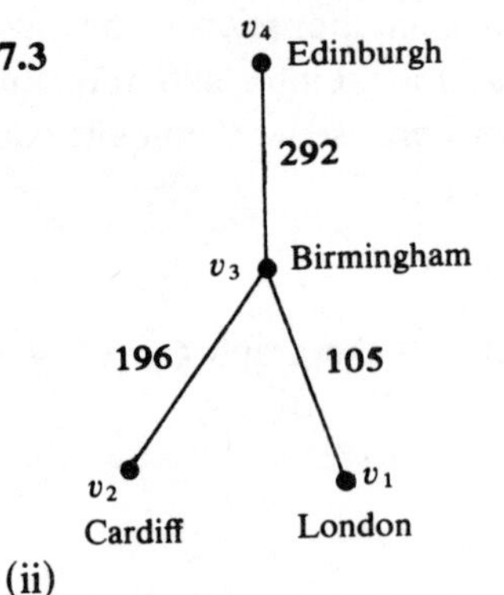

(ii)

$G_1=(\{v_1, v_2, v_3\}, \{[v_1, v_2], [v_1, v_3], [v_2, v_3]\})$

$G_2=(\{w_1, w_2, w_3\}, \{[w_1, w_2], [w_1, w_3], [w_2, w_3]\})$

labelled as in Figure 7.4; G_1 and G_2 are equivalent labelled graphs (no vertex labelling). //

Fig. 7.4

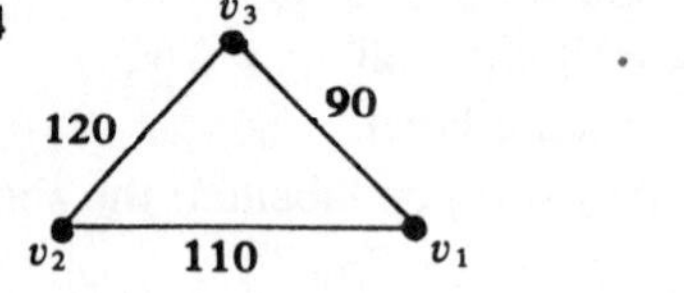

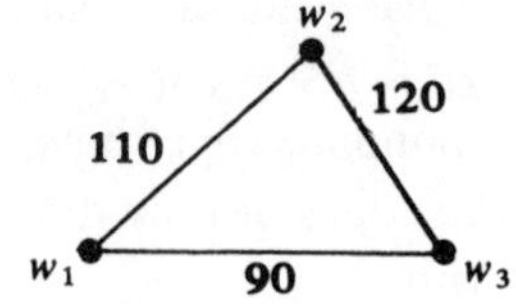

Exercises 7.1

1. Construct a proof of the first proposition of this section.
2. Draw the graphs represented by the following adjacency matrices:

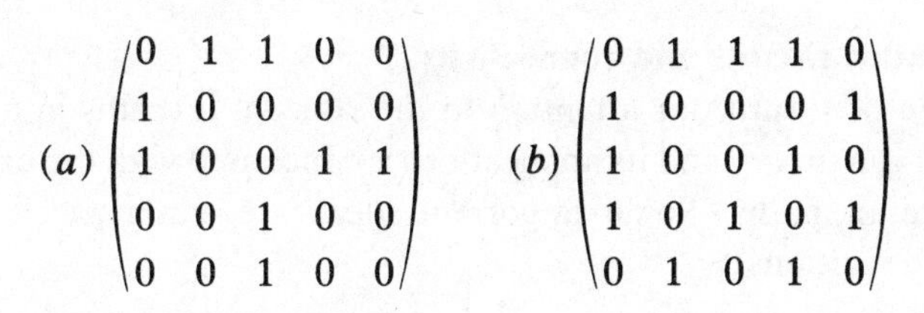

3. Determine the adjacency matrices of the graphs represented by the pictures below.

Fig. 7.5

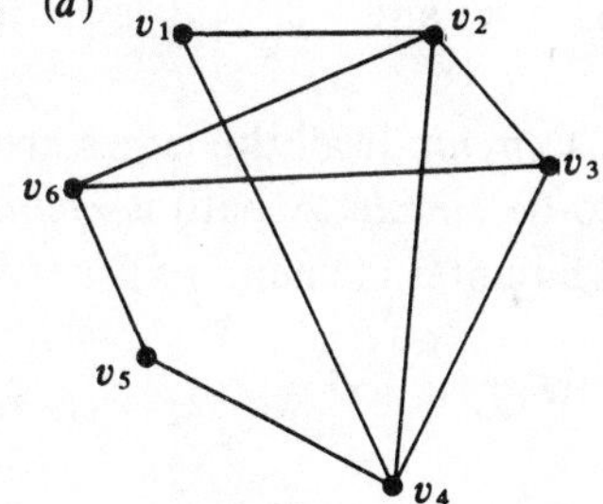

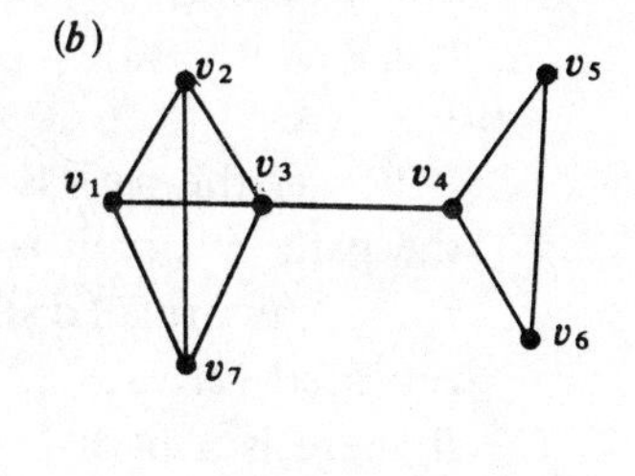

4. Draw the subgraph generated by the vertices $\{v_2, v_3, v_4, v_6\}$ of the graph in Question 3(a) above.
5. If $G=(V, E)$ is a graph with $|V|=n$ what is the maximum possible value for $|E|$?
6. How many distinct graphs are there on n vertices?

The remaining exercises of this section require some additional definitions.

Definition.

(i) A graph $G=(V, E)$ is said to be *complete* if for all $v_1, v_2 \in V$ we have $[v_1, v_2] \in E$. The complete graph on n vertices is denoted K_n.

(ii) A graph $G=(V, E)$ is said to be *bipartite* if there is a partition $\{V_1, V_2\}$ of V such that no two vertices of V_1 are adjacent and no two vertices of V_2 are adjacent. A bipartite graph is said to be *complete* if for any pair $v_1 \in V_1$ and $v_2 \in V_2$ we have $[v_1, v_2] \in E$. If $|V_1|=m$ and $|V_2|=n$ then the complete bipartite graph (V, E) is written $K_{m,n}$. //

7. Draw the graph K_5.
8. Construct an example of a bipartite graph.
9. (i) Draw the graph $K_{3,3}$.
   (ii) How many edges does the graph $K_{m,n}$ have?

7.2 Paths, circuits and connectivity

We now turn our attention to the concept of paths in a graph. Much of graph theory and its applications is concerned with the existence and nature of paths. Some important ideas are summarized in the following definitions.

Definitions.

(i) If $G=(V, E)$ is a graph then a *path of length* k from v to w is a sequence $\langle v_0, v_1, \ldots, v_k\rangle$ of not necessarily distinct vertices $v_i \in V$ with $v_0 = v$ and $v_k = w$ and $[v_{i-1}, v_i] \in E$ for all $i = 1, \ldots, k$.
If $v_0 = v_k$ the path is called a *circuit.* If all the edges are distinct the path or circuit is said to be *simple.* A path is *elementary* if $v_0, \ldots, v_k$ are all distinct; a circuit is elementary if v_0 is the only repeated vertex.

(ii) If there is a path from $v \in V$ to $w \in V$ we say w is *reachable* from v.

(iii) A *cycle* is a simple circuit.

(iv) A graph without cycles is said to be *acyclic.* //

Cycles and length of cycles were defined for permutations in Section 3.4 and we note that the idea of a circuit in a graph is similar in essence as well as name.

Definitions.

(i) A graph $G=(V, E)$ is *connected* if each pair of distinct vertices may be joined by a path.

(ii) A *tree* is a connected acyclic graph.

(iii) A *rooted tree* is a tree with a distinguished vertex called the *root.*

(iv) A *spanning tree* for $G=(V, E)$ is a spanning subgraph that is also a tree. //

In 7.1 we remarked that calculations with the adjacency matrix reveal important information about the nature of a graph. The following results are examples of this interrelationship between algebra and topology in graph theory.

In the following theorem and its corollary the powers A^k of A are computed in $\mathcal{M}(n, \mathbb{Z})$ *not* in $\mathcal{M}(n, \mathbb{B})$. Hence the arithmetic in $\mathbb{Z}$ can give rise to integer values greater than 1 in the matrix A^k.

Theorem. Let A be the adjacency matrix of a graph $G=(V, E)$ with $|V| = n$ then $(A^k)_{ij}$ is the number of paths of length k from v_i to v_j.

Proof. (By induction on k.)

For $k=1$, a path of length 1 is just an edge of G. Hence the result is true for $k=1$ from the definition of A. Assume that it is true for $k-1$ and consider A^k. Let

$$(A^{k-1})_{ij}=\alpha_{ij} \quad \text{and} \quad A_{ij}=a_{ij}$$

then

$$(A^k)_{ij}=(A^{k-1}A)_{ij}=\sum_{q=1}^{n} \alpha_{iq}a_{qj}.$$

By the induction hypothesis

α_{iq} = number of paths of length $(k-1)$ from v_i to v_q

and

a_{qj} = number of paths of length 1 from v_q to v_j.

Hence

$\alpha_{iq}a_{qj}$ = number of paths of length k from v_i to v_j where v_q is the next to last vertex in the path

and therefore

$\sum_{q=1}^{n} \alpha_{iq}a_{qj}$ = number of paths of length k from v_i to v_j.

The result is now true by induction. //

Corollary.

(i) There is a path from v_i to $v_j (i \neq j)$ in $G=(V, E)$ if and only if the (i, j)th element of the $n \times n$ matrix (where $n=|V|$)

$A+A^2+\ldots+A^{n-1}$

is non-zero.

(ii) With the $i \neq j$ condition removed the required matrix is

$A+A^2+\ldots+A^{n-1}+A^n$.

Proof.

(i) Let $\langle v_i, v_1, v_2, \ldots, v_j\rangle$ be a path from v_i to v_j in G. If there are no repeated vertices then as $|V|=n$ there are at most $n-1$ edges and so by the theorem the corollary is true. If there is a repeated vertex, the path looks like

$$\langle v_i, \ldots, \underbrace{v_r, \ldots, v_r}_{\text{this is a circuit}}, \ldots, v_j\rangle.$$

When all such circuits are removed this results in a path of length at most $n-1$ from v_i to v_j.

The existence of a path from v_i to $v_j(i \neq j) \Rightarrow (i, j)$th entry of $\sum_{k=1}^{n-1} A^k$ is non-zero. The argument in the reverse direction is obvious.

(ii) If $i = j$ is allowed then the existence of a path from v_i to v_j implies that there is a sequence $\langle v_i, v_1, \ldots, v_j \rangle$. If there are no repeated vertices (except possibly $v_i = v_j$) then the path consists of at most $n+1$ vertices (no more than n edges) therefore the (i, j)th entry of $\sum_{k=1}^{n} A^k$ is non-zero. It may be deduced from this result that when $|V| = n$,

$$A(R^+(E)) = A(R(E)) \vee A(R^2(E)) \vee \ldots \vee A(R^n(E))$$
$$= \bigvee_{k=1}^{n} A(R^k(E))$$

and

$$A(R^*(E)) = I \vee A(R(E)) \vee \ldots \vee A(R^{n-1}(E))$$
$$= \bigvee_{k=0}^{n-1} A(R^k(E)). \quad //$$

The reader will recall that for any binary relation R, R^+ is defined by

$$R^+ = \bigcup_{k=1}^{\infty} R^k$$

and when $R \subseteq V \times V$ where $|V| = n$ then it follows from 6.1 that

$$A(R^+) = A^+(R) = \bigvee_{k=1}^{n} A(R^k)$$

and similarly,

$$A(R^*) = A^*(R) = \bigvee_{k=0}^{n-1} A(R^k).$$

In the remainder of this section in order to simplify the notation we shall write $A(R^k(E))$ as $A(R^k)$, $A(R^+(E))$ as $A(R^+)$ and $A(R^*(E))$ as $A(R^*)$. Warshall's algorithm requires $4n^3$ operations to determine $A(R^+)$ whereas the above closed form requires $4n^4 - 7n^3$. Other still more efficient algorithms may be used for large values of n.

The matrix $C = A(R^*)$ is called the *connection*, *connectivity* or *reachability* matrix of $G = (V, E)$. There is a path from v_i to v_j $(i \neq j)$ in G if and only if the (i, j)th element of C is 1; G is connected if and only if $C_{ij} = 1$ for all $1 \leqslant i, j \leqslant n$.

The important properties of the relation R^* may be summarized as follows:

Proposition. R^* is an equivalence relation on V.

Proof. Since by definition R^* is the reflexive closure of R we need to verify only that R^* is symmetric. Now vR^*w implies that there is a path $\langle v, v_1, \ldots, v_k, w\rangle$ from v to w in G. That is

$$[v, v_1] \in E, [v_1, v_2] \in E, \ldots, [v_k, w] \in E.$$

Therefore

$$[w, v_k] \in E, [v_k, v_{k-1}] \in E, \ldots, [v_2, v_1] \in E, [v_1, v] \in E.$$

Thus

$$\langle w, v_k, v_{k-1}, \ldots, v_2, v_1, v\rangle$$

is a path from w to v in G and wR^*v. //

The relation R^* determines an important class of subgraphs which we now define. Some related concepts are also given; these will be of importance later when we discuss graph 'traversal'.

Definition. If $\{V_i: 1 \leq i \leq p\}$ is the partition of a graph defined by R^* then p is called the *connectivity number* of G. The subgraphs (V_i, E_i) generated by the equivalence classes are called the *connected components* of G. //

A forest is a graph for which every connected component is a tree. A *spanning forest* for a graph $G = (V, E)$ is a collection of vertex disjoint trees $T_i = (V_i, E_i)$ such that

$$V = \bigcup_i V_i \quad \text{and} \quad E_i \subset E \quad \text{for all } i.$$

(Vertex disjoint means that $V_i \cap V_j = \varnothing$ when $i \neq j$.)

Figure 7.6 illustrates the aforementioned concepts for a graph with $p = 2$.

Fig. 7.6

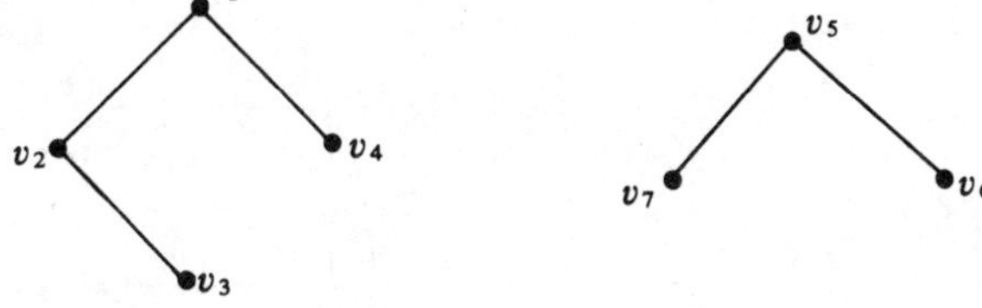

is a spanning forest for the graph

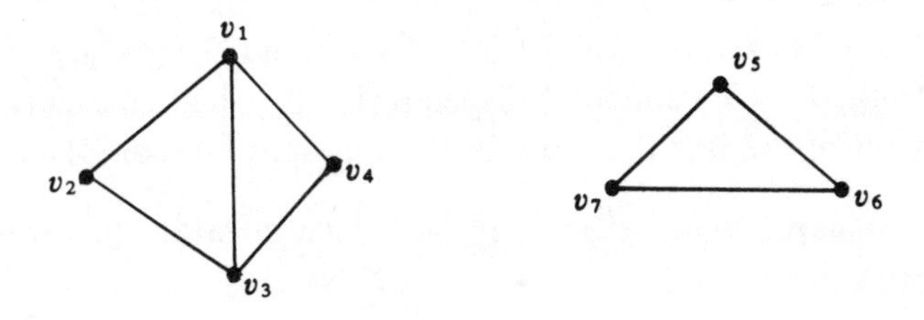

Exercises 7.2

1. Let $G=(V, E)$ be defined by $V=\{v_1, v_2, v_3, v_4\}$ and $E=\{[v_1, v_2], [v_1, v_3], [v_2, v_4], [v_3, v_4]\}$. Use the adjacency matrix of G to determine
   (*a*) the number of paths of length 2 from v_3 to v_2,
   (*b*) the number of paths of length 3 from v_1 to v_2 and
   (*c*) whether G is connected.
2. Draw spanning trees for the graphs in Question 3 of Exercises 7.1.
3. Give a matrix characterization of acyclicity in a graph.

7.3 Planar graphs

Historically, much work in graph theory has been concerned with the particular class of graphs that may be accurately represented by pictures in the plane $\mathbb{R}^2$. In this section three major results on such graphs are discussed.

7.3.1 Euler's theorem and Kuratowski's theorem

Definition.

(i) A graph G which can be pictured in the plane so that its edges do not cross is said to be *planar*. Such a picture is called a *map* of G.

(ii) A map of G is said to be *connected* if G is connected. //

Example 7.3.1

(1) $G=(\{v_1, v_2, v_3, v_4\}, \{[v_1, v_2], [v_1, v_3], [v_1, v_4], [v_2, v_3], [v_2, v_4], [v_3, v_4]\})$. A picture and a suitable map of G are shown in Figure 7.7; G is therefore a planar graph.

Fig. 7.7

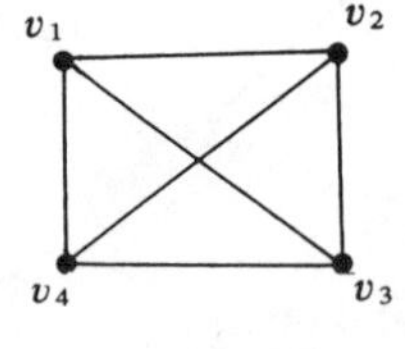

Picture of G

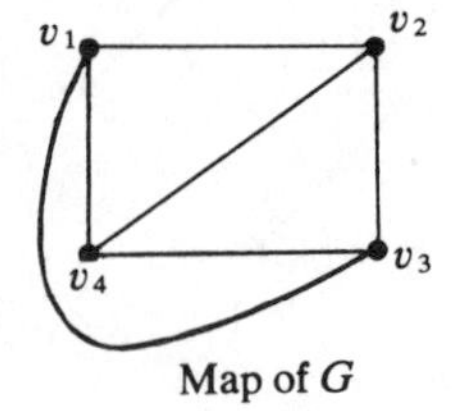

Map of G

(2) The graphs shown in Figure 7.8(*a*) and 7.8(*b*) are usually denoted by K_5 and $K_{3,3}$ respectively. We shall consider, later, proof of the fact that both of these graphs are non-planar. //

A map divides $\mathbb{R}^2$ into 'regions'; we illustrate this with the following example.

Fig. 7.8

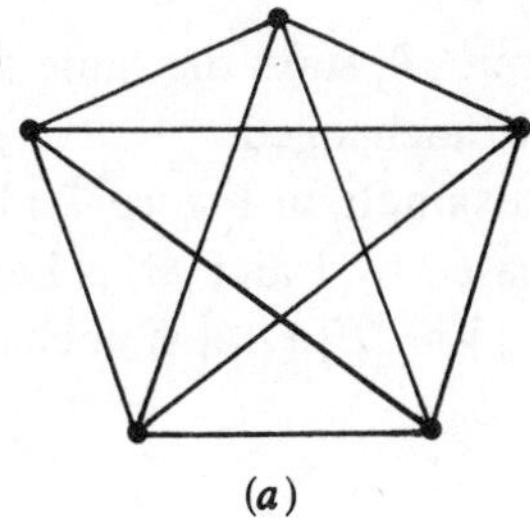

(a)

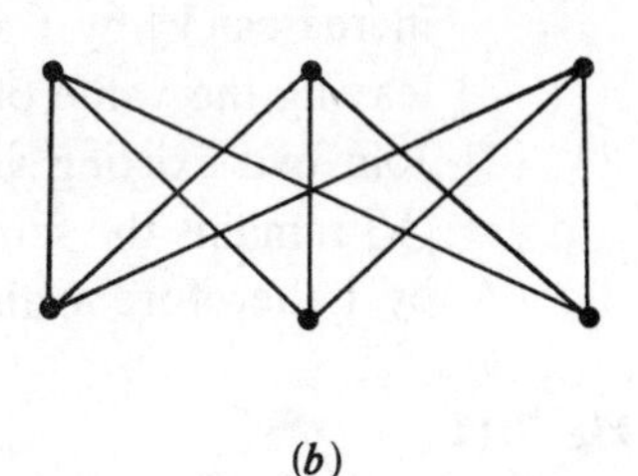

(b)

Example 7.3.2

Consider the map shown in Figure 7.9. This divides $\mathbb{R}^2$ into four regions; r_1, r_2 and r_3 being bounded regions and r_4 an 'unbounded' region. //

Fig. 7.9

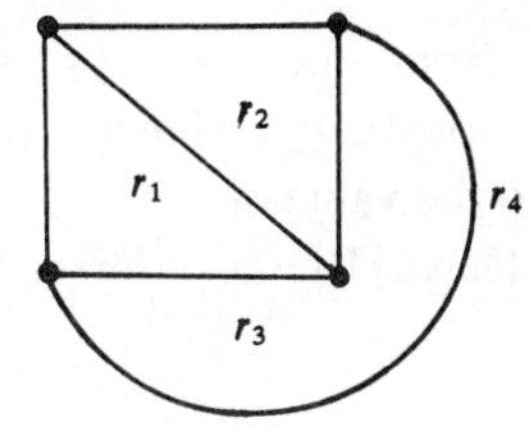

In the remaining sections we shall use $\mathscr{R}$ to denote the set of regions of a given map.

Euler's theorem. For any connected map

$$|V|-|E|+|\mathscr{R}|=2.$$

Proof. If $|V|=1$ then clearly $|E|=0$ and $|\mathscr{R}|=1$; the formula therefore holds for $|V|=1$. Now consider the following two ways of extending a given map:

(1) Add a new vertex and join it to an existing one.

(2) Join two existing vertices.

We show that the value of $|V|-|E|+|\mathscr{R}|$ is invariant under (1) and (2).

(1) Add a new vertex as for example in Figure 7.10. This process

Fig. 7.10

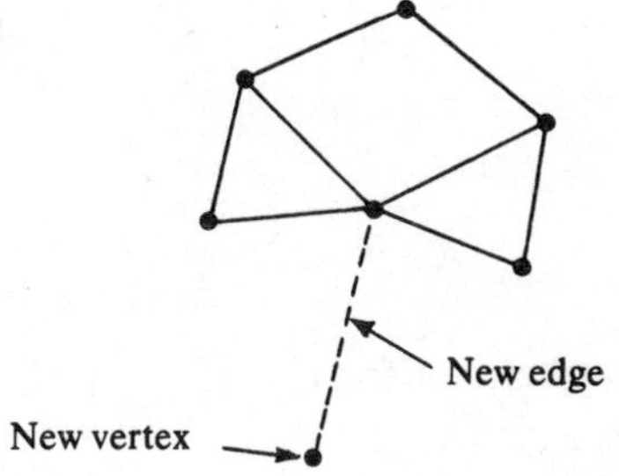

increases $|V|$ by 1 and $|E|$ by 1 but $|\mathcal{R}|$ stays the same thereby leaving the value of $|V|-|E|+|\mathcal{R}|$ unchanged.

(2) Join two existing vertices as for example in Figure 7.11. Now $|V|$ remains the same, $|E|$ is increased by 1 and $|\mathcal{R}|$ is increased by 1 therefore again the value of $|V|-|E|+|\mathcal{R}|$ is unchanged.

Fig. 7.11

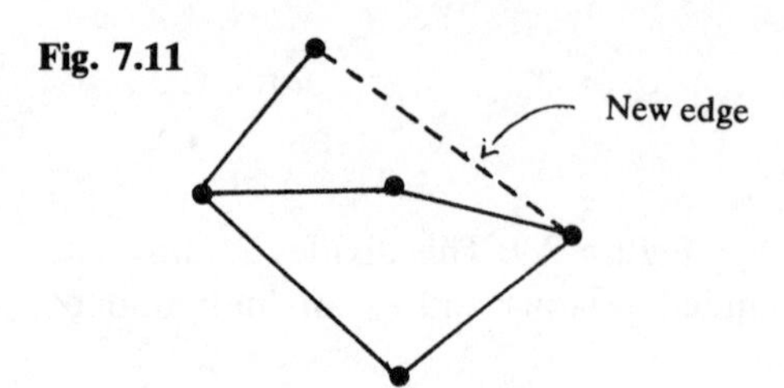

All maps may be obtained from the case $|V|=1$ by doing either (1) and/or (2) repeatedly. Hence, because $|V|-|E|+|\mathcal{R}|=2$ when $|V|=1$ and the value of $|V|-|E|+|\mathcal{R}|$ is invariant under (1) and (2), we must have $|V|-|E|+|\mathcal{R}|=2$ for all connected maps. //

Clearly, each region of a map has a bounding circuit. Two further simple results, one concerning these bounding circuits, are all that is required to demonstrate that the graphs K_5 and $K_{3,3}$ are non-planar. The following leads to a proof of these results.

Definition. If $G=(V, E)$ is a planar graph then for a map of G we define the *degree* Δ_r of a region r to be the length of a circuit bounding r. //

Example 7.3.3

For the map shown in Figure 7.12

$\langle v_1, v_3, v_4, v_3, v_2, v_1\rangle$ is a boundary circuit for r_1

Fig. 7.12

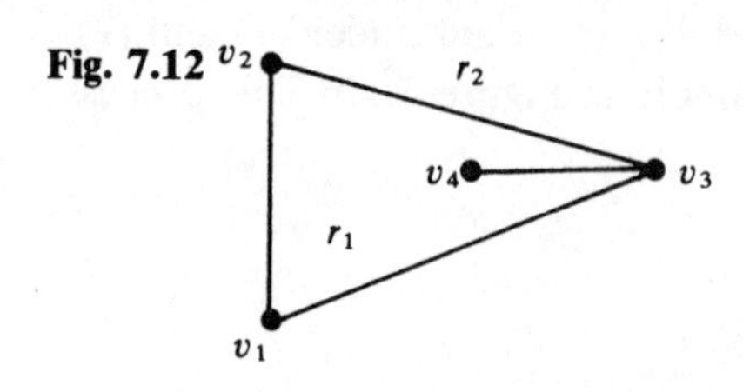

and

$\langle v_1, v_3, v_2, v_1\rangle$ is a boundary circuit for r_2.

Therefore $\Delta_{r_1}=5$, and $\Delta_{r_2}=3$. //

Proposition. If $G=(V, E)$ is a planar graph then
(i) $\sum_{r\in\mathcal{R}} \Delta_r = 2|E|$ for any map of G and
(ii) If $|V|\geqslant 3$ then $|E|\leqslant 3|V|-6$.

Proof. The proof is left as an exercise to the reader. //

Proposition. The graphs K_5 and $K_{3,3}$ are non-planar.

Proof.
(i) K_5 is non-planar.
K_5 planar $\Rightarrow |E|\leqslant 3|V|-6$ by the previous result. For K_5, $|E|=10$ and $|V|=5$ which implies $10\leqslant 15-6=9$. Since this is false K_5 is non-planar.
(ii) $K_{3,3}$ is non-planar.
$K_{3,3}$ planar $\Rightarrow \sum_{r\in\mathcal{R}} \Delta_r = 2|E|$. However, in $K_{3,3}$ no three vertices are connected to each other, therefore $\Delta_r \geqslant 4$ for all $r\in\mathcal{R}$. From Euler's formula

$$|\mathcal{R}|=2+|E|-|V|=2+9-6=5,$$

therefore

$$\sum_{r\in\mathcal{R}} \Delta_r \geqslant 5*4=20.$$

Thus $2|E|\geqslant 20$ so that $|E|\geqslant 10$ which is false, hence $K_{3,3}$ is non-planar. //

What is interesting about the graphs K_5 and $K_{3,3}$ is that these are essentially the only non-planar graphs there are as all non-planar graphs have subgraphs 'similar' to either K_5 or $K_{3,3}$. Before making this precise with our second theorem we need two further definitions.

Definition.
(i) If $G=(V, E)$ is a graph, an *elementary contraction* of G is formed by deleting an edge $[v_i, v_j]$ from E, replacing every occurrence of v_i and v_j in E by a new symbol w, deleting v_i and v_j from V and adding w to V. Pictorially, an elementary contraction of G is obtained by merging two adjacent vertices, after deleting the edge between them, and calling the 'composite' vertex w.
(ii) A graph G is said to be *contractible* to the graph G' if G' may be obtained from G by a series of elementary contractions. //

Example 7.3.4
For the graphs shown in Figure 7.13 G is contractible to G'. //

Fig. 7.13

G

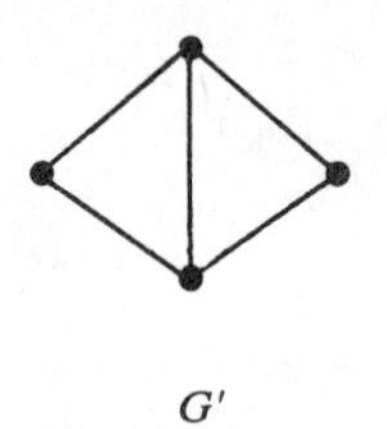

Theorem. (Due to Kuratowski circa 1930.) A graph is planar if and only if none of its subgraphs is contractible to K_5 or $K_{3,3}$.

Proof. The proof of this theorem is beyond the scope of this text and is therefore omitted. //

We deduce from Kuratowski's theorem that K_5 and $K_{3,3}$ are essentially the only non-planar graphs. Algorithms based on this theorem have been devised for deciding whether a given graph is planar or not.

7.3.2 Colouring of maps and graphs

Definition.

(i) A *colouring* of $G=(V, E)$ is an assignment of a colour to the vertices of G such that if $[v, w] \in E$ then v and w have different colours.

(ii) The *chromatic number* $\chi(G)$ of a graph G is the minimum number of colours required to colour G. //

Theorem. $\chi(G) \leqslant 4$ for all planar graphs G. //

This theorem was first 'proved' in 1976 by exhaustive computer examination of all possible cases.

Definition. If M is a map we define a second map M', called the *dual* of M, in the following way: select an interior point in each region of M; if two regions share a common edge we draw an arc connecting the selected interior points. This process determines M'. //

Example 7.3.5
Figure 7.14 shows a map M together with its dual M'. //

A colouring of M' corresponds to colouring the regions of M such that no regions sharing a common edge are coloured the same. We may therefore restate the four colour theorem as follows:

Fig. 7.14

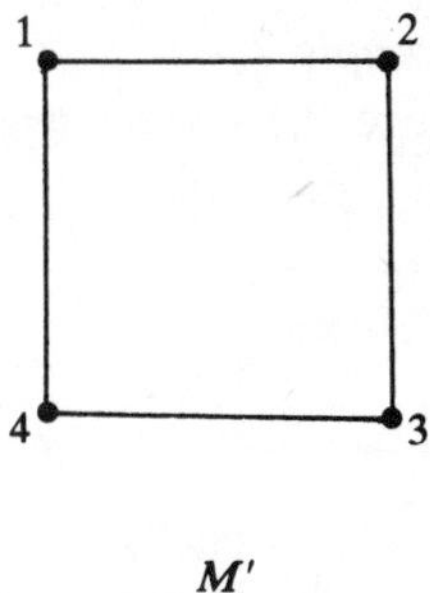

Theorem. If the regions of a map M are to be coloured so that adjacent regions have different colours then no more than four colours are required. //

Exercises 7.3

1. If $T=(V, E)$ is a tree with $|V|=n$, show that $|E|=n-1$.
2. Verify Euler's formula for the graphs of Question 3 of Exercises 7.1.
3. If $G=(V, E)$ is a planar graph and Δ_r denotes the degree of a region r in a map of G, show that

   $$\sum_{r\in\mathcal{R}} \Delta_r = 2|E|.$$

4. If $G=(V, E)$ is a connected planar graph with $|V|\geq 3$ show that $|E|\leq 3|V|-6$.
5. Give examples to show that the inequality in Question 4 above is not valid when $|V|<3$.
6. Determine a graph G for which $\chi(G)=4$.
7. If T is a tree what is the value of $\chi(T)$?

7.4 Data structures for graph representation

The adjacency matrix suggests an obvious method for representing a graph on a machine – in a high-level language we could use an array to hold the matrix entries. The symmetry and zero diagonal imply a minimum storage requirement, for a graph on n vertices, of $\frac{1}{2}n(n-1)$ cells. Normally many entries in the matrix will be zero and their storage tends to be wasteful, but despite this the adjacency matrix is sometimes the most convenient representation. For many problems the adjacency list representation is preferable. In this case we associate a linked list L_v with each vertex $v\in V$; L_v is the list of vertices adjacent to v.

Example 7.4.1

Figure 7.15 gives a graph with the lists that may be used to represent it. //

Fig. 7.15

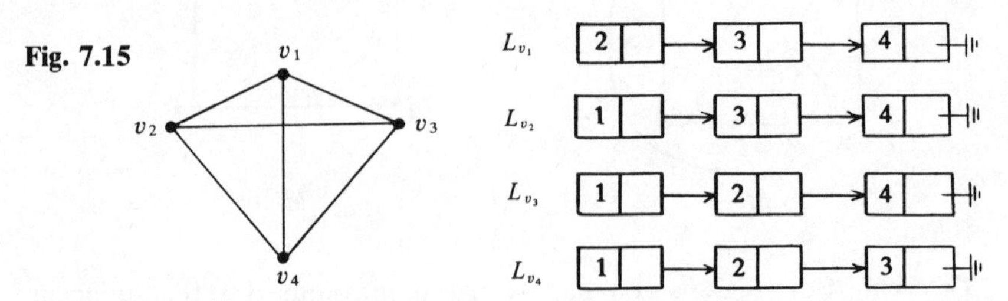

One way of implementing these lists is to use arrays

$$E(j, k) \quad 1 \leqslant k \leqslant 2$$

and

$$P(i) \quad 1 \leqslant i \leqslant n$$

where $E(j, 1)$ stores a vertex number, $E(j, 2)$ stores a link and $P(i)$ points to the start of the list L_{v_i} in E.

Example 7.4.2

For the graph of Example 7.4.1 we could have the following:

i	$E(i, 1)$	$E(i, 2)$	$P(i)$
1			5
2	1	6	2
3			11
4			35
5	2	9	
6	3	10	
7			
8			
9	3	13	
10	4	0	
11	1	19	
12	4	0	
13	4	0	
⋮	⋮	⋮	
19	2	12	
⋮	⋮	⋮	
35	1	36	
36	2	37	
37	3	0	

//

If v_i has no adjacent vertices we set $P(i)=0$ otherwise $E(P(i), 1)$ is adjacent to v_i and $[i, E(P(i), 1)]$ is an edge of the graph; similarly, if $E(P(i), 2) \neq 0$ then $E(E(P(i), 2), 1)$ is adjacent to v_i. A program segment to list the vertices adjacent to each vertex looks like the following:

```
for 1 ≤ i ≤ n do
  begin
    write 'vertices adjacent to vertex', i
    k ← P(i)
      while k ≠ 0 do
        begin
          write E(k, 1)
          k ← E(k, 2)
        end
  end
```

A similar program may be used to create the adjacency matrix M of a graph from the linked list representation E, P as follows:

```
for 1 ≤ i,j ≤ n do M(i, j) ← 0
for 1 ≤ i ≤ n do
  begin
    k ← P(i)
      while k ≠ 0 do
        begin
          M(i, E(k, 1)) ← 1
          k ← E(k, 2)
        end
  end
```

To create the E, P structure from M we may use the following code:

```
for 1 ≤ i ≤ n do P(i) ← i
for 1 ≤ j ≤ n do E(i, j) ← 0
for 1 ≤ i ≤ n do
  begin
    ifree ← i
    for 1 ≤ j ≤ n do
      begin
        if M(i, j) = 1 then do
          begin
            E(ifree, 1) ← j
            if j ≠ n do
              begin
```

```
                E(ifree, 2) ← ifree + n
                ifree ← ifree + n
            end
        end
    end
end
```

Alternative linked list representations of graphs are possible. The choice of representation depends largely on the algorithms to be employed.

A linked list representation L_v of the vertices adjacent to v defines an 'ordering' of the edges leaving v. Consider this last statement in relation to the graph and the linked lists given at the beginning of this section; the implied edge ordering is as shown in Figure 7.16. A graph with edge orderings of this type is called an ordered graph which we now define formally.

Fig. 7.16

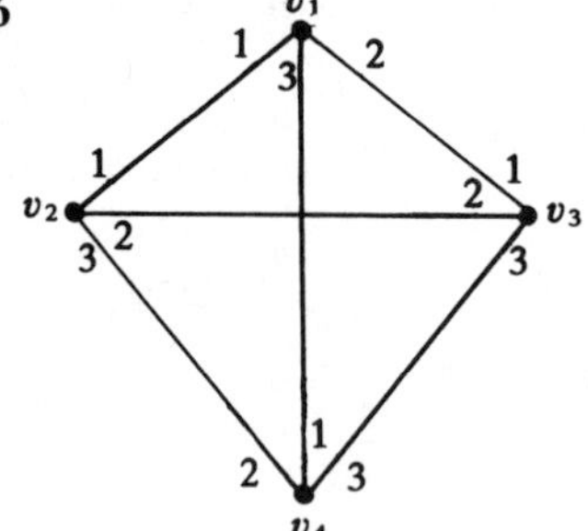

Definition. A set $V=\{v_1,\ldots,v_n\}$ of vertices together with a set $\{L_{v_1}, L_{v_2},\ldots, L_{v_n}\}$ of ordered lists of ordered pairs of vertices is an *ordered graph.*

Some conditions must be imposed on the lists L_v so that the underlying structure is still a graph. These are

(i) $(v, v)\notin L_v$ for any $v\in V$ and
(ii) $(w, u)\in L_w \Rightarrow (u, w)\in L_u$. //

The graph in Figure 7.16 may be written in ordered graph notation as

$$\begin{aligned}(\{v_1, v_2, v_3, v_4\}, \{&((v_1, v_2), (v_1, v_3), (v_1, v_4)),\\ &((v_2, v_1), (v_2, v_3), (v_2, v_4)),\\ &((v_3, v_1), (v_3, v_2), (v_3, v_4)),\\ &((v_4, v_1), (v_4, v_2), (v_4, v_3))\}).\end{aligned}$$

An ordered graph defines a unique graph without ordering. The converse of this statement is not true since in general many orderings of a graph will be possible, some of which are regarded as essentially the same according to the following definition.

Definition. Two ordered graphs G_1 and G_2 are *equivalent* if there is a bijection $f: V_1 \to V_2$ between the sets of vertices that also preserves the list structures. In other words if

$$L_v = ((v, w_1), \ldots, (v, w_k))$$

is an edge list of G_1 then

$$L_{f(v)} = ((f(v), f(w_1)), \ldots, (f(v), f(w_k)))$$

is an edge list of G_2. //

Example 7.4.3

Although the graphs shown in Figure 7.17 are equivalent they are *not* equivalent as ordered graphs. //

Fig. 7.17

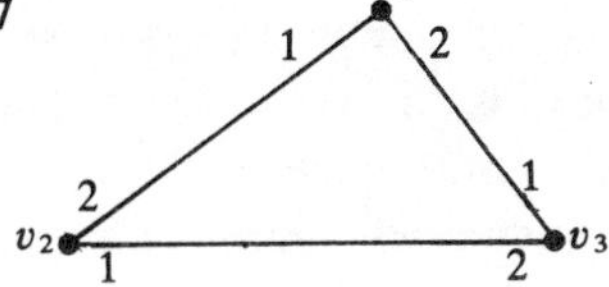

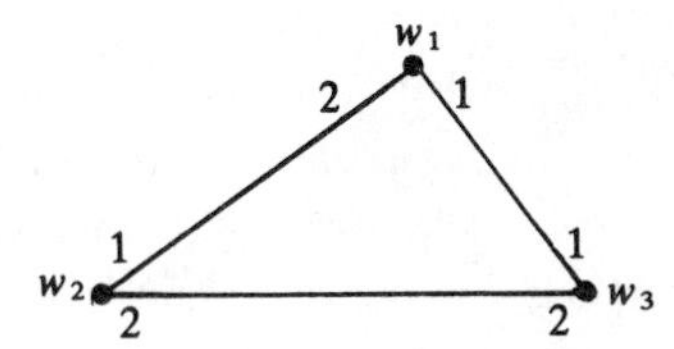

All the concepts of Section 7.2 (paths, circuits, connectedness and the concept of a tree) extend in an obvious way to ordered graphs.

Exercises 7.4

1. Rewrite the algorithms of Section 7.4 in a programming language of your choice and test them with suitable data.

7.5 Graph traversal

7.5.1 Introduction

Many problems involving graphs require the graph to be 'traversed', i.e. each vertex of the graph is 'visited' or 'processed' once only. A graph traversal may thus be represented by writing the vertices in sequence corresponding to the order in which they are processed.

If $G = (\{v_1, v_2, \ldots, v_n\}, E)$ and $\sigma: \mathbb{N}_n \to \mathbb{N}_n$ is a permutation, then the sequence

$$t = v_{\sigma(1)}, v_{\sigma(2)}, \ldots, v_{\sigma(n)}$$

defines a traversal of G. As there are $n!$ distinct permutations of $\mathbb{N}_n$ there must also be $n!$ distinct ways of traversing a graph on n vertices, or put another way there are $n!$ total orderings of the vertices. The vertex $v_{\sigma(1)}$ is called the *start vertex* for the traversal defined by σ.

From the $n!$ possibilities we describe just two methods of traversal which have proved useful in applications. Both require the graph to be ordered and use the edge ordering to define a permutation or total ordering of vertices.

7.5.2 Depth first traversal of a graph (dft)

Let $G = (\{v_1, \ldots, v_n\}, \{L_{v_1}, \ldots, L_{v_n}\})$ be an ordered graph. Select a start vertex $v_s (1 \leq s \leq n)$ and define $\sigma(1) = s$; $v_{\sigma(2)}$ is the vertex adjacent to $v_{\sigma(1)}$ occurring first on the adjacency list $L_{v_{\sigma(1)}}$, $v_{\sigma(3)}$ is defined to be the first vertex on $L_{v_{\sigma(2)}}$ that is not already on t, to determine $v_{\sigma(4)}$ the process is repeated from $v_{\sigma(3)}$ and similarly for $v_{\sigma(k)}$ where $k > 4$. When a vertex u is reached such that all vertices on L_u are already on t then the process is restarted from a vertex $w \in t$ where w is the last vertex on t such that L_w contains vertices not yet on t. The traversal terminates when all vertices in $V \backslash t$ cannot be reached from any vertex on t.

If G is connected the process described above defines a traversal of G, otherwise only the component of G containing $v_{\sigma(1)}$ is traversed. To obtain a complete traversal of G when G is not connected it is necessary to restart the process in each connected component, and this provides a method for the determination of the connected components of a graph. For connected graphs a unique traversal is obtained for each choice of start vertex so that a total of n depth first traversals of an ordered connected graph are possible. If G has connected components V_i, $1 \leq i \leq p$ with $|V_i| = n_i$, then $n_1 * n_2 * \ldots * n_p$ depth first traversals are defined.

The following recursive procedure may be used to perform a depth first traversal. Here t is an array of length $n = |V|$ initialized to zero and $t(v_i)$ is set to 1 to indicate that vertex v_i has been processed.

procedure dft(v)
  $t(v) \leftarrow 1$
  process vertex v
  for each $w \in L_v$ **do if** $t(w) = 0$ **then** dft (w)
endproc

Example 7.5.1

Consider the ordered graph shown in Figure 7.18. The depth first traversal with start vertex v_1 is given by

$$v_1, v_2, v_4, v_8, v_5, v_6, v_3, v_7. \quad //$$

7.5.3 Breadth first traversal (bft)

Let $G = (\{v_1, \ldots, v_n\}, \{L_{v_1}, \ldots, L_{v_n}\})$ be an ordered graph. Select a start vertex v_s and suppose $L_{v_s} = ((v_s, w_1), (v_s, w_2), \ldots, (v_s, w_k))$.

Fig. 7.18

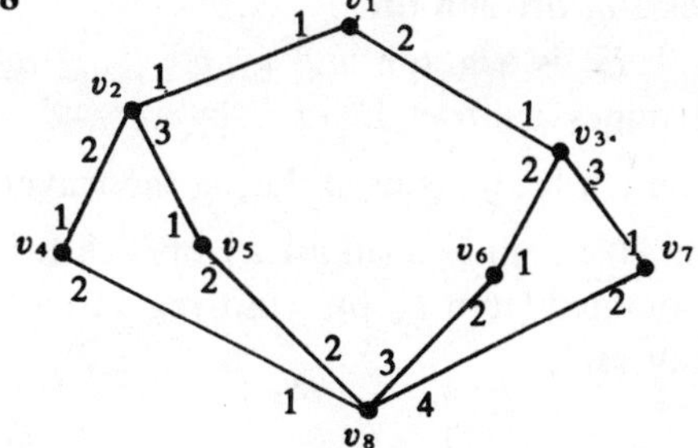

The first $k+1$ members of t are defined by $v_{\sigma(1)} = v_s$, $v_{\sigma(2)} = w_1, \ldots, v_{\sigma(k+1)} = w_k$ and $v_{\sigma(k+1+i)}$ is the ith vertex on L_{w_1} not already on t. This exhausts L_{w_1} and the process is restarted on L_{w_2}, and so on. The traversal terminates when all vertices reachable from $v_{\sigma(1)}$ are on t. The remarks of Section 7.5.2 on uniqueness, connectedness and number of possible traversals apply for this traversal also. The following procedure may be used to perform a breadth first traversal of a graph; t plays the same role as in dft and q is a queue with add and delete procedures add$q$ and delete$q$ respectively.

```
procedure bft(v)
  t(v) ← 1
  process vertex v
  initialize q with v
    while q ≠ ∅ do
      begin
        deleteq(v, q)
        for w ∈ L_v do
          begin
            if t(w) = 0 then do
              begin
                addq(w, q)
                t(w) ← 1
                process vertex w
              end
          end
      end
  endproc
```

Example 7.5.2

The breadth first traversal of the graph in Example 7.5.1 with start vertex v_1 is:

$v_1, v_2, v_3, v_4, v_5, v_6, v_7, v_8$. //

7.5.4 The spanning forests of dft and bft

If $G=(\{v_1,\ldots,v_n\},E)$ is a graph and $t=v_{\sigma(1)},\ldots,v_{\sigma(n)}$ is a traversal of G then t determines a subset E^t of E by

$$[v,w]\in E^t \text{ if and only if } [v,w] \text{ is used during the traversal.}$$

For ordered graphs a traversal t defines a sublist L_v^t of each list L_v in a similar way, that is L_v^t is obtained from L_v by removing all pairs (v,w) that are not used in the traversal.

Proposition. If $G=(\{v_1,\ldots,v_n\},\{L_{v_1},\ldots,L_{v_n}\})$ is an ordered connected graph and t is a bft or a dft of G then

$$G^t=(\{v_1,\ldots,v_n\},\{L_{v_1}^t,\ldots,L_{v_n}^t\})$$

is an ordered spanning tree for G.

Proof. G is connected hence the subgraph G^t spans G and is connected. If G^t has a circuit then some vertex appears more than one in t but since t is a traversal this is impossible and G^t is acyclic; hence G^t is a tree. //

Corollary. Every connected graph has a spanning tree. //

Example 7.5.3

For the graph of Example 7.5.1 the spanning trees defined by dft and bft with start vertex v_1 are shown respectively in Figures 7.19(a) and (b). //

Fig. 7.19 (a) (b)

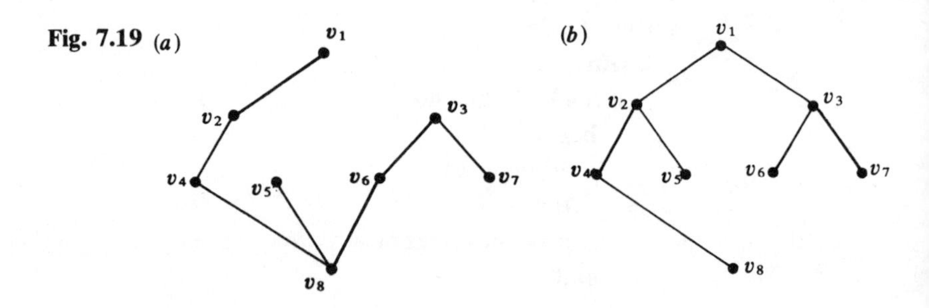

For graphs that are not connected a complete dft or bft defines a spanning forest for the graph.

Exercises 7.5

1. If $G=(\{v_1,\ldots,v_5\},\{L_{v_1},\ldots,L_{v_5}\})$ is the ordered graph defined by the lists

$L_{v_1} = ((v_1, v_2)),$
$L_{v_2} = ((v_2, v_5), (v_2, v_4), (v_2, v_3), (v_2, v_1)),$
$L_{v_3} = ((v_3, v_2)),$
$L_{v_4} = ((v_4, v_2), (v_4, v_5)),$
$L_{v_5} = ((v_5, v_4), (v_5, v_2)),$

determine

(a) the depth first traversal with start vertex v_2

(b) the breadth first traversal with start vertex v_4.

2. Draw spanning trees corresponding to the traversals of Question 1.
3. If the adjacency matrix of a graph G has the block structure

$$\begin{pmatrix} A_1 & & 0 \\ & A_2 & \\ 0 & & \ddots \quad A_p \end{pmatrix}$$

where each A_i is a square matrix with Boolean entries and all other entries are zero, what can be deduced about the nature of G?

Explain how a depth first or breadth first traversal may be used to block structure the adjacency matrix of any graph in this way.

4. Implement the dft and bft procedures in a suitable programming language.

7.6 Directed graphs

7.6.1 Introduction

Many applications of graph theory in computer science require a direction to be associated with each edge. For example, flow through a program is directed.

Definition. A *directed graph* or *digraph G*, is a pair $G = (V, E)$ where V is a finite set of vertices and E is any subset of $V \times V$. //

Proposition.

(i) A directed graph $G = (V, E)$ defines a relation on V.

(ii) If V is a finite set then a relation on V defines a directed graph with V as vertex set.

Proof.

(i) As in Section 7.1 we define $R(E)$ by $vR(E)w$ if and only if $(v, w) \in E$. Clearly, $R(E)$ is a relation.

(ii) If R is a relation on V the directed graph $G=(V, E)$ defined by R has edge set E where $(v, w) \in E$ iff vRw. //

The direction of an edge is indicated by the order in $V \times V$; for example if $(v, w) \in E$ the edge is said to *leave* v and *enter* w. Arrows are used in the picture of a digraph to indicate direction.

Examples 7.6.1

If $V=\{v_1, v_2, v_3\}$ and $E_1=\{(v_1, v_2), (v_2, v_3), (v_3, v_1)\}$ then the adjacency matrix and a picture for the digraph $G_1=(V, E_1)$ are as shown in Figure 7.20. Similarly, Figure 7.21 illustrates the adjacency matrix and a picture for $G_2=(V, E_2)$ where $E_2=\{(v_1, v_1), (v_1, v_2), (v_1, v_3), (v_2, v_3), (v_3, v_1)\}$. //

Fig. 7.20

$$\begin{pmatrix} 0 & 1 & 0 \\ 0 & 0 & 1 \\ 1 & 0 & 0 \end{pmatrix}$$

Adjacency matrix

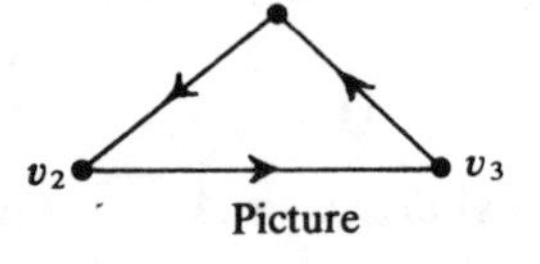

Picture

Fig. 7.21

$$\begin{pmatrix} 1 & 1 & 1 \\ 0 & 0 & 1 \\ 1 & 0 & 0 \end{pmatrix}$$

Adjacency matrix

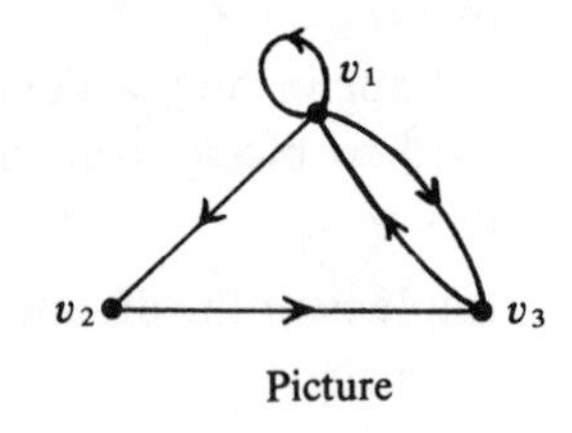

Picture

Because the edge relation for a digraph is not necessarily symmetric or irreflexive we do not have $A=A^T$ or $A_{ii}=0$ in general. An edge of type (v, v) is called a *loop*. The degree $\delta(v)$ of a vertex $v \in V$ may be written as a sum

$$\delta(v)=\delta^-(v)+\delta^+(v)$$

where $\delta^-(v)$ is the number of edges entering v, the *indegree*, and $\delta^+(v)$ is the number of edges leaving v, the *outdegree*. The sets $\{w: (w, v) \in E\}$ and $\{w: (v, w) \in E\}$ are called respectively the *inbundle* and *outbundle* of the vertex $v \in V$. The concepts of equivalence and labelling extend to digraphs in an obvious way.

7.6.2 Paths and connectivity in a digraph

Definition. A *path of length k* from v to w in a digraph $G=(V, E)$ is a sequence of edges of the form

$$(v, w_1), (w_1, w_2), (w_2, w_3), \ldots, (w_{k-2}, w_{k-1}), (w_{k-1}, w),$$

i.e. the second vertex of each edge is equal to the first vertex of the following edge. //

It is often convenient to represent a path by the sequence

$$v, w_1, w_2, w_3, \ldots, w_{k-2}, w_{k-1}, w$$

of vertices it defines. If $v = w$ the path is called a *circuit* or a *cycle.* A digraph without cycles is said to be *acyclic.* A *dag* is a directed acyclic graph.

The theorem of Section 7.2 also holds for digraphs, with identical proof, and we define the connection or reachability matrix in the same way. We note, however, that for digraphs the relation R^* is *not* an equivalence relation on V and consequently does *not* partition V.

If $\mathscr{D}$ denotes the set of all digraphs and $\mathscr{G}$ the set of all graphs, we may define a mapping $\mathscr{F}: \mathscr{D} \to \mathscr{G}$ in the following way.

Definition. If $G = (V, E) \in \mathscr{D}$ then $\mathscr{F}(G) \in \mathscr{G}$ has vertex set V and edge set defined by performing the following operations on E:

(i) delete all loops from E,
(ii) (v, w) is replaced by $[v, w]$ for all $(v, w) \in E$, then $\mathscr{F}(G)$ is the graph *associated* with the digraph G. //

For digraphs the notion of 'connectedness' is more involved than for graphs and is related to the problem of traversal. Three important kinds of digraph connectivity are now identified.

Definition. If $G = (V, E)$ is a digraph we say that

(i) G is *weakly connected* if the graph $\mathscr{F}(G)$ is connected,
(ii) G is *unilaterally connected* if for each pair of distinct vertices $v, w \in V$ there is a path from v to w or a path from w to v,
(iii) G is *strongly connected* if for each pair of distinct vertices $v, w \in V$ there is a path from v to w and a path from w to v. //

Clearly,

$$G \text{ strongly connected} \Rightarrow G \text{ unilaterally connected}$$
$$\Rightarrow G \text{ weakly connected.}$$

Examples 7.6.2

For the digraphs in Figure 7.22 we see that

(i) is weakly connected only,
(ii) is unilaterally connected but *not* strongly connected and
(iii) is strongly connected. //

Fig. 7.22

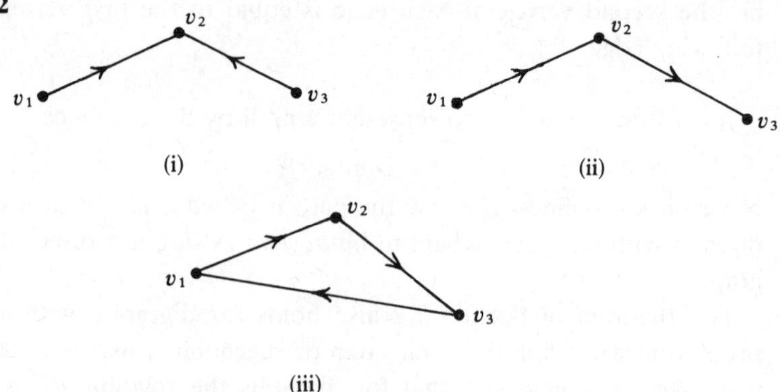

In terms of the connection matrix $C = A(R^*)$ we have

G strongly connected iff $C_{ij} = 1$ for all $1 \leq i,j \leq n$,

G unilaterally connected iff $C_{ij} \vee C_{ji} = 1$ for all $1 \leq i,j \leq n$.

Example 7.6.3

Consider the digraph represented by the picture in Figure 7.23.

Fig. 7.23

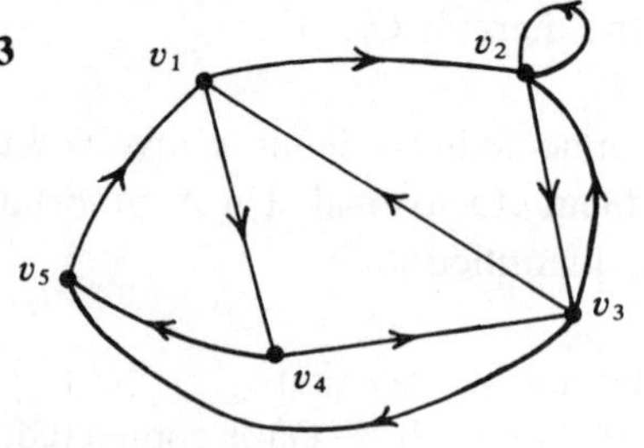

For this digraph we have

$$A(R) = \begin{pmatrix} 0 & 1 & 0 & 1 & 0 \\ 0 & 1 & 1 & 0 & 0 \\ 1 & 1 & 0 & 0 & 1 \\ 0 & 0 & 1 & 0 & 1 \\ 1 & 0 & 0 & 0 & 0 \end{pmatrix}, \qquad A(R^2) = \begin{pmatrix} 0 & 1 & 1 & 0 & 1 \\ 1 & 1 & 1 & 0 & 1 \\ 1 & 1 & 1 & 1 & 0 \\ 1 & 1 & 0 & 0 & 1 \\ 0 & 1 & 0 & 1 & 0 \end{pmatrix},$$

$$A(R^3) = \begin{pmatrix} 1 & 1 & 1 & 0 & 1 \\ 1 & 1 & 1 & 1 & 1 \\ 1 & 1 & 1 & 1 & 1 \\ 1 & 1 & 1 & 1 & 0 \\ 0 & 1 & 1 & 0 & 1 \end{pmatrix} \text{ and } A(R^4) = \begin{pmatrix} 1 & 1 & 1 & 1 & 1 \\ 1 & 1 & 1 & 1 & 1 \\ 1 & 1 & 1 & 1 & 1 \\ 1 & 1 & 1 & 1 & 1 \\ 1 & 1 & 1 & 0 & 1 \end{pmatrix},$$

so that for

$$C = \bigvee_{k=0}^{4} A(R^k) = I \vee A(R) \vee A(R^2) \vee A(R^3) \vee A(R^4)$$

we have $C_{ij} = 1$ for all $1 \leqslant i,j \leqslant 5$, and the graph is strongly connected. For more efficient machine computation of C we could use Warshall's algorithm. //

If $G = (V, E)$ is a digraph we may partition V by defining an equivalence relation ρ as follows:

$v \rho w$ if $v = w$ *or* there is a path from v to w and a path from w to v.

If $\{V_i: 1 \leqslant i \leqslant p\}$ is the partition and $\{E_i: 1 \leqslant i \leqslant p, E_i = (V_i \times V_i) \cap E\}$ are the associated sets of edges then the subgraphs $G_i = (V_i, E_i)$, $1 \leqslant i \leqslant p$ are called the *strongly connected components* of G.

Clearly, $\rho \subseteq R^*$, and $A(\rho)$ may be determined from $A(R^*)$ as $A(\rho)_{ij} = A(R^*)_{ij} \wedge A(R^*)_{ji}$; G is strongly connected iff G has only one strongly connected component, i.e. if $p = 1$.

Example 7.6.4

For the digraph in Figure 7.24 we have

$$A(R^*) = \begin{pmatrix} 1 & 1 & 1 & 1 \\ 0 & 1 & 1 & 0 \\ 0 & 0 & 1 & 0 \\ 0 & 0 & 1 & 1 \end{pmatrix} \quad \text{and} \quad A(\rho) = \begin{pmatrix} 1 & 0 & 0 & 0 \\ 0 & 1 & 0 & 0 \\ 0 & 0 & 1 & 0 \\ 0 & 0 & 0 & 1 \end{pmatrix}.$$

Fig. 7.24

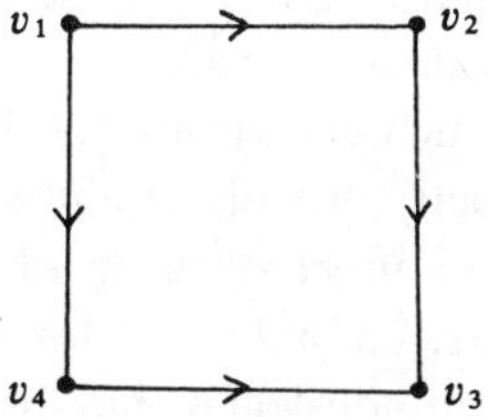

The strongly connected components of the graph are thus $G_i = (\{v_i\}, \varnothing)$, $1 \leqslant i \leqslant 4$. //

If $G = (V, E)$ is a dag, a vertex $v \in V$ is called a *leaf* if $\delta^+(v) = 0$. If $(v, w) \in E$ then v is a *direct ancestor* of w and w is a *direct descendant* of v. If there is a path from v to w in G, v is said to be an *ancestor* of w and w a *descendant* of v.

These concepts are not meaningful for a digraph with cycles, for within such graphs a vertex may be descended from itself.

Example 7.6.5

For the dag defined by the picture in Figure 7.25 vertices v_2, v_4 and v_5 are leaves, v_1 is an ancestor of v_5, v_5 is a direct descendant of v_3, and so on. //

Fig. 7.25

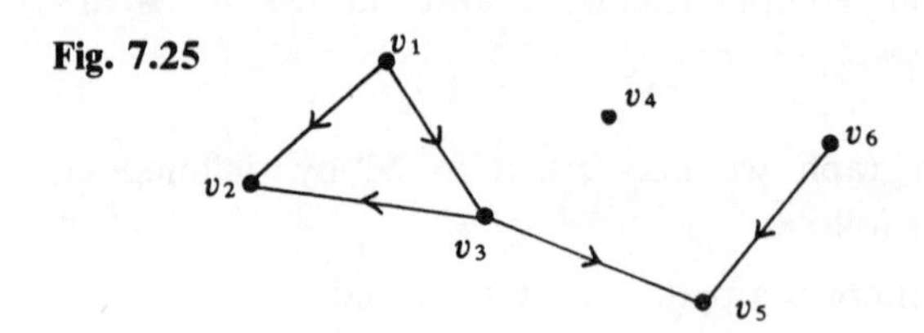

There is a strong connection between dags and partial order relations; in particular we have the following result, the proof of which is left as an exercise. Notice that in order to shorten some of the proofs that follow, partial orders will be based on $<$ rather than $\leqslant$ and hence are transitive and irreflexive (see p. 43).

Proposition.

(i) If $<$ is a partial order relation on a finite set V, then if $E = \{(v, w): v < w\}$ the pair $G = (V, E)$ is a dag.

(ii) If $G = (V, E)$ is a dag and we define a relation $<$ by $v < w$ if v is an ancestor of w then $<$ is a partial order relation on V. //

The terminology of digraphs can be used to give a precise definition of the familiar data structure known as a *directed* tree.

A *directed tree* $T = (V, E)$ is a dag in which one vertex $v_r \in V$ has no ancestors and every other vertex has exactly one direct ancestor; v_r is called the *root* of the tree. A *binary tree* is a directed tree in which every vertex has at most two direct descendants, i.e. $\delta^+(v) \leqslant 2$ for all $v \in V$. A binary tree is said to be *complete* if each non-leaf vertex has exactly two direct descendants.

Proposition. The following are equivalent for a digraph $G = (V, E)$.

(i) G is a tree.

(ii) The associated graph $\mathscr{F}(G)$ is connected and there is a vertex v_r with no ancestors, all other vertices having exactly one direct ancestor.

(iii) G has a vertex v_r which is joined to every other vertex by a unique path.

(iv) G has a vertex v_r that has no ancestors, all other vertices have exactly one direct ancestor and there is a path to each vertex from v_r.

Proof. The proof is left as an exercise to the reader. //

7.6.3 Ordered digraphs and traversals

Adjacency lists are an alternative to the adjacency matrix representation for digraphs. A given adjacency list representation defines an ordering of the edges leaving each vertex.

Definition. An *ordered directed graph* is a pair $G=(V, E)$ where V is a finite set of vertices and E is a set of ordered lists of directed edges. Elements of E have the form

$$L_v=((v, w_1), \ldots, (v, w_k))$$

for $v, w_i \in V$. //

Example 7.6.6

The ordered digraph

$$G=(\{v_1, v_2, v_3, v_4\}, \{((v_1, v_2), (v_1, v_4), (v_1, v_1)), ((v_2, v_4), (v_2, v_3))\})$$

may be represented by the adjacency lists

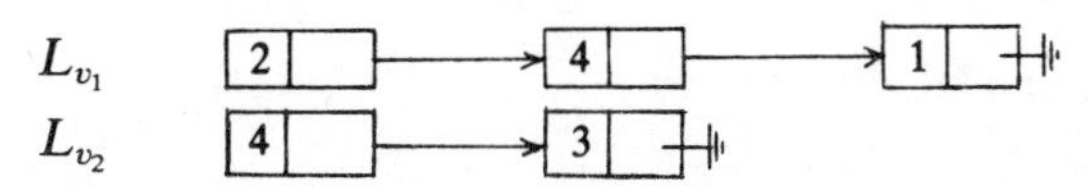

and may be pictured as in Figure 7.26. //

Fig. 7.26

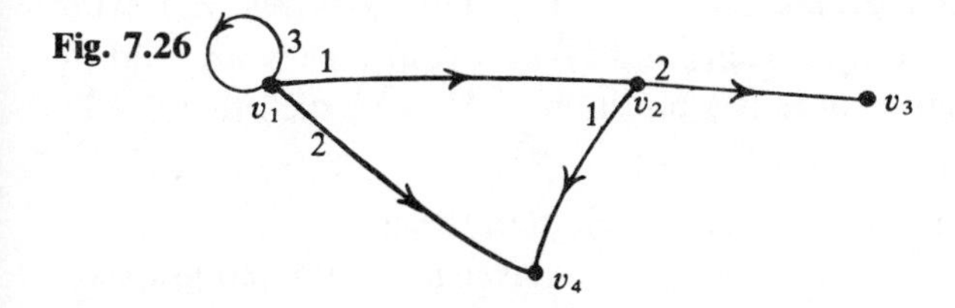

An ordered digraph G determines a unique digraph without ordering, we just replace each *list* $((v, w_1), \ldots, (v, w_k))$ by the *set* $\{(v, w_1), \ldots, (v, w_k)\}$. The digraph so determined is called the *digraph underlying G*. An *ordered dag* is an ordered graph whose underlying digraph is a dag. An *ordered directed tree* is an ordered digraph whose underlying digraph is a directed tree.

Example 7.6.7

$$T = (\{v_1, \ldots, v_6\}, \{((v_1, v_2), (v_1, v_3)), ((v_3, v_4), (v_3, v_5), (v_3, v_6))\})$$

is an ordered directed tree with v_1 as root; it may be pictured as in Figure 7.27. //

Fig. 7.27

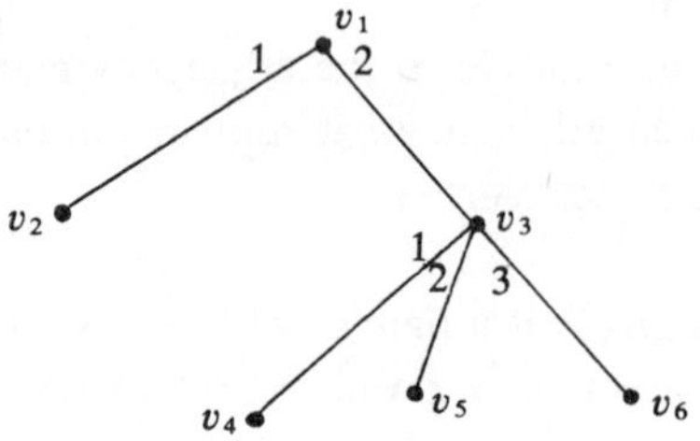

Convention. Ordered directed trees are pictured with the direct descendants of a vertex ordered from the left on the paper as in the example given in Figure 7.27. In this way edge ordering integers may be omitted.

Definition.

(i) If S_V and S_E are sets, a *labelling* of an ordered digraph $G = (V, E)$ is a pair of mappings (f, g)

$f\colon V \to S_V$ vertex labelling

and

$g\colon E \to \bigcup_{k=1}^{\infty} S_E^k$ edge labelling;

g takes the form

$g(((v, w_1), \ldots, (v, w_k))) = (\alpha_1, \ldots, \alpha_k) \in S_E^k$.

(ii) Two labelled graphs $G_1 = (V_1, E_1)$ and $G_2 = (V_2, E_2)$ with labelling functions (f_1, g_1) and (f_2, g_2) respectively are said to be *equivalent* if there is a bijection $h\colon V_1 \to V_2$ such that

(a) $((v, w_1), \ldots, (v, w_k)) \in E_1$
iff $((h(v), h(w_1)), \ldots, (h(v), h(w_k))) \in E_2$
(equivalent as ordered graphs),

(b) $f_1(v) = f_2(h(v))$ for all $v \in V$ (equal vertex labels) and

(c) for all $((v, w_1), (v, w_2), \ldots, (v, w_k)) \in E_1$ we have

$g_1(((v, w_1), \ldots, (v, w_k))) = g_2(((h(v), h(w_1)), \ldots, (h(v), h(w_k))))$
(equal edge labels). //

Following Section 7.5 we define a traversal of a digraph to be a permutation or total ordering of the vertices. For ordered digraphs dft and bft are defined in exactly the same way. For ordered directed trees other traversals are often useful and we describe some of them below.

Definition. If $T = (\{v_1, \ldots v_n\}, E)$ is an ordered directed tree and $L_v = ((v, w_1), \ldots, (v, w_k)) \in E$ we define a relation $<$ on the set

$$\{w_1, \ldots, w_k\}$$

by

$$w_i < w_j \quad \text{iff } i < j.$$

We extend $<$ by defining it on each list of E. //

Proposition. $<$ is a partial order relation on V.

Proof.

(i) $v < v$ implies a list of the form

$((q, w_1), \ldots, (q, v), \ldots, (q, v), \ldots, (q, w_k))$

which is impossible as there are no cycles in a tree, therefore $v \not< v$ for any $v \in V$.

(ii) $v < w$ and $w < u$ implies that there exist $x, y \in V$ such that

$L_x = ((x, w_1), \ldots, (x, v), \ldots, (x, w), \ldots, (x, w_k))$

and

$L_y = ((y, u_1), \ldots, (y, w), \ldots, (y, u), \ldots, (y, u_l)).$

If $x \neq y$ then $\delta^-(w) = 2$ which is impossible as T is a tree, therefore $x = y$ and

$L_x = ((x, w_1), \ldots, (x, v), \ldots, (x, w), \ldots, (x, u), \ldots, (x, w_k)),$

i.e.

$v < u$

so $<$ is a partial order relation on V. //

Under $<$ only vertices directly descended from the same vertex are comparable.

Example 7.6.8

For the ordered directed tree in Figure 7.28 we have

$$< = \{(v_2, v_3), (v_4, v_5), (v_4, v_6), (v_5, v_6)\}$$

or $v_2 < v_3$, $v_4 < v_5$, $v_4 < v_6$ and $v_5 < v_6$. //

Notation. We denote the set of all descendants of a vertex $v \in V$ by $\Gamma^+(v)$; similarly $\Gamma^-(v)$ denotes the set of all ascendants of v.

Fig. 7.28

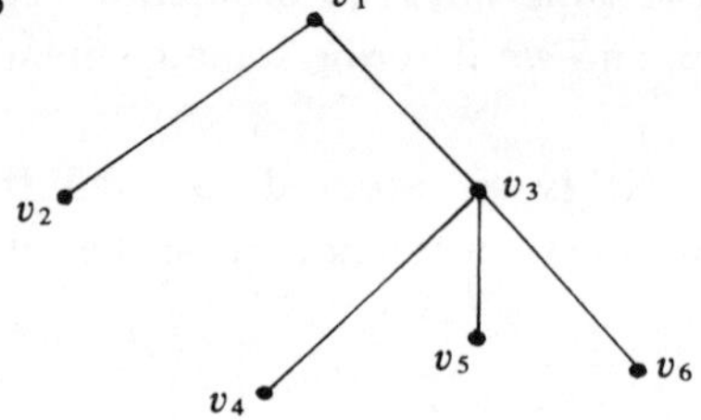

Definition. The relation $<$ is called the *transverse ordering* of the vertices of an ordered directed tree $T=(V, E)$. //

Our goal in the remainder of this section is to derive various useful tree traversal methods in a systematic way by extending the transverse ordering. Two further results are necessary before defining the traversals explicitly. Let $<$ define a further relation $<_e$ on V as follows: if $w_i < w_j$ then

$$w'_i <_e w'_j \quad \text{for all } w'_i \in \Gamma^+(w_i) \cup \{w_i\} \quad \text{and}$$
$$\text{for all } w'_j \in \Gamma^+(w_j) \cup \{w_j\}.$$

Proposition.

(i) $< \subseteq <_e$

(ii) $<_e$ is a partial order relation on V.

Proof.

(i) This is obvious.

(ii) (a) $v <_e v$ implies either $v < v$ or there exist $x, y \in V$ such that $x < y$ and $v \in \Gamma^+(x) \cup \{x\}$ and $v \in \Gamma^+(y) \cup \{y\}$. But $v \not< v$ as $<$ is a partial order relation. Also $x < y$ implies $x \neq y$ therefore

$v \in \Gamma^+(x) \cup \{x\}$

and

$v \in \Gamma^+(y) \cup \{y\}$,

but this means $\delta^-(v) = 2$ or $\delta^-(w) = 2$, for some $w \in \Gamma^-(v)$. This is impossible because $T=(V, E)$ is a tree, thus $v \not<_e v$ for all $v \in V$.

(b) $v <_e w$ means $v < w$ or there exist $x, y \in V$ such that $x < y$ and

$v \in \Gamma^+(x) \cup \{x\}$

and

$w \in \Gamma^+(y) \cup \{y\}$;

$w <_e u$ means $w < u$ or there exist $r, s \in V$ such that $r < s$ and

$w \in \Gamma^+(r) \cup \{r\}$

and

$u \in \Gamma^+(s) \cup \{s\}$;

$w \in \Gamma^+(r) \cup \{r\}$ and $w \in \Gamma^+(y) \cup \{y\}$ gives either

$r \in \Gamma^+(y)$

or

$y \in \Gamma^+(r)$.

Hence the tree has one of the forms shown in Figure 7.29. If $r \in \Gamma^+(y)$ then $s \in \Gamma^+(y)$, and $x < y$ gives $x <_e s$, but $u \in \Gamma^+(s)$ therefore $x <_e u$ and $v \in \Gamma^+(x)$ therefore $v <_e u$. If $y \in \Gamma^+(r)$ then $x \in \Gamma^+(r)$ and $r < s$ gives $x <_e s$, but $u \in \Gamma^+(s) \cup \{s\}$ therefore $x <_e u$, and $v \in \Gamma^+(x)$ gives $v <_e u$.

Hence $<_e$ is a partial order relation on V. //

Sometimes $v <_e w$ is read 'v is to the left of w'.

Fig. 7.29

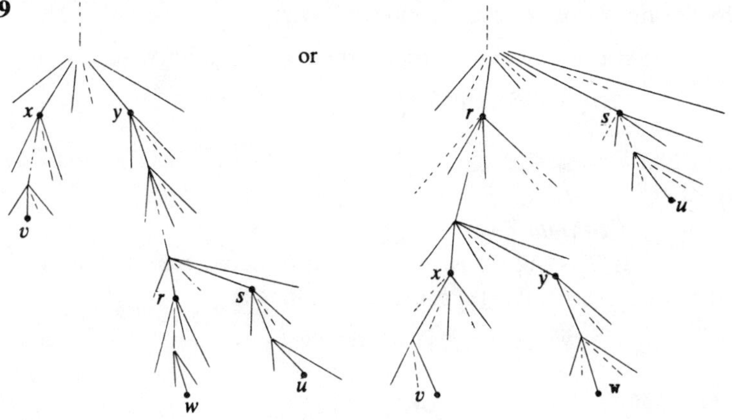

Proposition. If $T = (V, E)$ is an ordered directed tree then for $v_i, v_j \in V$ $(i \neq j)$ either

$$v_i <_e v_j \quad \text{or} \quad v_j <_e v_i$$

or v_i and v_j are on a path.

Proof. Let $v_i, v_j \in V$ with $v_i \neq v_j$ then there is a vertex v_a with

$$v_i \in \Gamma^+(v_a) \cup \{v_a\}$$

and

$$v_j \in \Gamma^+(v_a) \cup \{v_a\}.$$

If $v_i = v_a$ or $v_j = v_a$ then v_i and v_j are on a path, otherwise consider the direct descendants $w_1, \ldots, w_k$ of v_a. Then either

(1) $v_i \in \Gamma^+(w_l)$ and $v_j \in \Gamma^+(w_m)$ for $w_l \neq w_m$, or
(2) $v_i, v_j \in \Gamma^+(w_k)$.

If (1) then we have either $v_i <_e v_j$ or $v_j <_e v_i$ depending on whether $w_l < w_m$ or $w_m < w_l$. If (2) then repeat the process from w_k until (1) holds or we get

$$v_i = v_a \quad \text{or} \quad v_j = v_a$$

in which case v_i and v_j are on a path. //

Many useful tree traversals are defined by extending the relation $<_e$ to a total ordering of V. By the above result we need only extend $<_e$ to compare vertices that are on a path to define total orderings of V with $<_e$ as a subset.

Definition. If $T = (V, E)$ is an ordered directed tree we define a total ordering $<_1$ of V in the following way:

if v_j is descended from v_i then $v_i <_1 v_j$ otherwise $v_i <_1 v_j$ if $v_i <_e v_j$;

$<_1$ is called a *preorder* of V. //

Clearly $<_e \subseteq <_1$.

Example 7.6.9

If $T = (V, E)$ is the ordered directed tree in Figure 7.30 then the preorder of $V = \{a, b, c, d, e, f, g, h, i\}$ is

$$a <_1 b <_1 e <_1 f <_1 c <_1 d <_1 g <_1 i <_1 h.$$

Fig. 7.30

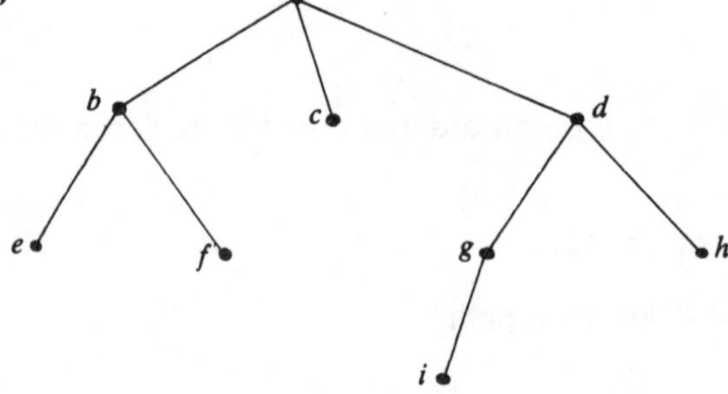

The corresponding traversal processes the vertices in the order

$a, b, e, f, c, d, g, i, h.$ //

Definition. If $T=(V, E)$ is an ordered directed tree we define a total order relation $<_2$ on V in the following way:

if v_i is descended from v_j then $v_i <_2 v_j$

otherwise $v_i <_2 v_j$ if $v_i <_e v_j$;

$<_2$ is called a *postorder* of V. //

Clearly $<_e \subseteq <_2$.

Example 7.6.10

With T as in the previous example the postorder of V is

$e <_2 f <_2 b <_2 c <_2 i <_2 g <_2 h <_2 d <_2 a$

with traversal sequence

e, f, b, c, i, g, h, d, a. //

Definition. If $T=(V, E)$ is a complete binary tree we define the *symmetric ordering* $<_s$ of V in the following way: for each non-leaf vertex $v \in V$ having direct descendants w_1 and w_2, with $w_1 < w_2$ then

$w_1' <_s v$ for all $w_1' \in \Gamma^+(w_1) \cup \{w_1\}$

and

$v <_s w_2'$ for all $w_2' \in \Gamma^+(w_2) \cup \{w_2\}$. //

The ordering $<_s$ also extends $<_e$.

Example 7.6.11

If T is the tree in Figure 7.31 then $<_s$ is defined by

$d <_s b <_s e <_s a <_s f <_s c <_s h <_s g <_s i$ //

Fig. 7.31

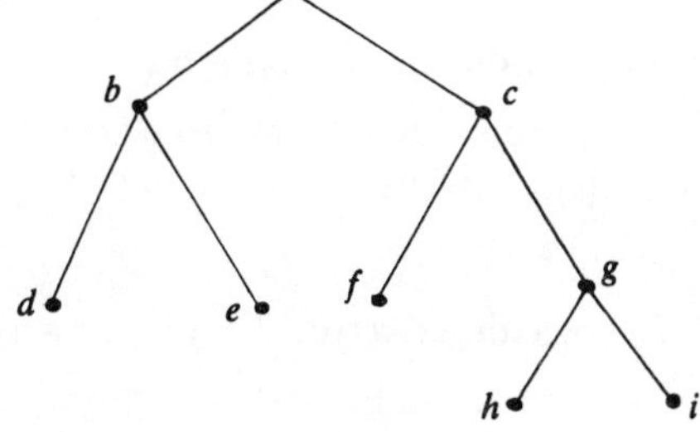

The preorder traversal of an ordered directed tree corresponds to the dft with the root as start vertex.

If $T = (V, E)$ is an ordered directed tree the vertices of T may be processed in preorder using the following algorithm, starting at the root.

```
procedure pre(v)
process v
if L_v ≠ ∅ do pre(w) for each w ∈ L_v
endproc
```

A corresponding algorithm for postorder processing is

```
procedure post(v)
if L_v = ∅ then process v
           else begin
                     post(w) for each w ∈ L_v
                     process v
                 end
```

Exercises 7.6

1. If $A \in \mathcal{M}(n, \mathbb{B})$ is the adjacency matrix of a digraph write down expressions for the functions δ^+ and δ^- in terms of A.
2. How many distinct digraphs are there on n vertices?
3. If $G = (V, E)$ is a digraph, what is the maximum possible value of $|E|$?
4. A digraph $G = (V, E)$ is defined by $V = \{v_1, v_2, v_3, v_4, v_5\}$ and $E = \{(v_1, v_2), (v_2, v_3), (v_3, v_4), (v_4, v_5), (v_5, v_3)\}$. Determine the matrix of
   (*a*) $R(E^+)$,
   (*b*) $R(E^*)$.
5. Give matrix characterizations of weak, unilateral and strong connectivity in a digraph.
6. If $G = (V, E)$ is an acyclic digraph what is the maximum possible value of $|E|$?
7. Prove the last two propositions in Section 7.6.2.
8. Show that if $T = (V, E)$ is a complete binary tree and V^l denotes the set of leaves of T then
   $|V^l| = |V \backslash V^l| + 1.$
9. A *subtree* $T' = (V', E')$ of a directed tree $T = (V, E)$ is a directed tree such that
   (i) $\varnothing \neq V' \subseteq V$,
   (ii) $E' = V' \times V' \cap E$,
   (iii) no vertex of $V \backslash V'$ is a descendant of a vertex in V'.

Draw all the subtrees of

$$T = (\{v_1, v_2, \ldots, v_6\}, \{(v_1, v_2), (v_1, v_3), (v_3, v_4), (v_3, v_5), (v_3, v_6)\}).$$

10. ***Definition.*** If T is a directed tree the *level* of a vertex is defined to be the maximum length of a path from that vertex to a leaf. The *depth* of a vertex is the length of the path from the root to that vertex. The *depth of T* is the length of the longest path in T. The *height of a vertex of T* is the depth of T minus the depth of the vertex. The *height of T* is the height of the root. //

    If T is the directed tree $(\{v_1, \ldots, v_9\}, \{(v_1, v_2), (v_1, v_3), (v_1, v_4), (v_3, v_5), (v_3, v_6), (v_3, v_7), (v_5, v_8), (v_5, v_9)\})$,
    (i) draw T with level numbers as vertex labels;
    (ii) draw T with depth numbers as vertex labels;
    (iii) draw T with height numbers as vertex labels;
    (iv) what is the depth of T?
    (v) what is the height of T?

11. If T is a directed tree a *cut*, C, of T is a subset of vertices of T such that
    (i) no two vertices of C are on a path in T,
    (ii) no vertex can be added to C without violating (i).
    Determine all the cuts of the directed tree defined in Figure 7.32.

Fig. 7.32

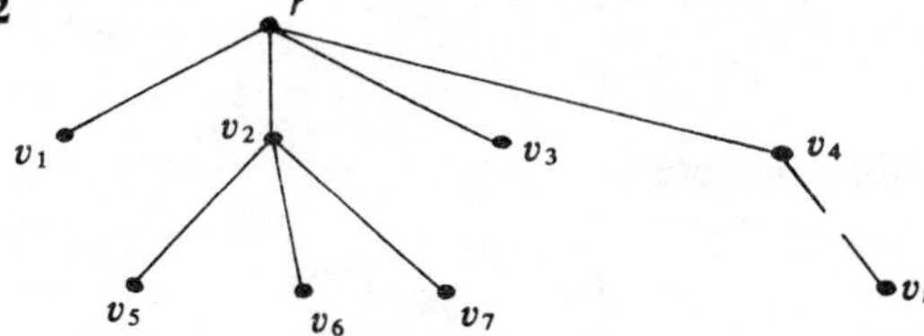

12. If $T = (V, E)$ is a complete binary tree with $|V| = n$ show that there are

    $n!/2^{(n-1)/2}$

    total orderings of V that extend the transverse ordering $<$.

13. Interpret $<_1$, $<_2$ and $<_s$ for binary trees associated with the structure of arithmetic expressions, as in Section 3.6.

8 LANGUAGES AND GRAMMARS

All means of communication involve language. Currently, we normally 'talk' to computers by written language (via punched cards, teletypes, VDUs, etc.) and hence the sentences of the language consist of strings of symbols. Indeed the whole computing process may be viewed as the transmutation of one set of strings into another. Such processes behave in very precise ways and consequently can be manipulated as mathematical objects – at the very least strings can be thought of as elements of a monoid.

Throughout this chapter we shall approach the study of language from a mathematical rather than a literary standpoint. In Section 8.1 strings are introduced and some related problems discussed and in Section 8.2 language structure is introduced. This is followed by a more detailed treatment of certain important classes of language and an introduction to parsing.

8.1 Fundamental concepts

8.1.1 Strings

A *letter* (or *character*) is a single indivisible token or symbol and an *alphabet* is a set of letters.

Examples 8.1.1

$A=\{a, b, c\}$,

$B=\{0, 1\}$,

$C=\{\text{PERFORM}, \text{ADD}, \text{GIVING}, \text{TO}, \ldots\}$,

$E=\{a, b, c, d, e, f, \ldots, w, x, y, z\}$.

Here we may regard B as the binary alphabet, C as the Cobol alphabet (in which PERFORM etc. cannot be separated) and E as the English alphabet. //

Alphabets are sets and hence we can apply set-theoretic notation to them. In particular, if A and B are alphabets such that $A \subseteq B$, then we say that A is a *subalphabet* of B, or alternatively that B is an *extension* of A.

Strings are ordered collections of letters from an alphabet, A say, and hence look very much like elements of $A^n = A \times A \times \ldots \times A$ for some $n \in \mathbb{N}$ but would normally be written as $a_1 a_2 \ldots a_n$ rather than $(a_1, a_2, \ldots, a_n)$. So letters are themselves strings derived from the case where $n = 1$. We also allow the possibility of taking no letters and have an *empty string* which we shall write as Λ. Note that Λ is not the same as the space character and $\Lambda \notin A$ for any alphabet A.

Continuing the linguistic analogy, strings are also called *words.* The set of all words over the alphabet A is called the *closure* of A and is written A^*, so

$$\begin{aligned} A^* &= A^0 \cup A^1 \cup A^2 \cup A^3 \cup \ldots \\ &= \bigcup_{n=0}^{\infty} A^n \quad \text{where } A^0 = \{\Lambda\}. \end{aligned}$$

It is also convenient to define the set of all non-empty strings over A as

$$A^+ = A^* \backslash \{\Lambda\} = \bigcup_{n=1}^{\infty} A^n.$$

As already mentioned in Example 5.2.1 the basic string operation is called *concatenation* and may be defined formally as a binary operation $\odot$ on A^* in the following way

$$\odot : (\alpha, \beta) \mapsto \alpha\beta;$$

similarly for A^+. The operation $\odot$ is associative and, on A^*, has Λ as its identity but it is *not* commutative. We summarize the properties of $\odot$ on A^* and A^+ below.

Proposition. Under concatenation

(i) A^* is a monoid,

(ii) A^+ is a semigroup. //

The result of concatenating strings α and β, i.e. $\alpha\beta$, is called the *juxtaposition* of α and β – you place the string β immediately after the string α. Alternatively a string α over an alphabet A may be recursively defined as

$$\alpha = \Lambda$$

or

$$\alpha = a\beta \quad \text{where } a \in A \text{ and } \beta \text{ is a string over } A.$$

Here $a\beta$ is simply the letter a followed immediately by the string β.

All words in A^n for some $n \in \{0\} \cup \mathbb{N}$ are composed of exactly n letters and are said to have *length n*. The length of $\alpha \in A^*$ is denoted by $|\alpha|$ and

$$|\alpha| = n \quad \text{iff } \alpha \in A^n.$$

Trivially $|\Lambda| = 0$ and $|a| = 1$ iff $a \in A$.

When strings are made up of repetitions of letters it is usual to adopt a superscript notation to indicate what we might regard as a repeated product (wrt concatenation). So, if $a \in A$, we may write

$$\Lambda = a^0$$
$$aa = a^2,$$

and in general

$$aa^{n-1} = a^n \quad \text{for } n \in \mathbb{N}.$$

The same convention can be applied to repeated strings; thus *ababa* can be written $(ab)^2a$ or $a(ba)^2$.

This is our first encounter with one of the main difficulties in talking about strings. We are using strings to describe strings and hence we need to be able to distinguish between the alphabets used. If the expressions above are over the alphabet A and the symbols '(' and ')' are not in A then the meaning is clear; if, on the other hand, the closed bracket is in A the expression $(ab)^2a$ could mean $(ab))a$. Provided that we are aware of the possibility of such problems and we know which alphabets we are dealing with these pit-falls can be avoided by using a different alphabet and a bijection between the two alphabets. In some cases it may be more desirable to construct a monomorphism between sets of strings so that one set is over a particularly simple alphabet.

Example 8.1.2

Let $B = \{0, 1\}$ and $A = \{a, b, c\}$. Then $\phi : A^* \to B^*$ defined by

$$\phi(xy) = \phi(x)\phi(y)$$

with

$$x, y \in A^*$$

and

$$\phi(a) = 0,$$
$$\phi(b) = 10,$$
$$\phi(c) = 110$$

is a monoid homomorphism (we need to preserve combinations) from A^* to the set $\{0, 10, 110\}^* \subseteq B^*$. For example

$$\phi : abbca \mapsto 010101100$$

and

$$\phi^{-1} : 01011010110 \mapsto abcbc.$$

In fact this kind of construction can be used to map any finite alphabet into $\{0, 1\}^*$. //

From the definition of length it follows that if $\alpha, \beta \in A^*$ for some alphabet A then

$$|\alpha\beta| = |\alpha| + |\beta|$$

and

$$|\alpha^n| = n|\alpha|.$$

Moreover, if $a \in A$ then

$$|a^n| = n.$$

In transforming one string into another it is unlikely that the entire input string will be changed in a single operation; indeed if this were so the process would possibly be specified only by a set of input–output pairs. To do anything else we need the notion of substrings.

Given strings α and β over the alphabet A, β is a *substring* of α if

$$\alpha = \gamma\beta\delta \quad \text{with } \gamma, \delta \in A^*.$$

Example 8.1.3

Let $A = \{a, b, c\}$ and $\alpha = abac$ then the substrings of α are

$\Lambda, a, b, c,$

$ab, ba, ac,$

$aba, bac,$

$abac.$

Notice in particular that α is a substring of itself and Λ is a substring of α (and any other string) since

$$\begin{aligned}\alpha &= \Lambda abac\\ &= \Lambda a \Lambda bac\\ &= \Lambda a \Lambda b \Lambda ac\\ &= \Lambda a \Lambda b \Lambda a \Lambda c\\ &= \Lambda a \Lambda b \Lambda a \Lambda c \Lambda\\ &= a \Lambda bac\\ &= a \Lambda\Lambda bac\\ &\ \vdots\end{aligned}$$

//

Isolation of a substring leads naturally to the replacement of the substring by another string. However, we have not yet made matters watertight enough to be able to do this properly. Consider what happens if we have a 'function' f that replaces the string xy by yxx when the former occurs as a substring in the operand. So

$$f(pqxy) = pqyxx.$$

But what happens about $f(xypxy)$? Do we take $yxxpxy$ or $xypyxx$? Potentially still more dangerous, if we apply f several times to successive resultant strings then $f^2(xypxy) = yxxpyxx$ uniquely but applying f to $xxyy$ gives

$$f(xxyy) = xyxxy$$
$$f^2(xxyy) = yxxxxy \text{ or } xyxyxx,$$

so the 'function' is ill-defined and we cannot correct matters by requiring the operation to change all substrings since they may overlap!

The situation gets even worse when several replacement functions are involved. What is needed is a means of selecting a specific substring whenever there is a choice – in particular we shall be concerned with the one which occurs first, counting from left to right. To formalize this we use the order properties of integers and the integers derived from the lengths of strings.

Suppose α and β are strings over A and $|\alpha| \leqslant |\beta|$ such that α is a substring of β. Assume that α occurs in β in m different ways and that $|\beta| - |\alpha| = n$. Therefore we can write β in the following m different ways:

$$\begin{aligned} \beta &= \gamma_1 \alpha \delta_1 \\ &= \gamma_2 \alpha \delta_2 \\ &\vdots \\ &= \gamma_m \alpha \delta_m \end{aligned}$$

where $\gamma_i, \delta_i \in A^*$ for all $i : 1 \leqslant i \leqslant m$,

$$|\gamma_1| < |\gamma_2| < \ldots < |\gamma_m|$$

and $m \leqslant n+1$ (if $\alpha = \beta$ then $n = 0$ so there is only one possibility). We say that γ_1 to γ_m *specify* the different occurrences of α in β, that γ_1 gives the *leading occurrence* and that the occurrence of α immediately after γ_1 *is* the leading occurrence.

Example 8.1.4

If g is a function $A^* \to A^*$ such that $\{x, y, p, q\} \subseteq A$ and g replaces the leading occurrence of xy in a string by yxx then

$$g(pqxy) = pqyxx,$$
$$g(xypxy) = yxxpxy,$$
$$g^2(xypxy) = yxxpyxx.$$

Notice also that

$$\begin{aligned} g^6(x^2y^2) &= g^5(xyx^2y) \\ &= g^4(yx^4y) \\ &= g^3(yx^3yx^2) \\ &= g^2(yx^2yx^4) \\ &= g(yxyx^6) \\ &= y^2x^8 \end{aligned}$$

and

$$g(y^2x^8) = y^2x^8$$

so

$$g^6(x^2y^2) = g^7(x^2y^2) = \ldots. \quad //$$

Before going on we note an alternative collection of terms for letter, alphabet and word; these are *word, vocabulary* and *sentence* respectively. In some contexts these terms are more sensible but special care must be taken because of the dual usage of 'word'.

8.1.2 Languages

So much for strings as such. A collection of strings (or sentences) is a language. Formally a *language* L over the alphabet A is a set of strings in A^*, so $L \subseteq A^*$.

Consequently operations on strings induce operations on languages. Hence we obtain L^+ (the *transitive closure* of L) and L^* (the *reflexive closure* of L) by

(*a*) $L^0 = \{\Lambda\}$,

(*b*) if L_i and L_j are languages then

$L_iL_j = \{xy: x \in L_i, y \in L_j\}$,

(*c*) $L^n = L^{n-1}L \quad n \in \mathbb{N}$

(*d*) $L^+ = \bigcup_{n \geq 1} L^n$

(*e*) $L^* = \bigcup_{n \geq 0} L^n$.

We now turn our attention to how words may be assembled into sentences and the set of all *meaningful* sentences collected into a language. Our interest will primarily be in artificial languages (such as programming languages or languages describing valid mathematical expressions) but in the first instance it will be helpful to consider an example from English. This will enable us to give some definitions so that we can make the initial steps into language theory. Take the sentence

'The dog bit me.'

Regarding the sentence, as the reader does, devoid of any effect that it may convey we can view it at two levels. Firstly, if we treat it as a mere collection of words each of which is in turn an ordered collection of letters, then we consider its *syntax*; secondly, interpreting the sentence by using what we understand to be the meanings of the words and their interrelationships we obtain its *semantics* – its meaning. In addition, if we are the speaker of the sentence (or other person or entity affected by its meaning) we are influenced by its *pragmatics* – its effect. These three areas together comprise the *semiotic* of the language.

Example 8.1.5

In the computer languages Fortran and Cobol, the statements

$A = B + C$

and

ADD B TO C GIVING A

respectively both imply the semantic notions of addition and assignment but have differing syntax. Pragmatically they may be represented, on some machine, by the outcome of executing the code

```
LOAD    B
ADD     C
STORE   A.  //
```

Our main concern will be in the area of syntax.

To illustrate the kind of structure that we shall study consider the diagram in Figure 8.1. Although this needs formal definition, essentially

Fig. 8.1

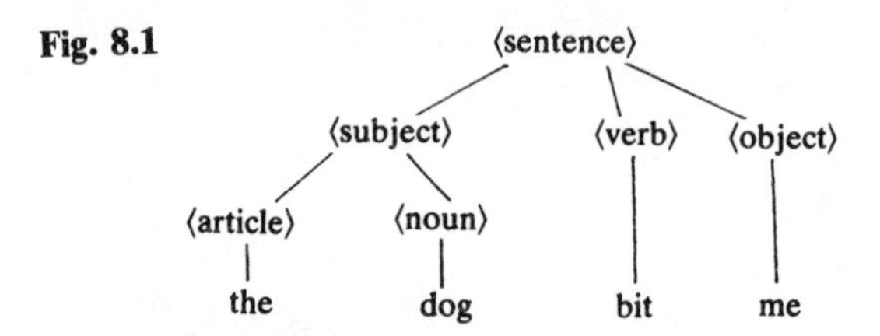

all that it means is that a ⟨sentence⟩ may be constructed by concatenating a ⟨subject⟩, a ⟨verb⟩ and an ⟨object⟩. Then a ⟨subject⟩ can be constructed by following an ⟨article⟩ with a ⟨noun⟩, and finally we choose the ⟨article⟩ 'the', the ⟨noun⟩ 'dog',... to give the sentence 'the dog bit me'.

Before giving the terminology and notation required to make precise the general concepts embodied in the specific situation depicted in Figure 8.1 we are now in a position to state the major aims of language theory and preview the rest of the chapter.

Recall that, for a given alphabet A, a language L is *any* subset of A^*; however, arbitrary subsets are of very little interest. We want to restrict attention to special languages containing strings which, because of *external* knowledge of the required semantics, we regard as meaningful or well-structured.

Most *interesting* languages are infinite and hence cannot be written out explicitly. In these cases we must devise ways of generating the language; a grammar G may be regarded as such a generative system.

The two main objectives of formal language theory may now be given:

(i) Given a grammar G (and an associated language L) how do we generate sentences $\alpha : \alpha \in L$?

(ii) Given $L \subseteq A^*$ and $\alpha \in A^*$ how do we analyse α to ascertain if $\alpha \in L$?

Analysis of strings to test for containment in L requires knowledge of how L is generated by G. In Sections 8.2 and 8.3 we describe the general principles of phrase-structure grammars and then look more closely at a certain subclass which has been found to be of great practical use.

If $L(G)$ denotes the language generated by G then determining the validity of $\alpha \in L(G)$, a process known as parsing, involves both α and G. Often the original grammar is not suitable for a particular parsing technique but can be modified into an equivalent form which is suitable. These related topics of parsing and grammar modification are addressed in Section 8.4.

Carrying further the idea of manipulating G before even looking at α, in certain very specialized (but not uncommon) situations it is possible to transfer almost all of the parsing effort (once and for all) into analysis of the *grammar* and hence greatly simplify the analysis of individual strings. Specification of the restrictions which must be satisfied by these grammars is necessarily more complex and Section 8.5 is devoted to the study of such matters.

Exercises 8.1

1. Suppose A and B are non-empty alphabets such that $|A| = p$ and $|B| = q$ and that

$$\phi : A \to \mathbb{N}_p$$

and

$$\psi : B \to \mathbb{N}_q$$

are bijections. Let $\chi_1 : A^* \to \mathbb{N}$ be defined by

$$\chi_1 : a_1 \ldots a_k \mapsto \sum_{i=1}^{k} \phi(a_i) * p^{(i-1)}$$

and $\chi_2 : B^* \to \mathbb{N}$ by

$$\chi_2 : b_1 \ldots b_k \mapsto \sum_{i=1}^{k} \psi(b_i) * q^{(i-1)}.$$

Show that $\chi_2^{-1} \circ \chi_1$ is a string bijection and demonstrate its application by finding the direct and inverse images of the strings x^2yxz in A^* and 145332 in B^* where $A = \{x, y, z\}$ and $B = \{1, 2, 3, 4, 5\}$.

8.2 Phrase-structure grammars

8.2.1 Basic definitions

Definition. A *phrase-structure grammar* (PSG) G is an algebraic structure consisting of an ordered 4-tuple (N, T, P, S) where

(i) N and T are non-empty finite alphabets of *non-terminal symbols* and *terminal symbols* respectively such that $N \cap T = \emptyset$,

(ii) P is a finite set of *productions*, $P \subseteq V^+ \times V^*$, where $V = N \cup T$ is called the *vocabulary* of G, and

(iii) $S \in N$ and S is called the *start symbol* or *source.* //

Assuming the symbol $\to$ not to be in V, $(\alpha, \beta) \in P$ is usually written as

$\alpha \to \beta$.

The notion behind productions, which are also referred to as *rewriting rules,* is that of being able to change one string of symbols into another. *Conceptually* terminal symbols are generally regarded as symbols that ought not to be changed and hence the definition of production used in PSGs is perhaps a little too general. In practical cases suitable restrictions will be imposed so as not to violate the permanence of terminal symbols but for now the definition will suffice.

As a first step refer to Figure 8.1 and try to see how it relates to the next example.

Example 8.2.1

The English sentence given as an illustration earlier can thus be specified by the grammar $G = (N, T, P, S)$ where

$N = \{$⟨sentence⟩, ⟨subject⟩, ⟨article⟩, ⟨noun⟩, ⟨verb⟩, ⟨object⟩$\}$,

$T = \{$the, dog, bit, me$\}$,

$P = \{$(⟨sentence⟩, ⟨subject⟩⟨verb⟩⟨object⟩),

(⟨subject⟩, ⟨article⟩⟨noun⟩),

(⟨article⟩, the),

$((\langle noun \rangle, dog),$

$((\langle verb \rangle, bit),$

$((\langle object \rangle, me)\},$

$S = \langle sentence \rangle.$

This particular system derives only the one sentence 'the dog bit me' and hence could be replaced by

$$N = \{\langle \text{sentence} \rangle\}$$

$$P = \{(\langle \text{sentence} \rangle, \text{the dog bit me})\}$$

or even

$$L = \{\text{the dog bit me}\}.$$

However, if now we wish to extend the language to include all sentences starting with say 'the lion', 'the rat' and 'the tiger', with the verbs 'ate' and 'attacked', and with objects 'you' and 'Napoleon' (so L would then have 35 more elements), this can be done by adding only 7 extra elements to each of the sets T and P. In this instance the size of the language is $4*3*3$ whereas the size of the set P is approximately $4+3+3$. Of even greater significance is the fact that we can include all the sentences of the form 'the dog bit (the son of)^n Napoleon' – an infinite set of them – by adding a handful of elements to T and P. //

Before describing the mechanics of generating sentences we must mention a notation due to Backus (*Backus normal form* or *Backus Naur form*, BNF). This is particularly useful when we wish to use elements in N that may be confused with those of T – such as ⟨sentence⟩ and 'sentence' – and the notation uses the four symbols

::= (meta-becomes),

< (meta-open),

> (meta-close),

| (meta-or).

The meta-open and meta-close are used to surround strings as elements of N, the meta-becomes replaces →, and if $(\alpha, \beta) \in P$ and $(\alpha, \gamma) \in P$ then this can be written as

$$\alpha ::= \beta \mid \gamma$$

which is read as 'α is β or γ'.

BNF was first used to define the syntax of Algol 60. The reader is encouraged to read the Algol 60 report (Naur, P. (editor) *et al.* 'Revised report on the Algorithmic Language ALGOL 60', *Computer Journal* Vol. 5, pp. 349–67 (1963)) if he has any doubts that BNF is capable of defining anything 'serious'. In formal language work, long strings are usually avoided in N and hence the Backus notation is not used except for the meta-or symbol. Typically, upper case letters will be used for elements of N and lower case for elements of T.

Example 8.2.2

$$G=(N, T, P, S)$$

where

$$N=\{S, T\},$$

$$T=\{a, b, c, d\},$$

$$P=\{S \to aTd, T \to bT|b|cT|c\}.$$

Notice that the dual use of T in this example causes no difficulties whatsoever. //

This grammar will generate all strings $a\{b, c\}^+d$ but as yet we have not said exactly how we can achieve this. We use the productions in the following way.

Let $\alpha, \beta \in V^*$; then α *directly derives* β if $\alpha = \gamma\sigma\delta$ and $\beta = \gamma\rho\delta$ where $\gamma, \delta, \rho \in V^*$, $\sigma \in V^+$ and $\sigma \to \rho \in P$. This is written as $\alpha \Rightarrow \beta$ and can be informally regarded as changing the string α into the string β by replacing a substring σ in α by ρ. (Notice that there is no compulsion to replace a specific instance of σ in α, nor to use a particular production whose left part is σ. All variations are possible.)

If now α and β are words over V and there is a finite sequence α_0, $\alpha_1, \alpha_2, \ldots, \alpha_r$ where $\alpha_0 = \alpha$, $\alpha_r = \beta$ and $\alpha_{i-1} \Rightarrow \alpha_i (i=1, \ldots, r)$ then we say that α *derives* β, written $\alpha \overset{*}{\Rightarrow} \beta$, and the derivation is said to be realized by $\alpha \Rightarrow \alpha_1 \Rightarrow \alpha_2 \Rightarrow \ldots \Rightarrow \alpha_{r-1} \Rightarrow \beta$. By analogy $\alpha \overset{+}{\Rightarrow} \beta$ if the derivation uses a non-empty sequence of direct derivations. Now if $\alpha \in V^*$ such that $S \overset{*}{\Rightarrow} \alpha$ then α is called a *sentential form*. Moreover if $\alpha \in T^*$ and $S \overset{*}{\Rightarrow} \alpha$ then α is a *sentence* generated by G. Thus the *language*, $L(G)$, generated by G is $\{\alpha: \alpha \in T^* \text{ and } S \overset{*}{\Rightarrow} \alpha\}$. Where G is understood we can define $L(X) = \{\alpha: \alpha \in T^*, X \in N \text{ and } X \overset{*}{\Rightarrow} \alpha\}$. By virtue of the degree of variability allowed in applying productions to sentential forms there may be many possible derivation sequences for a given sentence in $L(G)$ where G is a specific grammar. From these we select the one which at every stage acts on the leftmost possible substring that has to be changed by an element of P. This is called the (left) *canonical derivation sequence* for the sentence.

Example 8.2.3

Let $G = (\{B\}, \{(,)\}, P, B)$ where

$P = \{B \to (B) | BB | (\ )\}$.

The sentence ( )(( )( )) can then be derived in many different ways of which five are as follows:

(i) $B \Rightarrow BB$
$\Rightarrow (\ )B$
$\Rightarrow (\ )(B)$
$\Rightarrow (\ )(BB)$
$\Rightarrow (\ )((\ )B)$
$\Rightarrow (\ )((\ )(\ ))$

(ii) $B \Rightarrow BB$
$\Rightarrow (\ )B$
$\Rightarrow (\ )(B)$
$\Rightarrow (\ )(BB)$
$\Rightarrow (\ )(B(\ ))$
$\Rightarrow (\ )((\ )(\ ))$

(iii) $B \Rightarrow BB$
$\Rightarrow B(B)$
$\Rightarrow (\ )(B)$
$\Rightarrow (\ )(BB)$
$\Rightarrow (\ )((\ )B)$
$\Rightarrow (\ )((\ )(\ ))$

(iv) $B \Rightarrow BB$
$\Rightarrow B(B)$
$\Rightarrow (\ )(B)$
$\Rightarrow (\ )(BB)$
$\Rightarrow (\ )(B(\ ))$
$\Rightarrow (\ )((\ )(\ ))$

(v) $B \Rightarrow BB$
$\Rightarrow B(B)$
$\Rightarrow B(BB)$
$\Rightarrow B((\ )B)$
$\Rightarrow (\ )((\ )B)$
$\Rightarrow (\ )((\ )(\ ))$

Of these the first is the canonical derivation. //

8.2.2 The Chomsky hierachy

The system discussed so far is a powerful descriptive tool but as it stands it is too general. However, when restrictions are imposed we have a more interesting (yet still quite powerful) mathematical object. The initial restrictions that we shall impose on the PSGs qualify the elements of P and are due to Chomsky.

Definition. (The Chomsky hierachy.)

Let $G = (N, T, P, S)$ be a PSG described in Section 8.2.1. Such a grammar is called Chomsky type 0. If all elements of P are of the form $\alpha \to \beta$ where $\alpha = \gamma_1 x \gamma_2$ and $\beta = \gamma_1 \delta \gamma_2$ with $\gamma_1, \gamma_2 \in V^*$, $x \in N$ and $\delta \in V^+$ then G is said to be *context sensitive* (a CSG) or Chomsky type 1. (In this definition the strings γ_1 and γ_2 may be regarded as the context in which x may be replaced by δ.)

An alternative restriction for type 1 grammars is that in each production α and β should be such that $1 \leq |\alpha| \leq |\beta|$. (The equivalence of these two definitions is not obvious and is taken up below.) If the replacements may be carried out regardless of context then we may replace 'contexts' γ_1 and γ_2 by the empty string Λ and obtain the weaker restriction that if $x \to \delta \in P$ then $x \in N$ and $\delta \in V^+$. This restriction is satisfied by Chomsky type 2 grammars. Finally, if P consists only of productions of the form $x \to \delta$ where $x \in N$ and $\delta \in T \cup TN$ (so the right hand side is a single terminal or a single terminal followed by a single non-terminal) then G is said to be Chomsky type 3. //

Although *not* allowed in Chomsky's strict classification it is often useful to permit slightly more general forms within sets of productions. Essentially we would like to be able to include the empty string Λ as a right hand side of any production but, as will be seen later, Λs cause problems. Such Λ-productions are absolutely *necessary* from the generative point of view only if $\Lambda \in L$, in which case we can add $S \to \Lambda$ to P provided that S does not occur in the right hand side of any production. However, operationally, for certain parsing methods it is necessary to allow more general Λ-productions. To distinguish between Chomsky grammars and those in which Λ-productions are permitted we shall refer to the extended versions of Chomsky type 2 and 3 grammars as *context-free* (CFG) and *regular* (or right-linear) grammars respectively.

The languages generated by any of these kinds of grammar are similarly named. So a PSG generates a *phrase-structure language* (PSL), a CSG generates a CSL, a CFG generates a CFL and a regular grammar generates a *regular language* (or *regular set*). The majority of examples in this chapter will be of context-free languages and in Chapter 9 we shall concentrate on regular languages. However, most practical languages are to some extent context sensitive and in indicating the limitations of CFGs the following example is important.

Example 8.2.4

$\{x^n y^n z^n : n \in \mathbb{N}\}$ is a context sensitive language.

Suppose $G = (N, T, P, S)$ where

$N = \{S, X, Y, Z\}$,

$T = \{x, y, z\}$,

$P = \{P_1, \ldots, P_7\}$

and

$P_1 = S \to xSYZ$,

$P_2 = S \to xYZ$,

$$P_3 = xY \to xy,$$
$$P_4 = yY \to yy,$$
$$P_5 = yZ \to yz,$$
$$P_6 = ZY \to YZ,$$
$$P_7 = zZ \to zz.$$

First notice that for any value $n \in \mathbb{N}$ we can obtain

$$\begin{aligned} S &\overset{*}{\Rightarrow} x^{n-1}S(YZ)^{n-1} && \text{by } P_1 \\ &\Rightarrow x^n(YZ)^n && \text{by } P_2 \\ &\overset{*}{\Rightarrow} x^nY^nZ^n && \text{by } P_6 \\ &\Rightarrow x^nyY^{n-1}Z^n && \text{by } P_3 \\ &\overset{*}{\Rightarrow} x^ny^nZ^n && \text{by } P_4 \\ &\Rightarrow x^ny^nzZ^{n-1} && \text{by } P_5 \\ &\overset{*}{\Rightarrow} x^ny^nz^n && \text{by } P_7 \end{aligned}$$

so

$$\{x^ny^nz^n : n \in \mathbb{N}\} \subseteq L(G).$$

We must now show that no other strings can be generated by G.

Although some variation in the order of application of P_1, P_2 and P_6 is possible, any sentence must be derived via a sentential form such as $x^nYZ\alpha$ where α consists of $(n-1)$ Ys and Zs. In order to achieve a string over T we must eventually use P_4, P_5 and P_7 but P_7 can convert Z to z only in the context zZ and P_5 performs the same substitution in the context yZ. Similarly P_4 and P_3 require the contexts yY and xY respectively before changing Y to y. At this stage the only terminals are in the x^n substring so the next move must be $x^nyZ\alpha$, caused by P_3. But we know that a valid sentence can be generated following transformation of $Z\alpha$ into $Y^{n-1}Z^n$ by P_6, indeed that is the only way in which a successful generation can take place.

Assume that we had an intermediate substring of the form

$$yY^mZ^pY\beta$$

where β consists of the remaining Ys and Zs. By the preceding argument we must use m applications of P_4 to achieve $y^{m+1}Z^pY\beta$ but now if we use P_5 to give $y^{m+1}zZ^{p-1}Y\beta$ the following Y is condemned in that no rules exist to change it to y (or any other terminal). The only way out is to use P_6, p times, to move the Y to the left and hence achieve $x^ny^nz^n$. //

This is an example of a context-sensitive language which we shall subsequently show not to be context-free. Similarly, there exist context-free languages which are not regular (see Chapter 9). We now turn our attention to proving the equivalence of the alternative characterization of context-sensitive grammars.

Definition. Grammars G_1 and G_2 are *equivalent* if $L(G_1)=L(G_2)$. //

Proposition. L is a CSL if and only if it may be generated by a grammar whose productions $\sigma \to \mu$ satisfy $1 \leqslant |\sigma| \leqslant |\mu|$.

Proof.

(i) If L is a CSL then there is a grammar G with productions of the form

$\alpha A \beta \to \alpha \gamma \beta$ where $A \in N$, $\gamma \in V^+$ and $\alpha, \beta \in V^*$

such that

$L = L(G)$.

But

$$|\alpha A \beta| = |\alpha| + |A| + |\beta| = |\alpha| + 1 + |\beta| \geqslant 1$$

and

$$\begin{aligned} |\alpha \gamma \beta| &= |\alpha| + |\gamma| + |\beta| \\ &\geqslant |\alpha| + 1 + |\beta| = |\alpha A \beta|. \end{aligned}$$

Therefore

$1 \leqslant |\alpha A \beta| \leqslant |\alpha \gamma \beta|$ as required.

(ii) Let $G=(N, T, P, S)$ be a grammar whose productions $\sigma \to \mu$ satisfy $1 \leqslant |\sigma| \leqslant |\mu|$. We must produce a grammar G' equivalent to G with productions in the form

$\alpha A \beta \to \alpha \gamma \beta$.

Productions of G look like either

(1) $A \to \gamma_1 \ldots \gamma_p$, or

(2) $\alpha_1 \ldots \alpha_n \to \beta_1 \ldots \beta_q$ where $n \leqslant q$

and $A \in N$, $\alpha_i, \beta_i, \gamma_i \in V$.

In all productions replace each $a_i \in T$ that occurs by a new non-terminal A_i and include productions $A_i \to a_i$ in G'. Productions of type (1) are now in the right form and are included in G'. Productions of type (2) need to be modified however. They now have the form

$$W_1 \ldots W_n \to Y_1 \ldots Y_q \qquad n \leqslant q$$

where W_i and Y_j are non-terminal symbols of the new grammar. For each such production we introduce new non-terminals $\hat{Y}_1, \ldots, \hat{Y}_q$ and $n+q$ new productions,

n productions:

$$W_1 \ldots\ldots\ldots\ldots W_n \rightarrow \hat{Y}_1 W_2 \ldots W_n,$$
$$\hat{Y}_1 W_2 \ldots\ldots\ldots W_n \rightarrow \hat{Y}_1 \hat{Y}_2 W_3 \ldots W_n,$$
$$\vdots$$
$$\hat{Y}_1 \ldots \hat{Y}_{n-2} W_{n-1} W_n \rightarrow \hat{Y}_1 \ldots \hat{Y}_{n-2} \hat{Y}_{n-1} W_n,$$
$$\hat{Y}_1 \ldots \hat{Y}_{n-2} \hat{Y}_{n-1} W_n \rightarrow \hat{Y}_1 \ldots \hat{Y}_{n-2} \hat{Y}_{n-1} \hat{Y}_n \hat{Y}_{n+1} \ldots \hat{Y}_q,$$

q productions:

$$\hat{Y}_1 \hat{Y}_2 \ldots\ldots\ldots \hat{Y}_q \rightarrow Y_1 \hat{Y}_2 \ldots \hat{Y}_q,$$
$$Y_1 \hat{Y}_2 \ldots\ldots\ldots \hat{Y}_q \rightarrow Y_1 Y_2 \hat{Y}_3 \ldots \hat{Y}_q,$$
$$\vdots$$
$$Y_1 Y_2 \ldots\ldots Y_{q-1} \hat{Y}_q \rightarrow Y_1 Y_2 \ldots Y_{q-1} Y_q.$$

All of these $n+q$ productions are in the form $\alpha A \beta \rightarrow \alpha \gamma \beta$.

The new non-terminals $\hat{Y}_1, \ldots, \hat{Y}_q$ force application of these productions in the order written so that no sentences outside the original language may be generated. //

To conclude this section we discuss the concept of ambiguity. The classic case of an ambiguous sentence is

'They are flying planes.'

This has two interpretations depending on whether we take 'are flying' as the verb or 'flying planes' as the object, and leads us straight into a precise definition of ambiguity. A language is *ambiguous* if it contains an ambiguous sentence. In turn a sentence is *syntactically ambiguous* if it has more than one canonical derivation and is *semantically ambiguous* if, for a given canonical derivation, it has more than one interpretation. (Derivations are related not directly to a language but to a grammar that generates it. Hence we should more properly refer to an *ambiguous grammar*; however there do exist *inherently ambiguous* languages which can be generated only by ambiguous grammars.) For a discussion of semantic ambiguities the reader is referred to the more imaginative texts on computer languages but we can easily illustrate syntactic ambiguities by two examples.

Example 8.2.5

(1) From mathematics. Let

$G = (\{E\}, \{1, -\}, \{E \rightarrow E - E \mid 1\}, E),$

then

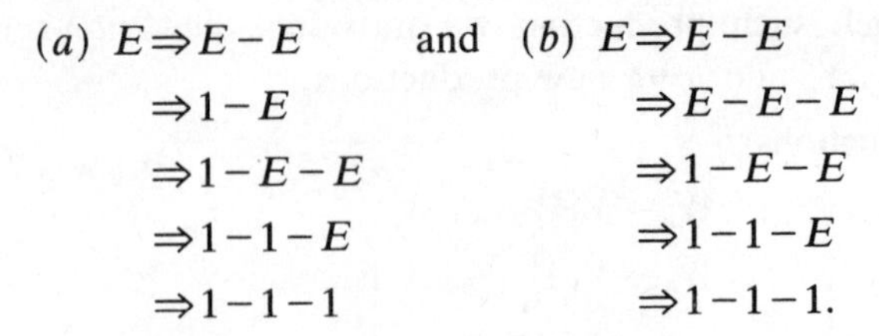

$$(a)\ E \Rightarrow E-E \quad \text{and} \quad (b)\ E \Rightarrow E-E$$
$$\Rightarrow 1-E \qquad\qquad \Rightarrow E-E-E$$
$$\Rightarrow 1-E-E \qquad\qquad \Rightarrow 1-E-E$$
$$\Rightarrow 1-1-E \qquad\qquad \Rightarrow 1-1-E$$
$$\Rightarrow 1-1-1 \qquad\qquad \Rightarrow 1-1-1.$$

From the derivation sequences these two derivations are obviously different and probably we would wish to attach different meanings to them. In (a) the second '$-$' is evaluated first giving the answer 1, in (b) the first '$-$' is executed first giving -1. (The diagrams in Figure 8.2 illustrate the different structures.)

Fig. 8.2

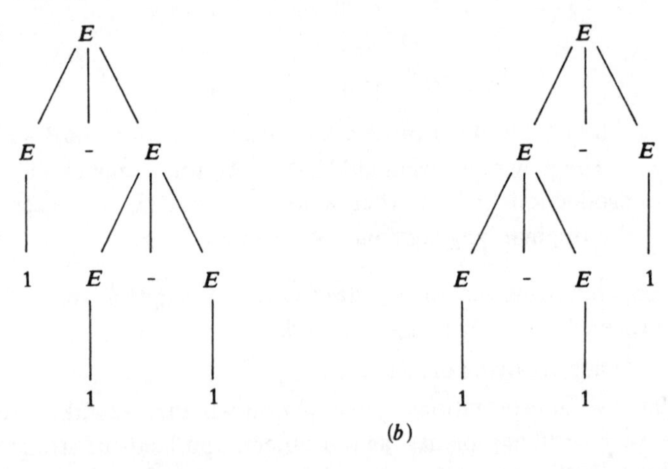

(2) A well-known example from the first specification of Algol 60, the 'dangling else'. Restricting the grammar to the relevant sublanguage we have productions

$S \rightarrow$ **if** B **then** S **else** $S\,|$
if B **then** $S\,|$
U

with S for statement, B for Boolean expression and U for unconditional statement. Now consider

if B_1 **then if** B_2 **then** U_1 **else** U_2.

We do not know whether the '**else** U_2' belongs to '**if** B_1' or '**if** B_2' (see Figures 8.3(a) and (b)). Formally, we can derive this sentence, regarding B and U as terminals, by

Fig. 8.3 (*a*)

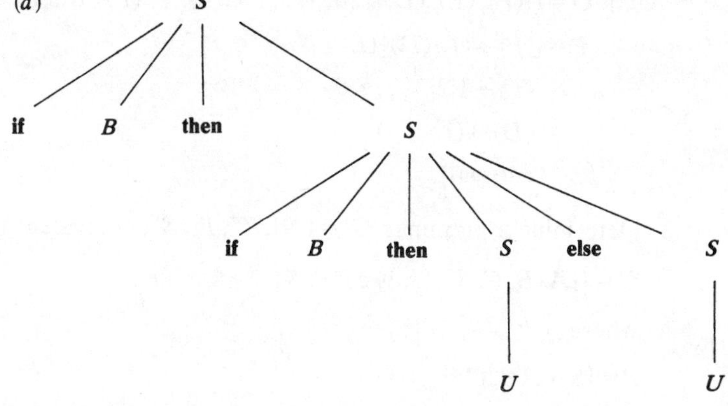

(*b*)

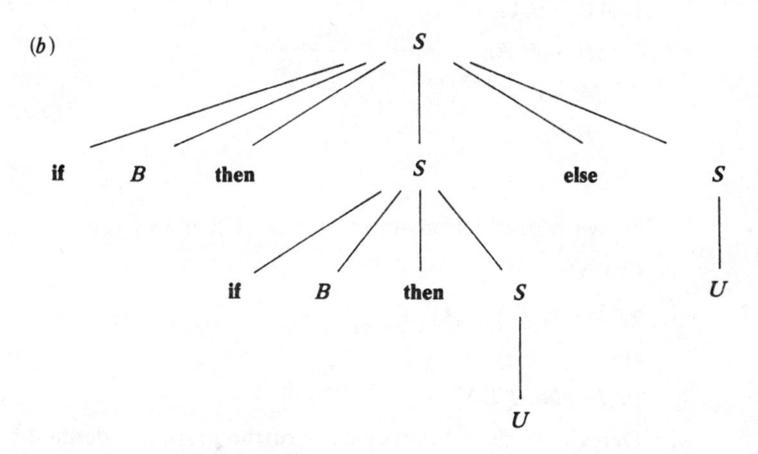

(*a*) $S \Rightarrow$ **if** B **then** S

$\Rightarrow$ **if** B **then if** B **then** S **else** S

$\Rightarrow$ **if** B **then if** B **then** U **else** S

$\Rightarrow$ **if** B **then if** B **then** U **else** U, or

(*b*) $S \Rightarrow$ **if** B **then** S **else** S

$\Rightarrow$ **if** B **then if** B **then** S **else** S

$\Rightarrow$ **if** B **then if** B **then** U **else** S

$\Rightarrow$ **if** B **then if** B **then** U **else** U. //

Exercises 8.2

1. State precisely the languages determined by the following grammars:
   (i) $G = (\{\langle\text{digit}\rangle\}, \{0, 1, 2, \ldots, 9\}, P, \langle\text{digit}\rangle)$
   where $P = \{\langle\text{digit}\rangle \to 0|1|2|3|\ldots|8|9\}$

(ii) $G = (\{\langle P\rangle, \langle L\rangle, \langle D\rangle\}, \{0, 1, 2, \ldots, 9\}, P, \langle P\rangle)$ where

$$P = \{\langle P\rangle \to \langle L\rangle\langle D\rangle | \langle L\rangle$$
$$\langle L\rangle \to 1|2|3 \ldots 8|9$$
$$\langle D\rangle \to \langle L\rangle$$
$$\langle D\rangle \to 0\}.$$

2. Determine a grammar $G' = (N', T', P', S')$ equivalent to

$G = (\{A, B, C, S\}, \{x, y, z\}, P, S)$

where

$$P = \{S \to AB^2C$$
$$AB \to BAz$$
$$zB \to A^2Bx$$
$$A \to x$$
$$B \to y$$
$$C \to z\}$$

but with productions in the form $\alpha Q\beta \to \alpha\gamma\beta$ for

$Q \in N'$,

$\gamma \in (N' \cup T')^+$,

and

$\alpha, \beta \in (N' \cup T')^*$.

3. Determine the Chomsky class of the grammar defined by

$G = (\{A, B, T, S\}, \{x, y, z\}, P, S)$

where

$$P = \{S \to xTB | xB$$
$$T \to xTA | xA$$
$$B \to yz$$
$$Ay \to yA$$
$$Az \to yzz\}.$$

Using a property of the class to which G belongs decide whether or not the strings

$(a)\ x^2yxz,\quad (b)\ x^2y^2z^2,\quad (c)\ xyxz,$

are in $L(G)$.

4. Determine the sequence of cuts of the derivation tree below that corresponds to the rightmost derivation of the sentence $x + x * x$.

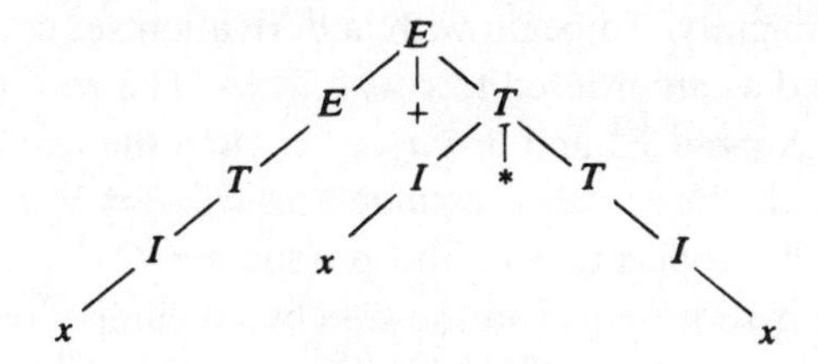

5. Assigning an order to the set P of productions in G enables canonical derivations to be specified by sequences of integers in $\mathbb{N}$. Demonstrate two such sequences for the sentence 'aza' in $L(G_1)$ where G_1 is as below. Also derive a string expression over $\mathbb{N}$ describing all derivations in the language $L(G_2)$ and hence show that $A \overset{+}{\Rightarrow} A$ implies ambiguity.

   $G_1 = (N, T, P, E)$

   $G_2 = (N, T, P, A)$

   where $N = \{A, B, C, E, R\}$, $T = \{a, d, e, x, z\}$ and

   $$P = \{A \to B \mid Cd$$
   $$B \to Bx \mid eC \mid C$$
   $$C \to A \mid xR$$
   $$E \to aE \mid Ea \mid R$$
   $$R \to z\}.$$

6. Decide whether or not the following grammars are ambiguous
   (i) $G = (\{A, B, S\}, \{a, b, c\}, P, S)$
   where $P = \{S \to AB,\ A \to a \mid ab,\ B \to c \mid bc\}$.
   (ii) $G = (\{\langle\text{unsigned integer}\rangle, \langle\text{digit}\rangle\}, D, P, \langle\text{unsigned integer}\rangle)$
   where $D = \{0, 1, \ldots, 9\}$ and

   $$P = \{\langle\text{unsigned integer}\rangle \to \langle\text{digit}\rangle$$
   $$\langle\text{digit}\rangle \to \langle\text{digit}\rangle\langle\text{digit}\rangle$$
   $$\langle\text{digit}\rangle \to 0 \mid 1 \mid 2 \ldots \mid 9\}.$$

8.3 Context-free languages

8.3.1 Basic definitions

CFGs and CFLs are of great practical importance in computing, for although most computing languages are to some extent context-sensitive it is usually easier to regard them as being CFLs and then reject some sentences on other (semantic) criteria. In these cases the exact CSG may be referred to as specifying the *tight* syntax and the CFG the *loose* syntax. CFGs also enable us to be more explicit about questions involving (syntactic) ambiguity. To begin with, a derivation sequence of $\alpha \in L(G)$ can be depicted as an ordered tree as follows. The root of the tree is labelled S and if $S \overset{*}{\Rightarrow} \alpha \in T^*$ and $\alpha = a_1 \ldots a_n$ then the leaves are labelled, in order, a_1 to a_n. In between, assume that $S \overset{*}{\Rightarrow} \beta \Rightarrow \gamma \overset{*}{\Rightarrow} \alpha$ and that $\beta \Rightarrow \gamma$ is achieved by application of the production $C \to \gamma_1 \ldots \gamma_m$ where each $\gamma_i \in V$. This is represented in the tree by labelling a node C and its m successors (the points arrived at by following the elements of the outbundle of C) γ_1 to γ_m in order. Consequently, the labels may not be distinct.

Example 8.3.1
Consider the grammar whose productions are

$$E \to T \mid E + T,$$

$$T \to I \mid I * T,$$

$$I \to (E) \mid x.$$

(Typically, in the CF case, we shall omit the other elements of the grammar; the first rule specifies the source and the non-terminals are exactly those symbols on the left hand sides of the productions.) Then the derivation of the sentence $x + x * x$ can be depicted as in Figure 8.4. //

Fig. 8.4

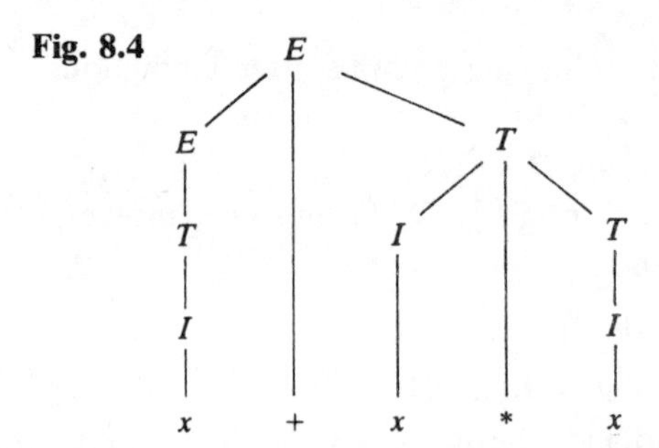

From the constructional (generative) point of view this is called a *derivation tree.* (When used as an aid to the analysis of a string which may or may not be in $L(G)$ it is called a *parse tree.*) It is now easy to see that a sentence is ambiguous if it has two derivation trees that are not isomorphic and that, for context-free grammars, the canonical parse is isomorphic to yet another tree traversal scheme.

We have already stated the obvious fact that CFGs are more limited in scope than CSGs but nevertheless they still have a very wide-ranging descriptive ability.

Example 8.3.2

Using CF rules we can generate

(1) all sequences of As

$S \to AS \mid \Lambda,$

(2) all non-empty lists of As separated by Bs

$S \to A \mid ABS,$

(3) as (2) but with the possibility of the empty list

$S \to T \mid \Lambda,$

$T \to A \mid ABT,$

(4) all strings beginning with a sequence of As and Bs and ending with matching Cs and Ds:

$S \to ASC \mid BSD \mid X,$

for example

$S \to [S] \mid (S) \mid X.$

In these examples A, B, C, D and X can be further defined. //

8.3.2 Characteristic properties

A characteristic of the above examples, about which we shall soon say more, is the ability to recurse as much as we like (viz one extra recursion for each $n \in \mathbb{N}$). The difficulties arise when it is required to put some restriction on the depth of recursion without devising individual rules for each allowable depth. Since N and P are finite, it should be apparent that if no recursion is allowed $L(G)$ will also be finite, and not very interesting. To go further we need some terminology.

(1) The grammar G is said to be *left-recursive* if it allows derivations of the form

$X \overset{*}{\Rightarrow} X\alpha$ where $X \in N$ and $\alpha \in V^+$.

(2) Similarly G is *right-recursive* if it has derivations of the form

$X \overset{*}{\Rightarrow} \alpha X$ (X and α as above)

and

(3) as a third possibility, G is *self-embedding* if it has derivations of the form

$X \overset{*}{\Rightarrow} \alpha X \beta$ where $X \in N$ and $\alpha, \beta \in V^+$.

A CFG is said to be *recursive* if any of the cases (1), (2) or (3) occur. From the comments above it is clearly desirable to have 'loops' within a grammar but not all loops. We shall return to this matter in the next section.

Now for a formal result about the capabilities of CFGs. This is probably the best known result in the theory of CFLs; its proof utilizes the recursive properties of CFGs and our knowledge of trees. It is known as the *pumping lemma* for CFLs or the *uvwxy* theorem.

Theorem. If L is a CFL then there is an $n \in \mathbb{N}$ such that if $z \in L$ and $|z| \geqslant n$ then z can be written as $uvwxy$ where $u, v, w, x, y \in T^*$ and $vx \neq \Lambda$, and for any $i \in \mathbb{N}$

$$uv^iwx^iy \in L.$$

Proof. Since L is a CFL, it can be generated by G, say, where $G = (N, T, P, S)$ and no production, except possibly $S \to \Lambda$ if $\Lambda \in L(G)$, reduces the length of sentential forms. (If $\Lambda \in L$, S is also excluded from right hand sides of productions thus preventing subsequent length reductions of sentential forms. That such a G can be found is shown in Section 8.4.)

If G is not recursive then, since N and P are finite, $L(G)$ is finite and hence the theorem is trivially satisfied if we take n greater than the length of the longest string in $L(G)$. On the other hand if G is recursive then there is a derivation tree in which a specific non-terminal, A say, occurs twice on a path from the root to a leaf. The situation is depicted in Figure 8.5 (only the salient features are included).

Fig. 8.5

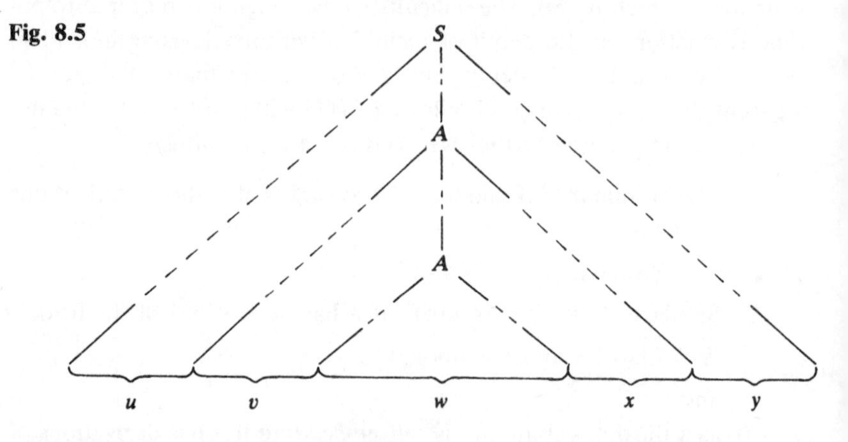

Moreover, since G is recursive we can choose this situation such that $|uvwxy| \geqslant n$ where n is greater than the length of the longest sentence obtained by a non-recursive derivation ($\leqslant k^m$ where k is the length of the longest right hand side of any production in P and $m = |N|$). Thus if $z \in L$ and $|z| \geqslant n$ then z must have this form for some five strings. Now

$$A \overset{+}{\Rightarrow} vAx \quad (vx \neq \Lambda \text{ so } |vAx| > |A|)$$

and

$$A \overset{*}{\Rightarrow} w.$$

Therefore

$$A \overset{i}{\Rightarrow} v^i A x^i \overset{*}{\Rightarrow} v^i w x^i \quad \text{for any } i \in \mathbb{N}.$$

Hence, since $S \overset{*}{\Rightarrow} uAy$, we have

$$S \overset{*}{\Rightarrow} uv^i wx^i y \quad \text{for any } i \in \mathbb{N}$$

and thus

$$uv^i wx^i y \in L \quad \text{for any } i \in \mathbb{N}. \quad //$$

This result can be used to verify that certain constructs in programming languages cannot be defined by a CFG. As a more tangible example which does not require knowledge about a specific language, we give the following.

Example 8.3.3

The grammar of Example 8.2.4 generates the language $\{x^n y^n z^n : n \in \mathbb{N}\}$. We are now in a position to show that this language is strictly CS and cannot be generated by a CFG. By the theorem there is some n large enough to ensure that $x^n y^n z^n$ can be written as $abcdf$ (an obvious change of symbols is required) for some strings $a, \ldots, f$. Because the xs and the zs are separated in the string, obviously a and b cannot both contain all the symbols x, y and z; similarly for all other pairs from $\{a, b, c, d, f\}$. In particular, at least one of x, y and z cannot be in both b and d thus the string ab^2cd^2f, which by the theorem is in L, contains an imbalance of x, y and zs (they are also shuffled but we shall not discuss this further). Hence we do not have the same language and the result follows. //

Similar contradictions can be achieved in many situations where information held in an earlier part of the string affects the required structure of a subsequent substring. The following is typical.

Example 8.3.4

The language $L = \{1^p : p \in \mathbb{N} \text{ and } p \text{ is prime}\}$ is not context free;

$$L = \{11, 111, 11111, \ldots\}.$$

Assume L is a CFL then, as there are infinitely many prime numbers, there is a prime number q such that

$$1^q = uvwxy \quad vx \neq \Lambda$$

and

$$uv^i wx^i y \in L \quad \text{for all } i \in \mathbb{N}$$

(this follows from the pumping lemma). Thus there are integers a, b, c, d, e such that

$$1^q = 1^a 1^b 1^c 1^d 1^e \quad \text{for } b+d>0$$

and

$$1^{q_i} = 1^a (1^b)^i 1^c (1^d)^i 1^e \in L \quad \text{for all } i \in \mathbb{N}$$

so that

$$q = a+b+c+d+e \quad \text{is prime and} > 1$$

and

$$q_i = a+c+e+(b+d)i \quad \text{is prime for all } i \in \mathbb{N}.$$

In particular q_i is prime for

$$i = a+b+c+d+e+1$$

and in this case

$$\begin{aligned} q_i &= (a+c+e)+(b+d)(a+b+c+d+e+1) \\ &= (a+c+e)+(b+d)((a+c+e)+(b+d)+1) \\ &= ((a+c+e)+(b+d))(1+(b+d)) \\ &= q(1+b+d). \end{aligned}$$

However $q>1$ and $b+d+1>1$ therefore q_i is not prime for all $i \in \mathbb{N}$ and we have a contradiction. Therefore

$$L = \{1^p : p \in \mathbb{N} \text{ and } p \text{ is prime}\} \text{ is not a CFL.} \quad //$$

This last example also demonstrates the practical importance of tight and loose syntax. It *may* be possible to devise a very rigid syntax that incorporates the *validation* semantics. However, even where this is possible, it is often much more convenient and efficient to allow a wider language to be generated by the (typically context-free) grammar and then, if necessary, restrict the set of sentences by further semantic checks.

In Example 8.3.4 we could use

$$S \to 1S \mid 11$$

to generate all strings $1^q : q>1$ and specify the checking of 'q to be prime' by a suitable *arithmetic* algorithm.

In short, CSGs are complicated and have not been studied in great depth. On the other hand CFGs have received very much attention and form the basis of almost all practical computer translation systems.

Exercises 8.3

1. Derive a CFG which generates the set of all strings over $\{a, b\}$ having equal numbers of as and bs.

2. Construct grammars to generate the following languages
   (i) $\{a^{3n}: n \geq 1\}$
   (ii) $\{a^n b^{2m-1}: n, m \geq 1\}$
   (iii) $\{a^n b^n: n \geq 1\}$ for $n, m \in \mathbb{N}$.
3. Use the pumping lemma to show that the language
   $L = \{a^{n^2}: n \in \mathbb{N}\}$
   is not context free.
4. Show that if L_1 and L_2 are CFLs then so is $L_1 \cup L_2$.
5. Prove that
   $\{x^n y^n z^m: n \geq 1, m \geq 1\}$
   and
   $\{x^m y^n z^n: n \geq 1, m \geq 1\}$
   are CFLs and hence show that L_1, L_2 CFLs does not imply that $L_1 \cap L_2$ is a CFL.

8.4 Parsing concepts and grammar modifications

The most immediate and obvious contact that the average computer user (and, indeed, computer scientist) has with translation processes is in the use of compiling systems for high-level programming languages such as Pascal, Fortran, Cobol, Algol, etc.

When using such a language, the program you write (called the *source program*) is translated into an *equivalent* machine code program (the *object program*) which can be deciphered and executed by the actual computer used. The general scheme of a compiler is shown in Figure 8.6.

Conceptually, the stages of the compilation process can be regarded as sequentially linked, as in the diagram; however, in practice, they are often carried out simultaneously. The code generator needs to know the semantic interpretations which are associated with each (syntactic) construct within the program; and the optimizer must have knowledge of the subtleties of the computer's architecture. Hence we shall not consider these stages but restrict our comments to the translation from source program to parse tree.

The source program is merely a string of characters. Within that string certain combinations of characters occur in common patterns in which individual characters are essentially meaningless but the composite pattern *does* convey meaning. (Refer back to Example 8.2.1; 'dog' has a meaning but the letter 'o' within 'dog' has no obvious individual meaning.) Such composite symbols, also called *lexical tokens*, are not absolutely necessary and may not be used in certain language translators but *are* usually sought for and encoded as single tokens so as to shorten the

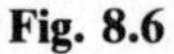
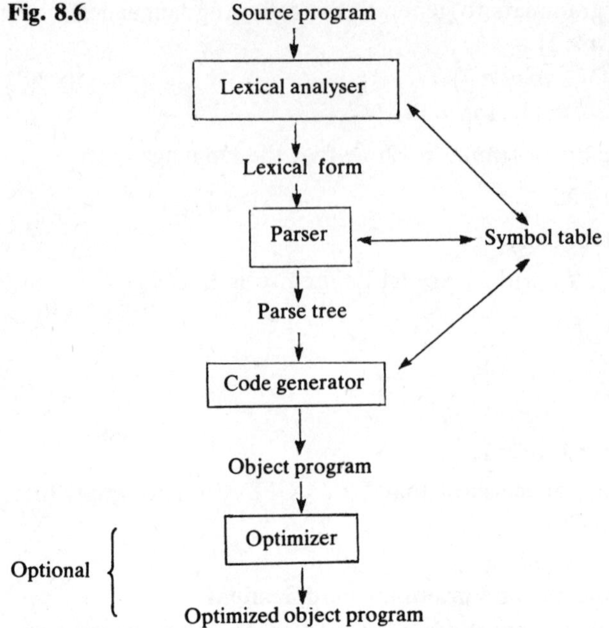

Fig. 8.6

length of the source program (now in its *lexical form*) and avoid the need to consider superfluous detail at subsequent stages.

Typical of lexical tokens are:

(1) keywords; i.e. those with constant meaning in the language; for example,

begin		GOTO	
end	Pascal	DO	Fortran
while		.OR.	

+, −, *, / in most languages,

(2) numbers; 52, 31.65, etc.,

(3) strings or sequences of characters,

(4) identifiers invented by the programmer.

Lexical tokens are usually specified by regular grammars.

We have therefore reduced the immediate problem to *parsing* a string of lexical tokens; graphically, we have to fill in the triangle of Figure 8.7 in a way that is consistent with the production rules of the grammar.

Fig. 8.7

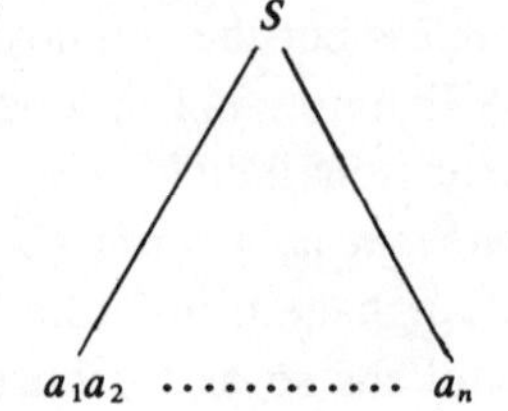

8.4.1 Tidying-up procedures

We are not generally allowed to tamper with the string $\alpha = a_1 a_2 \ldots a_n$ so all activity prior to carrying out the parsing process must be directed towards the grammar. Potentially, we shall need to perform quite complex transformations on the grammar so to begin with we check that all non-terminals can actually take part in some parse. There are two ways in which non-terminals may fail to contribute to any parse and these are now described formally.

Definition. If $G = (N, T, P, S)$ is a CFG then a non-terminal symbol $X \in N$ is said to be

(i) *inaccessible* if $X \neq S$ and there does not exist a derivation of the form

$$S \overset{+}{\Rightarrow} \alpha X \beta \quad \text{for } \alpha, \beta \in V^*,$$

(ii) *unproductive* if there is no string $\gamma \in T^*$ such that

$$X \overset{+}{\Rightarrow} \gamma,$$

(iii) *useless* if it is either inaccessible or unproductive.

A grammar with no useless non-terminals is said to be *reduced*. //

Clearly useless symbols play no role in sentence construction. Although useless symbols would not be deliberately included in a grammar, they can be introduced by algorithms designed to modify the grammar to suit a particular purpose (see later). Useless symbols increase the size of the parser unnecessarily and we now give an algorithm for their removal.

If $G = (N, T, P, S)$ is a CFG then we define a set N' by

$$N' = N \cup \{\tau\}$$

where τ is a new symbol ($\tau \notin V$), and a relation ρ on N' by

$$(A, B) \in \rho \quad \text{if } A \rightarrow \alpha B \beta \in P \text{ for } A, B \in N \text{ and } \alpha, \beta \in V^*.$$

$$(A, \tau) \in \rho \quad \text{if } A \rightarrow \gamma \in P \text{ for some } \gamma \in T^*.$$

Proposition.

(i) A is accessible iff $A = S$ or $(S, A) \in \rho^+$.

(ii) A is productive iff $(A, \tau) \in \rho^+$.

Proof.

(i) A is accessible iff there is a derivation of the form $S \overset{*}{\Rightarrow} \alpha A \beta$ for $\alpha, \beta \in V^*$, or equivalently iff there exists $i \geqslant 0$ such that

$$S \overbrace{\Rightarrow \ldots \Rightarrow}^{i} \alpha A \beta$$

and when $S \neq A$ this is true iff $S \rho^+ A$, so $(S, A) \in \rho^+$.

(ii) A is productive iff $A \overset{i}{\Rightarrow} \gamma$ for some $i > 0$ and $\gamma \in T^*$, i.e. iff there is a sequence of sentential forms $\alpha_0, \alpha_1, \ldots, \alpha_i$ such that

$A = \alpha_0 \Rightarrow \alpha_1 \Rightarrow \ldots \Rightarrow \alpha_i = \gamma,$

i.e. iff there is a sequence $A = A_0, A_1, \ldots, A_{i-1} \in N$ such that A_i is a substring of α_i and hence

$A_0 \rho A_1, A_1 \rho A_2, \ldots, A_{i-2} \rho A_{i-1}, A_{i-1} \rightarrow \beta$

where β is a substring of γ, i.e.

$A_0 \rho A_1, A_1 \rho A_2, \ldots, A_{i-1} \rho \tau$ so $A \rho^+ \tau$. //

In practice ρ^+ may be calculated using Warshall's algorithm. If $N_u \subset N$ is the set of useless symbols of G and $N' = N \backslash N_u$ and $P' = P \backslash P_u$ where P_u is the set of productions containing elements of N_u then

$G = (N', T', P', S),$

where T' is the set of terminal symbols appearing in the productions of P', is an equivalent CFG without useless symbols.

Algorithm. Useless symbol removal.

Input: CFG $G = (N, T, P, S)$,

Output: equivalent CFG $G' = (N', T', P', S)$ without useless symbols,

Method: construct N', T', P' as above. //

Example 8.4.1

Consider the grammar

$G = (\{A, B, C, D\}, \{x, y, p, q, w, a\}, P, A)$

where

$$\begin{aligned} P = \{ & A \rightarrow x \mid yDC \mid D \\ & B \rightarrow q \mid Bx \\ & C \rightarrow Cx \mid yC \\ & D \rightarrow Da \mid Cw \mid p\}. \end{aligned}$$

Using ρ, as defined above, encoded in matrix form:

$$M(\rho) = \begin{array}{c} \\ A \\ B \\ C \\ D \\ \tau \end{array} \begin{array}{c} \begin{array}{ccccc} A & B & C & D & \tau \end{array} \\ \begin{pmatrix} 0 & 0 & 1 & 1 & 1 \\ 0 & 1 & 0 & 0 & 1 \\ 0 & 0 & 1 & 0 & 0 \\ 0 & 0 & 1 & 1 & 1 \\ 0 & 0 & 0 & 0 & 0 \end{pmatrix} \end{array}$$

For this particular example we have

$$M(\rho^+) = M(\rho) = M.$$

Thus $M_{AB} = M_{C_T} = 0$ and so B is inaccessible and C is unproductive. Therefore the grammar reduces to

$$G' = (\{A, D\}, \{x, a, p\}, P', A)$$

where

$$P' = \{A \to x | D$$
$$D \to Da | p\}. \quad //$$

After useless symbol removal each remaining non-terminal symbol X occurs in at least one derivation tree (Figure 8.8) with X connected above to S and below to some terminal string $a_1 \ldots a_n$.

Fig. 8.8

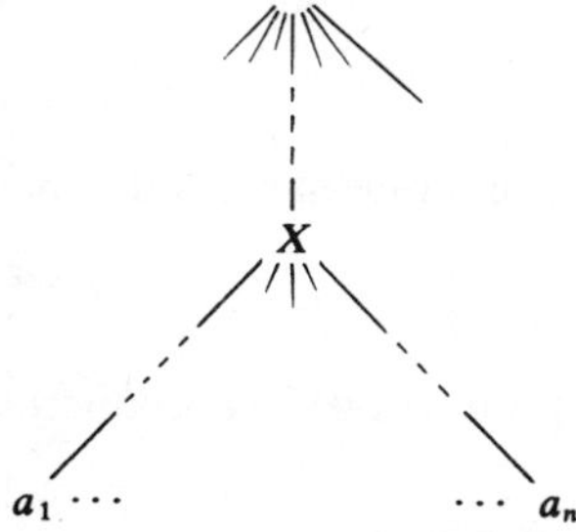

One 'obvious' way of parsing a string is to derive *all* strings, note their respective canonical derivation sequences, then check each sentence against the string and, if a match is found, use the derivation sequence to define the parse tree. Of course, in most instances, $|L|$ is infinite so this process is impossible; however, if the grammar has no awkward productions then this approach provides the basis of a technique which, at least in a local context, can be of practical use. Explicitly, when checking for $\alpha \in L(G)$ and $|\alpha| = n$, providing the length of sentential forms cannot be reduced by G, then these forms can be rejected (before reaching strings over T) whenever their length exceeds n. Sentences with lengths not exceeding n can be checked for equality with α in the usual way and, in a sensible grammar, all possible such strings should be generated in a finite number of steps. We now set about tidying up the grammar by producing an equivalent version which is 'easier to parse'.

To ensure that the generation process associated with a given sentence is finite we need to guarantee that all derivation sequences actually *do* get somewhere. Consequently, occurrences of situations such as $X \to X$ and $X \to \Lambda$ are of interest. When written directly as productions these are easy to detect; however, the more general situations $X \overset{+}{\Rightarrow} X$ and $X \overset{+}{\Rightarrow} \Lambda$ are more difficult to locate and, moreover, the two types of derivation are interlinked as demonstrated in the following example.

Example 8.4.2

Suppose that included within the productions of a certain grammar are

$$X \to Y, Y \to W, Z \to V, W \to Z, V \to X.$$

Following a possible derivation sequence from X we deduce that $X \overset{+}{\Rightarrow} X$ and hence whenever X occurs in a sentential form we could insert the progression $X \Rightarrow Y \Rightarrow W \Rightarrow Z \Rightarrow V \Rightarrow X$ (and again and again ...) thus achieving absolutely nothing but ambiguity.

This 'loop' within the set, N, of non-terminals is not difficult to see; however, notice that we could have essentially the same situation in a disguised form if instead of $X \to Y$ and $Y \to W$, we had, for instance, $X \to AY$, $Y \to AWA$ and $A \to \Lambda$. //

Before describing algorithms to perform grammar transformations we need to make our terminology precise.

Definitions. A Λ-*production* (lambda-production) is a production of the form $X \to \Lambda$ with $X \in N$.

A CFG $G = (N, T, P, S)$ is called Λ-*free* if either

(1) P has no Λ-productions, or

(2) there is exactly one Λ-production, namely $S \to \Lambda$ and S does not appear on the right hand side of any production in P.

A production is a *single production* if it is of the form $X \to Y$ where $X, Y \in N$.

A production of the form $X \to X$ for some $X \in N$ is called *trivial*.

A CFG $G = (N, T, P, S)$ is *cycle-free* if there are no derivations of the form $X \overset{+}{\Rightarrow} X$ for any $X \in N$. //

As already noted the detection and removal of cycles and of Λ-derivations are closely linked. We begin by locating all non-terminals from which Λ can be reached.

Notation. Given $G = (N, T, P, S)$ denote by N_Λ the set $\{X : X \overset{+}{\Rightarrow} \Lambda\} \subseteq N$. //

Algorithm. Computation of N_Λ.

Input: arbitrary CFG $G = (N, T, P, S)$.

Output: N_Λ.
Method: let $P=\{P_1,\ldots,P_{|P|}\}$ where each P_i is of the form
$\alpha_i \to \beta_i$: $\alpha_i \in N$, $\beta_i \in V^*$;

then treating N_Λ as a 'variable' of type set, we have

$N_\Lambda \leftarrow \varnothing$,
$i \leftarrow |P|$
repeat (**if** $(\alpha_i \notin N_\Lambda)$ **and** $(\beta_i \in N_\Lambda^*)$
  then $(N_\Lambda \leftarrow N_\Lambda \cup \{\alpha_i\}, i \leftarrow |P|)$
  else $i \leftarrow i-1$)
until $i=0$. //

Algorithm. Conversion to Λ-free grammar.
Input: arbitrary CFG $G=(N, T, P, S)$.
Output: equivalent Λ-free CFG $G'=(N', T, P', S')$.
Method:
(1) Determine N_Λ.
(2) Construct P' as follows.
(*a*) If $A \to \alpha_0 B_1 \alpha_1 B_2 \alpha_2 \ldots B_k \alpha_k \in P$, where $k \geq 0$ and for $1 \leq i \leq k$ each B_i is in N_Λ but no symbols in any $\alpha_j \in V^*$ are in N_Λ (for $0 \leq j \leq k$), then add to P' all productions of the form
$A \to \alpha_0 X_1 \alpha_1 X_2 \alpha_2 \ldots X_k \alpha_k$,
where X_i is either B_i or Λ, without adding $A \to \Lambda$ to P' (this could happen if all α_i were Λ).
(b) If $S \in N_\Lambda$ then add to P' the productions
$S' \to \Lambda | S$
where S' is a new symbol and let
$N' = N \cup \{S'\}$
otherwise $N' = N$ and $S' = S$. //

We can now set about removing cycles from a CFG. Location of cycles can easily be achieved by extracting the relation $\rho = \{(A, B): A \to B \in P\}$ and forming the closure ρ^+; any $X : X\rho^+X$ must then clearly be in a cycle. We incorporate this together with a 'back-substitution' scheme that removes all but one non-terminal within any cycle. (It also removes any trivial productions.)

Algorithm. Conversion of a CFG into an equivalent cycle-free grammar.
Input: a Λ-free CFG $G=(N, T, P, S)$.
Output: equivalent cycle-free grammar $G'=(N', T', P', S')$.

(The non-terminals of G are renamed, A_1 to A_n where $n = |N|$, S is renamed as A_1 and each production P_i is expressed as $\alpha_i \to \beta_i$. Additionally, we use a set INCYCLES and (potentially) n non-terminals called REPLACEi with $1 \leq i \leq n$.) The algorithm is as follows:

1. define ρ over $\mathbb{N}_n$ such that

   $i\rho j$ iff $A_i \to A_j \in P$,

2. let $\sigma = \rho^+$.
3. INCYCLES $\leftarrow \varnothing$,
4. **for** i **from** 1 to $n-1$

   do (**for** j **from** $i+1$ to n

       do if ($j \notin$ INCYCLES **and**

           $i\sigma j$ **and**

           $j\sigma i$)

         then (INCYCLES $\leftarrow$ INCYCLES $\cup \{j\}$,

             REPLACE$j \leftarrow A_i$))

5. $j \leftarrow 0$

   for i **from** 1 to $|P|$

   do (**for all** $k \in$ INCYCLES

       in P_i replace A_k by REPLACEk giving *new*P_i

       if (*new*$P_i \notin \{P'_1, \ldots, P'_j\}$

         and $\alpha_i \neq \beta_i$ in *new*P_i)

       then ($j \leftarrow j+1$,

           $P'_j \leftarrow$ *new*P_i))

6. $G' = (N \backslash \text{INCYCLES}, T, P', S)$. //

So much for general grammar tidying procedures.

A CFG which is Λ-free, cycle-free and reduced is said to be *proper.* Having obtained a proper CFG, G, how do we use it to check whether $\alpha \in L(G)$ for some given $\alpha \in T^*$?

8.4.2 Parsing modifications

As stated in the introduction to Section 8.4, the task of parsing a string is essentially that of filling the derivation triangle (Figure 8.7) with a suitable tree. Of course in most cases this cannot sensibly be done in a single step and the usual approach is via sequences of subtrees. These sequences of subtrees can be generated in many ways, three of the most common being as depicted in Figure 8.9.

Fig. 8.9

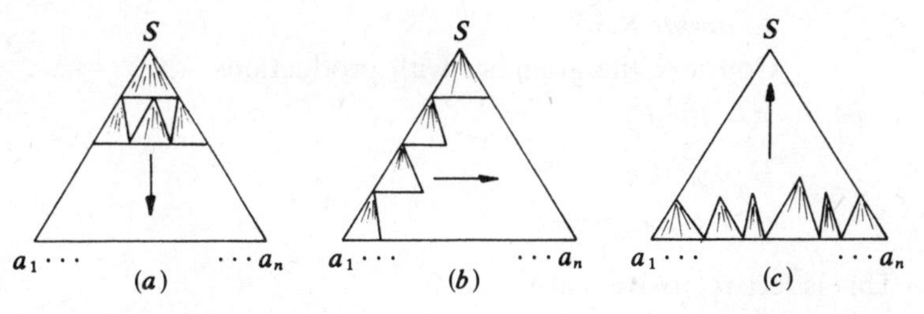

The parsing strategy indicated in Figure 8.9(a) is referred to as *top-down parsing* in which productions are applied (in certain chosen orders) to sentential forms in an attempt to expand S into the string $a_1 \ldots a_n$. It is normal in such a parse to use part of the target string ($a_1 \ldots a_n$) as a guide to steer the derivation. For practical considerations, consistent with reading $a_1 \ldots a_n$ from left to right and the desirability of beginning the parse *before* inputting the entire string, it is conventional to use the start of the string. Hence, in seeking sentential forms which begin with $a_1 \in T$ (and, correspondingly, searching for terminal strings as initial substrings of subsequent portions of the input), we must reject the possibility of derivations of the form $X \overset{+}{\Rightarrow} X\beta$ ($X \in N$, $\beta \in V^+$). Therefore, in order to *begin* a top-down parse, we need to remove left recursion.

In general this can be done by use of a process analogous to the algebraic solution of systems of linear equations. Having said this it is often possible to consider one recursion at a time and remove it using a very simple identity. Justification of this technique is via the language generated.

Consider

$$X \overset{+}{\Rightarrow} X\alpha.$$

By 'back substitution' of right hand sides of productions for left hand sides we can achieve a direct recursion as an element of P (modified) thus

$$X \to X\alpha \mid \beta.$$

Regarding this as the complete grammar over $\{\alpha, \beta\}$ (β representing all the other, non-left-recursive possibilities for X), trivially

$$L(X) = \{\beta\alpha^n \colon n = 0 \text{ or } n \in \mathbb{N}\}$$

and this can also be generated by

$$X \to \beta Y,$$
$$Y \to \alpha Y \mid \Lambda$$

which is not left-recursive (it is right-recursive). To complete the transformation, $X \to \beta Y$ must then be expanded, if necessary, into the correct number of terms.

Example 8.4.3

Consider the grammar with productions

$$A \to Bc|dC,$$
$$B \to xA|Ce,$$
$$C \to Ab|w.$$

This is left-recursive since

$$\begin{aligned} A &\Rightarrow Bc \\ &\Rightarrow Cec \\ &\Rightarrow Abec. \end{aligned}$$

Back substituting for C in B and B in A, we get

$$B \to xA|(Ab|w)e \equiv B \to xA|Abe|we$$

and

$$\begin{aligned} A \to Bc|dC &\equiv A \to (xA|Abe|we)c|dC \\ &\equiv A \to xAc|Abec|wec|dC \\ &\equiv A \to A\underbrace{bec}_{\alpha}|\underbrace{xAc|wec|dC}_{\beta}. \end{aligned}$$

So, using the transformation

$$A \to \beta Y,$$
$$Y \to \alpha Y|\Lambda,$$

we have

$$\begin{aligned} & A \to (xAc|wec|dC)Y \\ \equiv\ & A \to xAcY|wecY|dCY \\ & Y \to becY|\Lambda \\ & B \to xA|Ce \\ & C \to Ab|w. \quad // \end{aligned}$$

Notice that in this example B has been 'cut off' and can play no part in *any* sentence generated from the root A (hence we would need to remove all traces of B from the grammar) and also that we have introduced a Λ-production. The fact that this production has been created as a side-effect of a 'parsing transformation' should cause us less of a problem than a similar production occurring naturally. B can be removed by the 'useless symbol' algorithm.

Although the piecemeal removal of individual left-recursive chains within N will usually suffice, we must use a matrix generalization to cope with interleaved left-recursions.

Recall that in the simple case we replace

$$X \to X\alpha|\beta \quad \text{by} \quad \begin{array}{l} X \to \beta Y \\ Y \to \alpha Y|\Lambda. \end{array}$$

If we now have n non-terminals $X_1, \ldots, X_n$ which are mutually left-recursive in a fashion that does not reduce (by a sequence of back-substitutions) to the simple case, then we can represent the associated productions schematically as

$$\begin{array}{l} X_1 \to X_1A_{11}|X_2A_{21}| \ldots |X_nA_{n1}|B_1 \\ X_2 \to X_1A_{12}|X_2A_{22}| \ldots |X_nA_{n2}|B_2 \\ \vdots \\ X_n \to X_1A_{1n}|X_2A_{2n}| \ldots |X_nA_{nn}|B_n \end{array}$$

where each A_{ij} represents the remainder of all options which can be derived from X_j and which begin with X_i and, similarly, each B_j represents all alternatives for X_j which do *not* begin with an element of $\{X_1, \ldots, X_n\}$.

Now since A_{ij} and B_j are sets of strings it follows that

(i) if $X_j \to X_i$ then $\Lambda \in A_{ij}$,
(ii) if $X_j \to \alpha$ and $\alpha \neq X_i\beta$ for some $\beta \in V^*$ then $\alpha \in B_j$, and
(iii) if $X_j \not\to X_i\gamma$ for any $\gamma \in V^*$ then $A_{ij} = \varnothing$.

Consequently, over the algebraic system $(V^*, \odot, |)$ we can reduce these productions to the matrix scheme

$$\mathbf{X} = \mathbf{XA}|\mathbf{B} \text{ in } \mathcal{M}(n, (V^*, \odot, |)),$$

or, writing the alternation operator | as +, we have

$$\mathbf{X} = \mathbf{XA} + \mathbf{B} \text{ in } \mathcal{M}(n, (V^*, \odot, +)).$$

By analogy with the simple (non-matrix) case, about which more will be said in Chapter 9, we claim that

$$\mathbf{X} = \mathbf{BY} \quad \text{where } \mathbf{Y} = \mathbf{AY} + \mathbf{I}$$

and $\mathbf{I}$ is defined over $(V^*, \odot, +)$ by

$$\begin{aligned} \mathbf{I}_{ij} &= \{\Lambda\} \text{ if } i = j \\ &= \varnothing \quad \text{otherwise.} \end{aligned}$$

Example 8.4.4

Suppose that G has the following productions and $S = D$;

$D \to Dx|Ey|Fz,$

$E \to Da \quad |Fc,$

$F \to Dp|Eq|Fr|w.$

Thus, using the general scheme, we have

$$X_j \to X_1 A_{1j} | X_2 A_{2j} | X_3 A_{3j} | B_j$$

$X_1 =$	$D \to$	Dx	$\mid Ey$	$\mid Fz$	
$X_2 =$	$E \to$	Da		$\mid Fc$	
$X_3 =$	$F \to$	Dp	$\mid Eq$	$\mid Fr$	$\mid w.$

Now

$$X_i = \sum_k B_k Y_{ki}$$

so

$$\begin{aligned} X_1 = D \to & \sum_k B_k Y_{k1} \\ & = B_3 Y_{31} \\ & = w Y_{31}, \\ E & \to w Y_{32}, \\ F & \to w Y_{33} \end{aligned}$$

and

$$Y_{ij} = \sum_k A_{ik} Y_{kj} + I_{ij},$$

therefore

$$\begin{aligned} & Y_{11} \to x Y_{11} | a Y_{21} | p Y_{31} | \Lambda, \\ & Y_{12} \to x Y_{12} | a Y_{22} | p Y_{32}, \\ & Y_{13} \to x Y_{13} | a Y_{23} | p Y_{33}, \\ & Y_{21} \to y Y_{11} | \quad q Y_{31}, \\ & Y_{22} \to y Y_{12} | \quad q Y_{32} | \Lambda, \\ & Y_{23} \to y Y_{13} | \quad q Y_{33}, \\ & Y_{31} \to z Y_{11} | c Y_{21} | r Y_{31}, \\ & Y_{32} \to z Y_{12} | c Y_{22} | r Y_{32}, \\ & Y_{33} \to z Y_{13} | c Y_{23} | r Y_{33} | \Lambda. \end{aligned}$$

In this instance the transformation produces superfluous non-terminals and removing those that are not usable we have:

$$\begin{aligned} & D \to w Y_{31}, \\ & Y_{31} \to z Y_{11} | c Y_{21} | r Y_{31}, \\ & Y_{11} \to x Y_{11} | a Y_{21} | p Y_{31} | \Lambda, \\ & Y_{21} \to y Y_{11} | q Y_{31} \end{aligned}$$

and with appropriate name changes:

$$D \to wJ,$$
$$J \to zK|cL|rJ,$$
$$K \to xK|aL|pJ|\Lambda,$$
$$L \to yK|qJ.$$

To illustrate the power of the transformation we give, in Figure 8.10 parse trees of the string '*wcqzayx*' for both (*a*) the original and (*b*) the modified grammar. //

Fig. 8.10

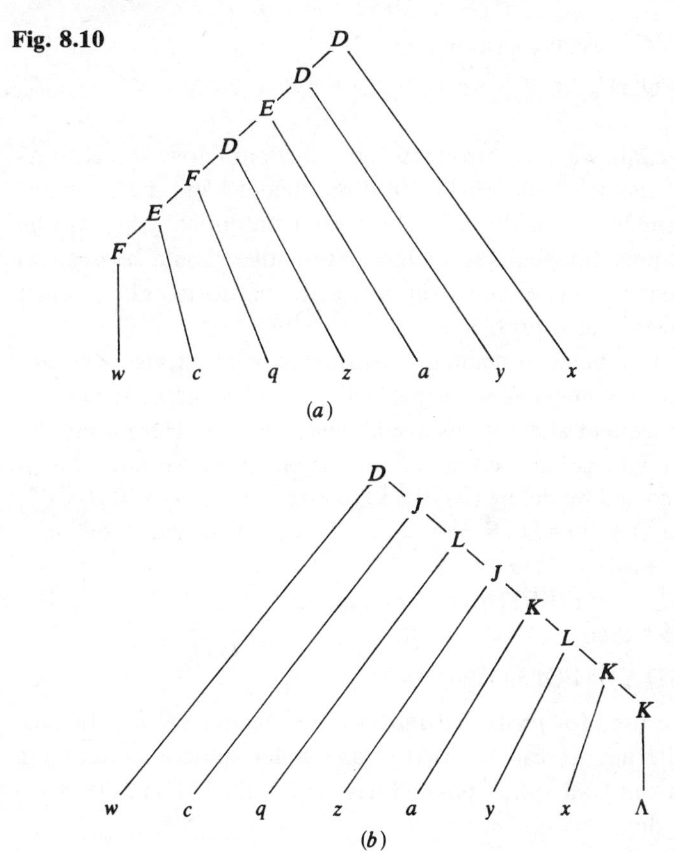

Removal of left-recursion ensures that, if possible, we actually do derive a string beginning with the required terminal symbol; it does not, however, guarantee that we get only one such string and hence it is possible to follow false trails.

In an attempt to avoid incorrect parsing sequences we can make explicit use of a manipulation already encountered within the previous transformation; this is *left-factorization.* The process requires all productions involving a given non-terminal on the left hand side to be examined; then all right hand sides are expanded and those which begin with a common substring over T are collected together.

Example 8.4.5

$$A \to xyB|xyBC|yPQ|yxV$$
$$\equiv A \to xyB(\Lambda|C)|y(PQ|xV)$$
$$\equiv A \to xyBA_1|yA_2$$

where

$A_1 \to \Lambda|C$ usually written $C|\Lambda$

$A_2 \to PQ|xV$. //

Notice that again we may introduce Λ-productions. However, if no Λs are present in the resultant left-factored productions and the grammar is not left-recursive, then the next symbol on the input string can be used to determine (immediately) which alternative should be used to expand the leading non-terminal in the sentential form. The explicit occurrence of Λs causes problems.

By virtue of Λ being a leading substring of *every* string over *any* alphabet, it always matches with the beginning of a target string and hence any subsequent alternatives would never be tried. Here is not the place to enter into a full analysis of the problem but we note that if $G=(N, T, P, S)$ and we define (a) FIRST$(\alpha)=\{x: \alpha \overset{*}{\Rightarrow} x\beta, x \in T, \beta \in V^*\}$ and (b) FOLLOW$(\alpha)=\{x: S \overset{*}{\Rightarrow} \gamma\alpha x\delta,\ x \in T,\ \gamma,\delta \in V^*\}$ and if for each production $X \to \alpha_1|\alpha_2|\ldots|\alpha_n$

(i) FIRST$(\alpha_i) \cap$ FIRST$(\alpha_j)=\varnothing$ $i \neq j$ and

(ii) if $X \overset{*}{\Rightarrow} \Lambda$ then

FIRST$(X) \cap$ FOLLOW$(X)=\varnothing$,

then G can be used for *predictive analysis*, (see Figure 8.9(b)). In such an analysis alternatives can be tried in any order subject to reserving Λ-derivations until all other possibilities have failed. Example 8.4.6 demonstrates the process.

Example 8.4.6

Suppose that the only productions of a grammar are

$$C \to xCx|\Lambda$$

and we are attempting to parse 'xx'. The string is rejected, even though legal, because we are forced to apply the first production twice, thus generating an erroneous move since $x \in$ FIRST$(C) \cap$ FOLLOW(C) and $C \to \Lambda$. Pictorially, this gives Figure 8.11.

Fig. 8.11

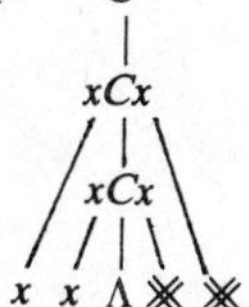

Using the grammar

$$C \to xxC|\Lambda$$

the parse (Figure 8.12) falls into no difficulty because now $x \notin \text{FOLLOW}(C)$. //

Fig. 8.12

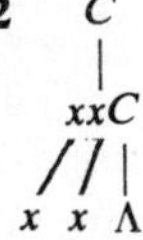

Before concluding this section we mention the other main parsing method, *bottom-up* parsing, in which productions are applied *backwards* in an attempt to reduce the target string to S (Figure 8.9(c)). This technique is more widely applicable than the top-down method and can be assisted by various grammar modifications to improve its efficiency.

Exercises 8.4

1. Modify the grammar whose productions are given below so that it is not left-recursive.

   $A \to Bx|Cz|w,$

   $B \to Ab|Bc,$

   $C \to Ax|By|Cp.$

2. If $G = (N, T, P, S)$ is a CFG and $A \in N$ is not a useless symbol of G show that the existence of any of the following derivations in G imply that G is ambiguous.

   (i) $A \overset{+}{\Rightarrow} A\gamma A,$

   (ii) $A \overset{+}{\Rightarrow} \alpha A|A\beta,$

   (iii) $A \overset{+}{\Rightarrow} \alpha A|\alpha A\beta A,$

   (iv) $A \overset{+}{\Rightarrow} A$

   for $\gamma \in (N \cup T)^*$ and $\alpha, \beta \in (N \cup T)^+$.

3. Determine a Λ-free CFG equivalent to the CFG defined by

   $G = (\{S\}, \{a, b\}, P, S)$

where

$P = \{S \to aSbS \mid bSaS \mid \Lambda\}$.

4. If $G = (\{A, B, C, D, E, S\}, \{a, b, c\}, P, S)$ where

$$
\begin{aligned}
P = \{S &\to A \mid B \\
A &\to C \mid D \\
B &\to D \mid E \\
C &\to S \mid a \mid \Lambda \\
D &\to S \mid b \\
E &\to S \mid c \mid \Lambda\}
\end{aligned}
$$

find a proper grammar equivalent to G.

8.5 Operator precedence grammars

An important subset of CFGs comprises the so called *operator grammars.* These are grammars in which all productions are such that no two non-terminals are adjacent in any right hand side and hence the intervening terminal may be thought of as an operator (though not necessarily in the arithmetic sense). We now attempt to define precedence relations on the set $T \cup \{\vdash, \dashv\}$, where $\vdash$ and $\dashv$ are new symbols, outside V, that delimit the 'sentence'. The rules are defined thus:

(1) $a \doteq b$ if $A \to \alpha a \beta b \gamma \in P$
where $\alpha, \gamma \in V^*$ and $\beta \in N \cup \{\Lambda\}$

(2) $a \lessdot b$ if $A \to \alpha a B \beta \in P$
where $B \overset{+}{\Rightarrow} \gamma b \delta$, $\gamma \in N \cup \{\Lambda\}$ and $\alpha, \beta, \delta \in V^*$

(3) $a \gtrdot b$ if $A \to \alpha B b \beta \in P$
where $B \overset{+}{\Rightarrow} \gamma a \delta$, $\delta \in N \cup \{\Lambda\}$ and $\alpha, \beta, \gamma \in V^*$

(4) $\vdash \lessdot a$ if $S \overset{+}{\Rightarrow} \alpha a \beta$
with $\alpha \in N \cup \{\Lambda\}$ and $\beta \in V^*$

and

(5) $a \gtrdot \dashv$ if $S \overset{+}{\Rightarrow} \alpha a \beta$
with $\beta \in N \cup \{\Lambda\}$ and $\alpha \in V^*$.

The symbols $\lessdot$, $\doteq$ and $\gtrdot$ denote precedence relations (read as 'has lower precedence than', 'is of equal precedence to' and 'is of higher precedence than') and provided that at most one such relation holds between any two elements of $T \cup \{\vdash, \dashv\}$ then the associated operator grammar is called an *operator precedence grammar.*

Although this is much more complicated than the other kinds of grammar so far encountered, the notion of precedence can be made to coincide with what we generally regard as the precedence of (arithmetic) operators and is extended to other operations, the actions of which are computationally important but usually taken for granted in a 'pencil and paper' calculation.

Example 8.5.1

$E \to E + T | T,$

$T \to T * P | P,$

$P \to (E) | x.$

For this grammar the precedence relations are as tabulated in Figure 8.13. //

Fig. 8.13

	+	*	(	)	x	$\dashv$
$\vdash$	$\lessdot$	$\lessdot$	$\lessdot$		$\lessdot$	
+	$\gtrdot$	$\lessdot$	$\lessdot$	$\gtrdot$	$\lessdot$	$\gtrdot$
*	$\gtrdot$	$\gtrdot$	$\lessdot$	$\gtrdot$	$\lessdot$	$\gtrdot$
(	$\lessdot$	$\lessdot$	$\lessdot$	$\doteq$	$\lessdot$	
)	$\gtrdot$	$\gtrdot$		$\gtrdot$		$\gtrdot$
x	$\gtrdot$	$\gtrdot$		$\gtrdot$		$\gtrdot$

To see what this is really all about consider a stage within the derivation of the sentence $\vdash x*(x+x)\dashv$. From rule 2 of the precedence definitions we see that for symbols $*$ and (

$$T \to T * P$$

and

$$P \stackrel{+}{\Rightarrow} (E).$$

Thus $* \lessdot ($, therefore the subtree for P must be evaluated before $T*P$ can be computed and hence the action associated with '(' – which is the removal of this bracket and its matching closing bracket – takes precedence over that indicated by '*'. (Graphically the situation can be illustrated by the three components of Figure 8.14. Here, the interpretations for rule 2 are $A \equiv T$, $\alpha \equiv T$, $a \equiv *$, $B \equiv P$, $\beta \equiv \Lambda$, $\gamma \equiv \Lambda$, $b \equiv ($ and $\delta \equiv E$ from which it can be seen that the basic structure of operator precedence grammars is fundamentally simple and natural but *looks* complicated when written down because of the generality present in the rules.) Replacing x by integers 2, 3 and 4 we get

$$\vdash 2*(3+4)\dashv.$$

Writing the precedence relations below this expression we see how the order of evaluation is determined.

Fig. 8.14

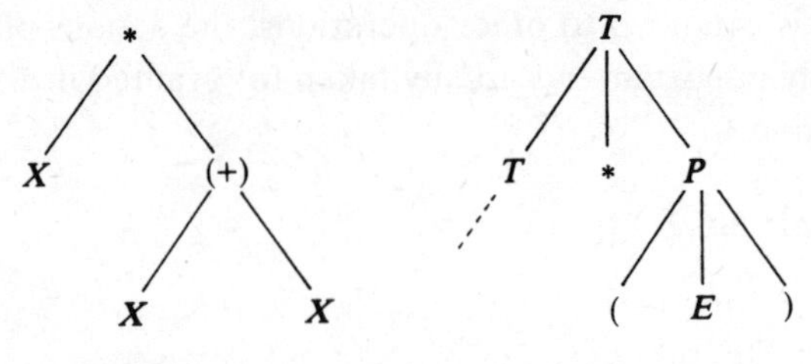

(*a*) Arithmetic structure (*b*) Syntax structure

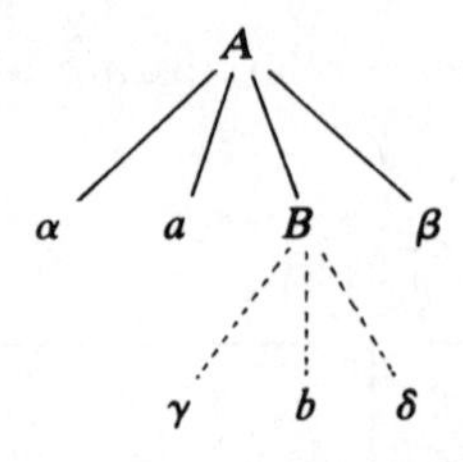

(*c*) General form of rule 2

$$\vdash \quad 2 \quad *$$
$$\lessdot \quad \gtrdot$$

(*a*) Get hold of 2 and save it on a stack.

$$\vdash \quad * \quad (\quad 3 \quad +$$
$$\lessdot \quad \lessdot \quad \lessdot \quad \gtrdot$$

(*b*) Similarly remove 3 from the expression and stack it.

$$\vdash \quad * \quad (\quad + \quad 4 \quad)$$
$$\lessdot \quad \lessdot \quad \lessdot \quad \lessdot \quad \gtrdot$$

(*c*) Likewise stack 4.

$$\vdash \quad * \quad (\quad + \quad)$$
$$\lessdot \quad \lessdot \quad \lessdot \quad \gtrdot$$

(*d*) Carry out addition with the top two items on the stack leaving the answer there. Remove + symbol.

$$\vdash \quad * \quad (\quad)$$
$$\lessdot \quad \lessdot \quad \doteq$$

(*e*) Throw away brackets.

$$\vdash \quad * \quad \dashv$$
$$\lessdot \quad \gtrdot$$

(*f*) Carry out multiplication with the top two items on the stack, leaving result on the stack. Remove * symbol.

$$\vdash \dashv$$

(*g*) No precedence relations so stop; the answer is on the stack.

Of course, instead of executing the arithmetic operations we could generate code to evaluate the expressions at a later time – this is what a compiler would do.

Exercises 8.5

1. Throughout the above discussion the semantics involved have been take for granted; however, they are intimately intertwined with the grammatical structure. Show that each of the following grammars is an operator precedence grammar and investigate how its inferred semantics differ from normal conventions.

$$
\begin{aligned}
P_1 = \{ & E \rightarrow E * T \mid T \\
& T \rightarrow T + P \mid P \\
& P \rightarrow (E) \mid x \}, \\
P_2 = \{ & E \rightarrow T + E \mid T - E \mid T \\
& T \rightarrow T * P \mid P \\
& P \rightarrow (E) \mid x \}.
\end{aligned}
$$

9 FINITE AUTOMATA

An automaton is a device that operates and controls itself. A conventional computer with a stored program is capable, subject to the availability of a suitable power supply, of controlling itself and hence *is* an automaton. (The plural of automaton is automata.) As such, computers have been studied for many years; however, it is becoming more usual to regard the program and the underlying machine as separate components.

Of course, to be able to carry out a computation we need not only a program but a machine on which it can run; but we are not intending here to embark on a detailed study of the theory of computation so, except where necessary for completeness, we shall restrict our attention to mathematical descriptions of certain finite machines.

Notwithstanding these comments we begin (in 9.1) with a general introduction which demonstrates important bounds on what machines can do. Subsequently (9.2), we study mathematical models of (typically small segments of) hardware devices and then, in 9.3, the associated algebra.

9.1 General concepts

All practical computing devices are, in some way, limited by the amount of information they can store; they are finite. The purpose of the current section is to introduce ways in which we can make statements about programs without becoming involved with tedious syntactic details and hence demonstrate that, even if we are *not* constrained by storage limits, there *are* problems which cannot be solved.

9.1.1 A universal machine

Despite the use of numerous different data types and character sets within *actual* programs, it suffices for *theoretical* purposes to restrict our consideration to programs that act on the set $V = \mathbb{N} \cup \{0\}$. Justification of the adequacy of studying programs which compute *number theoretic functions* is given in the following sections; our immediate task is to describe an idealized computer in which elements of V can be stored and manipulated and to give details of how programs for the machine can be represented.

We assume that the machine has a store consisting of an unlimited number of registers $R_1, R_2, R_3, \ldots$, etc. and that the contents of any register R_i is r_i where $r_i \in V$. Pictorially the store may be depicted as in Figure 9.1.

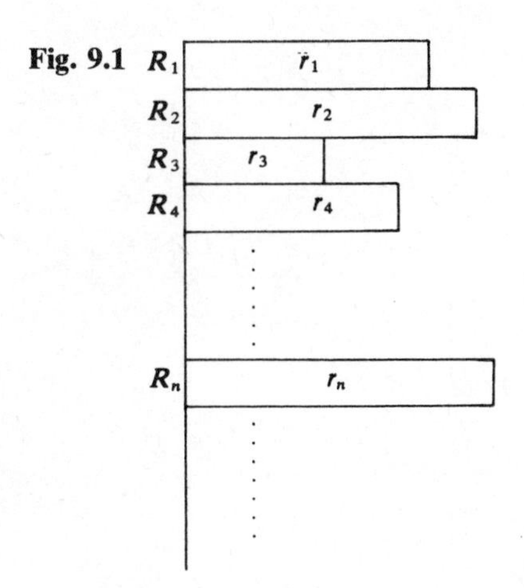

It will be convenient, for conciseness, to allow the use of many forms of operation acting on the store; however, only two (computing) operations are really necessary. These are:

$$R_n \leftarrow R_n + 1,$$
$$R_n \leftarrow R_n - 1.$$

Denoting the contents of register n before and after execution of a specific instruction by r_n and r'_n respectively we can describe the outcome of these operations as

$$R_n \leftarrow R_n + 1 \equiv r'_n = r_n + 1,$$

$$R_n \leftarrow R_n - 1 \equiv r'_n = \begin{cases} r_n - 1 & \text{if } r_n > 0 \\ 0 & \text{if } r_n = 0. \end{cases}$$

Apart from operations which change values within the store, there is a need for the machine to communicate *with* the program and thus influence the flow of control *through* the program. Only one test is absolutely necessary, namely $R_n = 0$, but again more general forms will be allowed. The result of $R_n = 0$ is true iff $r_n = 0$ and this can be used to control programs. The register set together with these operations constitute the *unlimited register machine*, URM.

Programs consist of finite collections of operations, numbered from 1 to n, for some $n \in \mathbb{N}$. We do not wish to make detailed pronouncements about these programs but rather to draw more general conclusions, and so we do not give a formal definition of their structure. Figures 9.2 and 9.3 will suffice to indicate the kind of constructions which are allowable.

Fig. 9.2

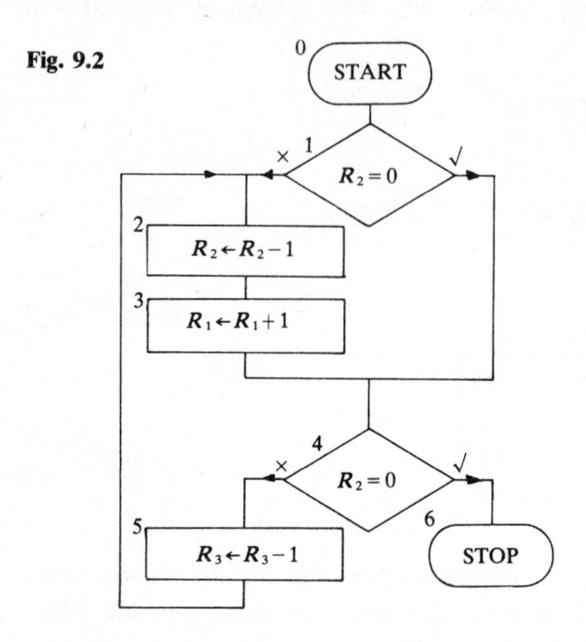

The program in Figure 9.2 is not intended to carry out a particularly sensible computation but to indicate the form of a flowchart program for our machine. Notice that it can be written down in a non-pictorial form as:

1: if $R_2=0$ then goto 4 else goto 2

2: $R_2 \leftarrow R_2-1$ (then goto 3)

3: $R_1 \leftarrow R_1+1$ (then goto 4)

4: if $R_2=0$ then goto 6 else goto 5

5: $R_3 \leftarrow R_3-1$ (then goto 2)

6: stop.

Figure 9.3 represents a more general form of a program in which *macro* instructions F_i and tests T_i are included. These may be standard instructions of the type already defined or they may represent sequences of basic instructions, which, for readability, have been given a name (much like a subroutine or procedure) and details of which are specified elsewhere. We shall call such sequences *macros*.

Fig. 9.3

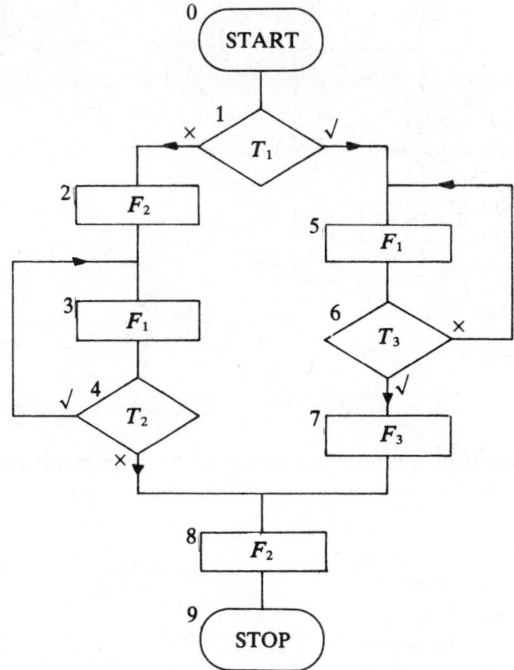

We can now detail some of these macros which, as well as providing a link to further topics, may also help convince the reader that our small repertoire of simple instructions is in fact quite powerful.

Example 9.1.1

$R_i \leftarrow 0$ can be realized by the following segment of program where the labels x, y and z are chosen so as not to interfere with other instructions in the program.

x: if $R_i = 0$ then goto z else goto y

y: $R_i \leftarrow R_i - 1$ then goto x

z: ?

In flowchart terms we have Figure 9.4. //

Example 9.1.2

Now, utilizing the macro instruction justified by the previous example, and including explicit 'goto' phrases only when deviating from

Fig. 9.4

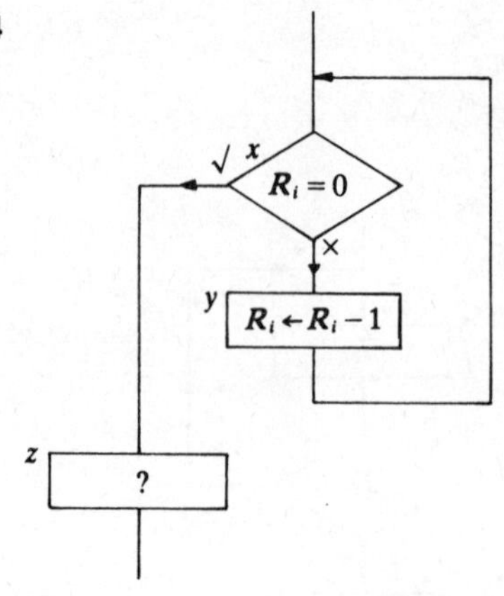

the ‘goto next instruction’ default, we can give an expansion of $R_i \leftarrow m$ (for some $m \in \mathbb{N}$);

$$\begin{aligned} &R_i \leftarrow 0 \\ &\left.\begin{aligned} &R_i \leftarrow R_i + 1 \\ &R_i \leftarrow R_i + 1 \\ &\vdots \\ &R_i \leftarrow R_i + 1 \end{aligned}\right\} \quad m \text{ copies.} \end{aligned}$$

//

The next complication in extending the instruction set requires the use of ‘working storage’. We presume that, at least in simple cases, the reader will be able to devise a suitable storage strategy and hence we shall make no specific mention of how these extra registers are chosen. Cases where unknown amounts of storage are required (such as stacks, etc.) will be discussed later.

Example 9.1.3

By use of R_k as a working register we can copy the contents of R_j into R_i, ($R_i \leftarrow R_j$). In doing this we destroy the contents of R_j which must subsequently be reset. The following program performs the required computation.

$R_k \leftarrow 0$

x: if $R_j = 0$ then goto y

$R_k \leftarrow R_k + 1$

$R_j \leftarrow R_j - 1$ then goto x

y: $R_i \leftarrow 0$

w: if $R_k = 0$ then goto z

$R_i \leftarrow R_i + 1$

$R_j \leftarrow R_j + 1$

$R_k \leftarrow R_k - 1$ then goto w

z: //

Now we can turn to 'proper' arithmetic calculations. Addition and subtraction are constructed in straightforward ways but, because of the limitations on the values held in registers, the action of the subtraction operation has to be slightly modified. Likewise we can perform multiplication and truncated division.

Example 9.1.4

Addition $R_i \leftarrow R_i + R_k$, having the effect $r_i' = r_i + r_k$, can be achieved by:

$R_j \leftarrow R_k$

x: if $R_j = 0$ then goto y

$R_i \leftarrow R_i + 1$

$R_j \leftarrow R_j - 1$ then goto x

y:

To obtain a full program, in terms of our basic instructions, we must expand the macro '$R_j \leftarrow R_k$' as in Example 9.1.3. Henceforth we shall take such expansions for granted.

In like manner, the 'bounded subtraction' $R_i \leftarrow R_i - R_k$ can be carried out by:

$R_j \leftarrow R_k$

x: if $R_j = 0$ then goto y

$R_j \leftarrow R_j - 1$

$R_i \leftarrow R_i - 1$ then goto x

y:

Notice that if the original values of R_i and R_k were such that $r_i < r_k$ then after r_i iterations the $R_i \leftarrow R_i - 1$ operation will have no effect.

Similarly, $R_i \leftarrow R_i * R_k$ can be expressed as:

$R_j \leftarrow 0$

$R_l \leftarrow R_k$

$$
\begin{aligned}
x:\quad & \text{if } R_l = 0 \text{ then goto } y \\
& R_l \leftarrow R_l - 1 \\
& R_j \leftarrow R_j + R_i \text{ then goto } x \\
y:\quad & R_i \leftarrow R_j. \qquad\qquad //
\end{aligned}
$$

In much the same way that the set of register operations can be extended by the definition of macros it is also possible to devise tests which are superficially more sophisticated but which in actual fact are constructed from sequences of the standard operations and tests.

Example 9.1.5

We may achieve the effect of 'if $R_i > R_k$ then goto x else goto y' by:

$$
\begin{aligned}
& R_j \leftarrow R_i \\
& R_j \leftarrow R_j - R_k \\
& \text{if } R_j = 0 \text{ then goto } y \text{ else goto } x.
\end{aligned}
$$

Checking for equality is slightly more involved and to model 'if $R_i = R_k$ then goto x else goto y' we need

$$
\begin{aligned}
& \text{if } R_i > R_k \text{ then goto } y \\
& \text{if } R_k > R_i \text{ then goto } y \text{ else goto } x.
\end{aligned}
$$

With these relational tests we can complete our set of primitive arithmetic operations with '$R_i \leftarrow R_i \div R_k$ where $r_i = 0$ if $r_k = 0$'. This corresponds to

$$
\begin{aligned}
& R_l \leftarrow 0 \\
& \text{if } R_k = 0 \text{ then goto } x \\
y:\quad & \text{if } R_i < R_k \text{ then goto } x \\
& R_l \leftarrow R_l + 1 \\
& \text{if } R_i = R_k \text{ then goto } x \\
& R_i \leftarrow R_i - R_k \text{ then goto } y \\
x:\quad & R_i \leftarrow R_l. \qquad\qquad //
\end{aligned}
$$

Of less general application but necessary in what will follow shortly are two tests associated with the exactness of our truncated integer division.

Example 9.1.6

In cases where $r_i \neq 0$, 'if R_i is a multiple of R_k then goto x else goto y' can be effected by:

if $R_k = 0$ then goto y

$R_j \leftarrow R_i$

$R_j \leftarrow R_j \div R_k$

$R_j \leftarrow R_j * R_k$

if $R_j = R_i$ then goto x else goto y.

With this, possibly rather strange, operation and a test of the form $R_i = m$ (left as an exercise) we can check whether r_i is prime.

Explicitly, we can model 'if R_i is prime then goto x else goto y' by:

if $R_i = 0$ then goto y

if $R_i = 1$ then goto y

$R_j \leftarrow R_i - 1$

z: if $R_j = 1$ then goto x

if R_i is a multiple of R_j then goto y

$R_j \leftarrow R_j - 1$ then goto z. //

Before our final example of this subsection we ought to mention how values are input to and output from the machine. Suppose we wish to compute a number-theoretic function, $f: V^n \to V^m$. The values of n and m are known before execution of the corresponding program commences so we can, in advance, nominate n registers as input registers (into which the initial values must be loaded before the program starts) and m registers, not necessarily distinct from the input registers, for output. When the program stops we assume that some external agent can retrieve the 'answers' from the relevant locations. This is, of course, a reasonable way to model transput since it is in effect exactly how input/output wells operate. With these conventions Example 9.1.7 can be regarded either as a complete program (in which R_i and R_j, for specific values of i and j, are designated as input and output registers) or as the scheme for a subprocedure.

Example 9.1.7

The sequence below deposits the nth prime number in R_j where n is the contents of R_i and n is assumed to be non-zero.

(Start)

$R_k \leftarrow R_i - 1$

$R_j \leftarrow 2$

x: if $R_k = 0$ then goto y

$R_j \leftarrow R_j + 1$

z: if R_j is prime then goto w

$R_j \leftarrow R_j + 1$ then goto z

w: $R_k \leftarrow R_k - 1$ then goto x

y: (stop). //

We can now explain our apparent pre-occupation with prime numbers.

9.1.2 Coding of programs

The programs in the previous section were only capable of manipulating elements of $V(=\mathbb{N} \cup \{0\})$; we must now explain how, in principle, *any* program can be regarded as a program of this type. This is done by describing ways in which different kinds of data may be *coded* into elements of V by regarding the data as sentences over suitable alphabets. Hence, almost as a side-effect of this process, we also have a method for encoding sentences of programming languages, namely the programs themselves.

The main mathematical tool used for this purpose is the *unique factorization theorem* (UFT), also known as the fundamental theorem of arithmetic. The UFT states that any element of V is either 0 or 1, or can be expressed uniquely as the product of ordered prime numbers. Explicitly, if $n \in V \backslash \{0, 1\}$ and

$$n = q_1 * q_2 * \ldots * q_i = s_1 * s_2 * \ldots * s_j$$

where $q_1, \ldots, q_i$ and $s_1, \ldots, s_j$ are all prime numbers such that

$$q_1 \leqslant q_2 \leqslant \ldots \leqslant q_i$$

and

$$s_1 \leqslant s_2 \leqslant \ldots \leqslant s_j,$$

then $i = j$ and $s_k = q_k$ for all k: $1 \leqslant k \leqslant i$. (Recall that prime numbers are elements of $\mathbb{N}$ which are multiples only of 1 and themselves.)

A full proof of the UFT is not difficult but its length would significantly distract us from the main task in hand. Therefore instead of giving a proof we suggest that the reader considers the construction of an algorithm (procedure/program) for extracting, in order, the prime factors $q_1, \ldots, q_i$ of any given n. Thus, side-stepping details of input and output, we may use the scheme in Figure 9.5.

Now presume that a certain program reads data that consists of sequences of symbols from the alphabet $A = \{x, y, z\}$. Arbitrarily choosing an order ϕ for the elements of A we may take $\phi(1) = x$, $\phi(2) = y$ and $\phi(3) = z$. If input α is of length n and α is expressed as $\alpha_1\alpha_2 \ldots \alpha_n$ where each $\alpha_i \in A$ then there is a (ϕ-) related sequence

$$\phi^{-1}(\alpha_1)\phi^{-1}(\alpha_2) \ldots \phi^{-1}(\alpha_n)$$

over $\{1, 2, 3\}$. Taking the first n prime numbers p_1 to p_n we can now

Fig. 9.5

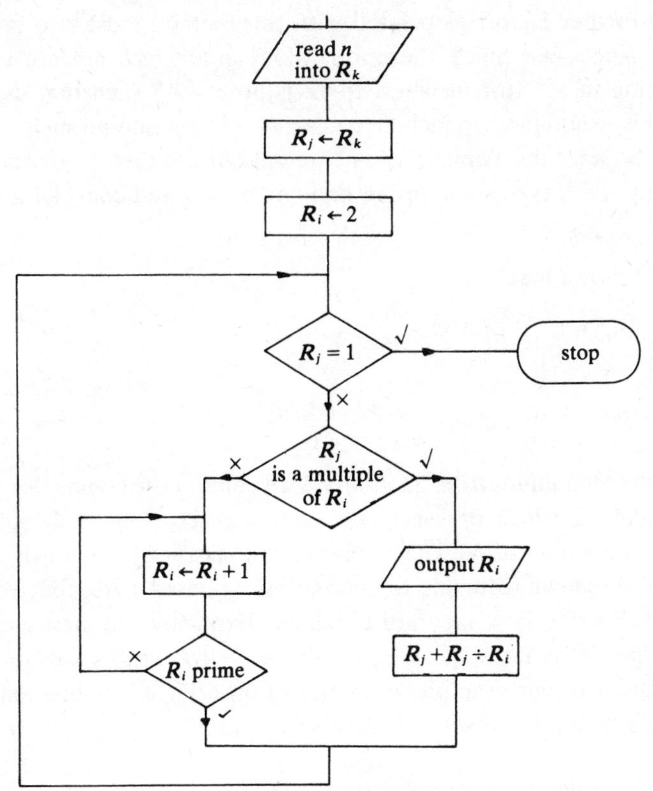

form the number

$$\prod_{i=1}^{n} p_i^{\phi^{-1}(\alpha_i)} = p_1^{\phi^{-1}(\alpha_1)} * \ldots * p_n^{\phi^{-1}(\alpha_n)}$$

– call this $\Phi(\alpha)$, and, by convention, let $\Phi(\Lambda) = 1$.

To illustrate the use of this general formula the string '*xyxz*' can be encoded via 1, 2, 1, 3 as $2^1 3^2 5^1 7^3 = 30870$. By the UFT, and the fact that Φ^{-1} is a bijection, '*xyxz*' is the only string over A that gives rise to this value; hence by the factorization procedure followed by application of Φ we can reverse the calculation to retrieve '*xyxz*'.

With this technique all strings over A can be encoded to give unique values in the set V which, by virtue of the ordering taken from $\mathbb{Z}$ (since $V \subseteq \mathbb{Z}$), implies an ordering of strings in A^*. For instance, '*xy*' > '*yx*' as indicated in Figure 9.6.

Fig. 9.6

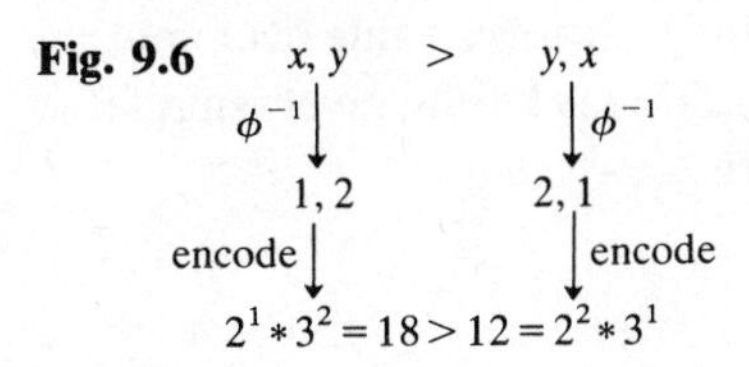

Two further factors associated with this method must also be noted. Firstly, since A is finite, there are values in $\mathbb{N}$ which are not codes of any string in A^* (for instance, there is no $\alpha \in A^*$ such that $\Phi(\alpha) = 2^5$ since this would have to include $a \in A$: $\phi^{-1}(5) = a$ and no such a exists). Secondly, since the elements of A^* are unbounded then so are the codes. (For any $n \in \mathbb{N}$ take some prime number $p_m > n$ and consider a string

$$\alpha = \alpha_1 \ldots \alpha_m \ldots \alpha_q \quad \text{so that } |\alpha| \geqslant m;$$

it then follows that

$$\begin{aligned} \Phi(\alpha) = \prod_{i=1}^{q} p_i^{\phi^{-1}(\alpha_i)} &\geqslant p_m^{\phi^{-1}(\alpha_m)} \\ &\geqslant p_m^1 \\ &= p_m > n.) \end{aligned}$$

Using the techniques from Section 3.3 we convert the coding Φ to obtain a new coding which preserves the ordering achieved by Φ but which utilizes the whole of V. This is one of the occasions when it is difficult to give *arithmetic* formulae to compute the revised codes, but yet quite easy to describe how they are obtained. Explicitly, the new coding Ψ, which has V as its range, is given by $\alpha \mapsto |\{\beta : \Phi(\beta) < \Phi(\alpha)\}|$. In the specific case of our example set A and its ordering ϕ, the first ten values of Φ and Ψ are as shown in Figure 9.7.

Fig. 9.7

string α	Λ	x	y	xx	z	yx	xy	zx	xxx	yy
$\Phi(\alpha)$	1	2	4	6	8	12	18	24	30	36
$\Psi(\alpha)$	0	1	2	3	4	5	6	7	8	9

To carry out these coding and encoding procedures we need a machine which is capable of understanding and creating symbols outside of the alphabet $D = \{0, 1, 2, \ldots, 8, 9\}$ over which elements of the set V are defined. However, all that is strictly necessary is a conversion device which could mimic the action of ϕ (and ϕ^{-1}) by a simple matching process acting on a table of input/output equivalents; the rest of the encoding procedure, involving Φ and Ψ, could then be handled by our standard URM.

Any input to a given computer system can therefore be regarded as a finite string (drawn from a potentially infinite set of strings over some finite alphabet) and, by the above process applied to the relevant alphabet A, this string can be transformed into a single element of V. The reverse procedure applied to the resultant value in V returns a value over another alphabet B, and hence a program $P: A^* \to B^*$ may be effectively simulated by the related program $P': V \to V$ where

$$P: \alpha \mapsto \Phi^{-1}(P'(\Phi(\alpha)))$$

and

$$P': n \mapsto \Phi(P(\Phi^{-1}(n))).$$

The magnitude of the values involved, even in 'small' situations, makes examples impracticable, but the associated diagrams in Figure 9.8 restate the essentials of the simulation.

Fig. 9.8

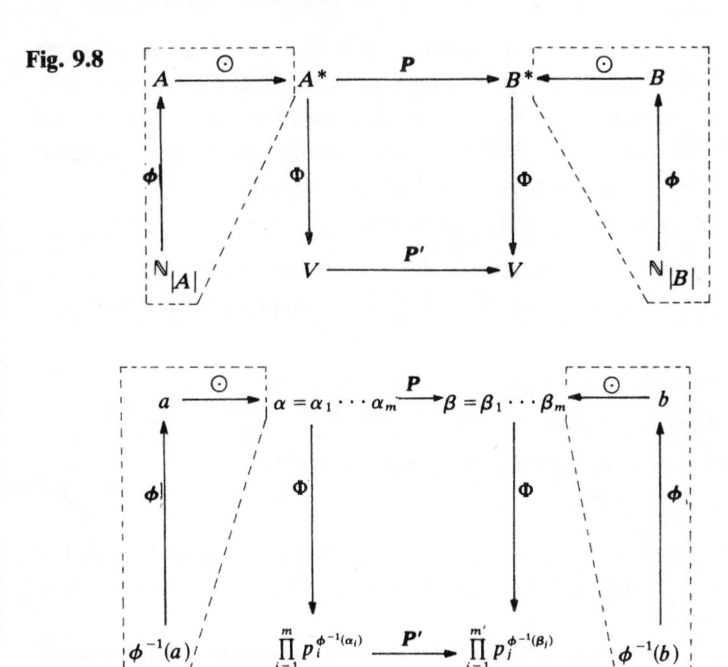

Because of the translations involved detailed specification of P' is complex and not necessarily related to P in any 'obvious' direct fashion; however, if P is computable then so is P'.

So much for strings over arbitrary alphabets.

As noted in the preamble we can also apply the same encoding procedures to programs and hence, having extended Ψ so as to ignore semantically invalid programs as well as those which include syntax violations, we can enumerate all programs for a given system using numbers in V. (We presume, so as to obtain an unbounded set of proper codes, that arbitrarily many null statements, such as $R_i \leftarrow R_i$, can be added to programs to exceed any given value in V.) However, in most programming languages there is much redundancy in the syntax so we need at most to take notice only of the lexical tokens. Rather than attempt to make general comments we consider what can be done with the simple URM flowchart language used earlier.

The four instruction types are:

x: $R_i \leftarrow R_i + 1$ then goto y

x: $R_i \leftarrow R_i - 1$ then goto y

x: if $R_i = 0$ then goto y else goto z

x: stop.

(Without loss of generality we can presume that all programs start at the instruction labelled 1.) All we need to know of the instruction at x is which type of operation is to be carried out, on what register, and which instruction, if any, to perform next. In order that the same scheme can be used for all instructions we adopt the format

(type, register, true exit, false exit).

Representing the 'stop', increment, decrement and 'is zero' instructions by 0, 1, 2 and 3 respectively gives a Φ-level coding of typical statements as follows.

$$\Phi(R_i \leftarrow R_i + 1 \text{ then goto } y) = 2^1 3^i 5^y 7^y$$

$$\Phi(R_i \leftarrow R_i - 1 \text{ then goto } y) = 2^2 3^i 5^y 7^y$$

$$\Phi(\text{if } R_i = 0 \text{ then goto } y \text{ else goto } z) = 2^3 3^i 5^y 7^z$$

$$\Phi(\text{stop}) = 2^0 3^0 5^0 7^0 = 1.$$

So Φ(instruction) contains all the information needed to carry out the instruction and, if appropriate, to proceed to the next step in the execution of the program.

Duplicating the argument that all strings in A^* are finite but unbounded to flowchart programs for the URM, any specific program has n statements for some value of $n \in \mathbb{N}$. It is then possible to extend Φ to programs by

$$\Phi(\text{prog}) = \prod_{i=1}^{n} p_i^{\Phi(s_i)}$$

where 'prog' consists of n labelled statements of which the ith is s_i.

From Φ(prog) we can obtain $\Phi(s_i)$ by extracting the p_i component and subsequently, by decomposing this value into primes, we know the details of the statement labelled by i.

To conclude, using a suitable 'squashing' function Ψ we have a coding which is a bijection between $\mathscr{P}$, the set of all programs for the URM, and V.

9.1.3 The halting problem

We are now in a position to prove that the construction of certain general programs is impossible – not just that we have not found a solution but that no solution can exist. Formal proofs will be limited to two fundamental cases. We shall prove one from first principles and then show how the other can be deduced from it, thus illustrating a commonly used reduction technique. After this we give a list of problems which, by further manipulation, can be shown to be unsolvable in a similar fashion.

Strictly speaking, all the statements below relate to flowchart programs for a URM with specific coding functions Ψ_1 and Ψ_2. However, given any 'universal' machine and a suitable programming language we are able to construct adequate coding functions and hence the results are of general applicability.

The first problem which we shall show to be unsolvable is the *self-applicability* problem.

Theorem. There is no URM flowchart program which, when given an arbitrary program A in the form of its coding $a = \Psi_1(A)$, will halt with output 0 if $A(a)$ halts (i.e. the program A halts given the input a) and with output 1 if $A(a)$ does not halt.

Proof. The proof is by contradiction.

Suppose such a program does exist; call it B. So

$$B(a) \text{ halts with result 0 if } A(a) \text{ halts}$$

and

$$B(a) \text{ halts with result 1 if } A(a) \text{ does not halt.}$$

Now we alter program B by replacing its stop instruction by a conditional loop such that if the output register has the value 0 then we loop on this instruction (indefinitely), otherwise stop. Call this program C. (See Figure 9.9.) Now

$$C(x) \text{ halts (and has the same output as } B(x)) \text{ iff the output is 1.}$$

Fig. 9.9

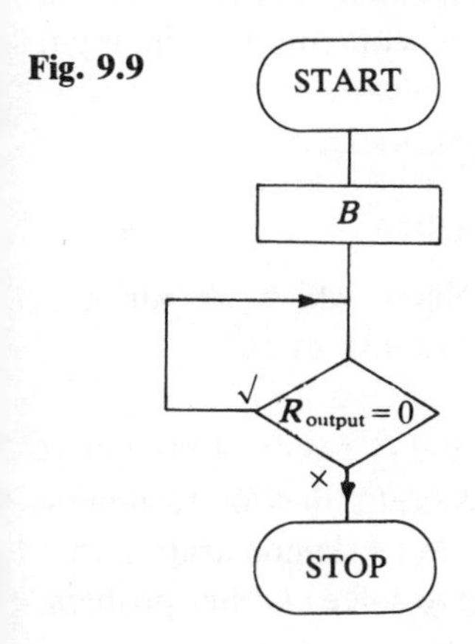

Presenting C with input $c = \Psi_1(C)$ then provides a contradiction since

$$C(c) \text{ halts iff } C(c) = B(c) = 1$$

but

$$B(c) = 1 \text{ iff } C(c) \text{ does not halt.}$$

In consequence the program C cannot exist and hence neither can B. //

The preceding proof is similar in style to that used to refute the existence of the Russell set in Examples 1.1. It is not difficult but, like many short mathematical arguments, is rather subtle and ought to be fully appreciated by the reader before he uses it in the proof of a more general result: the unsolvability of the *halting problem.*

Theorem. There is no URM flowchart program which, when presented with arbitrary input representing a program and its data, will determine whether or not execution of the program on the data halts.

Proof. We shall demonstrate how, if such a program existed, we could, by choosing the correct input, solve the self-applicability problem. But this is unsolvable and hence we deduce that no solution of the halting problem is possible.

The program seemingly requires two inputs, $a(=\Psi_1(A))$, the code of program A, and $x(=\Psi_2(X)$ where X is the input (data) for A). These can be encoded in the usual way by using two different prime numbers p and q. Let the input be $p^a q^x$ and suppose there is a program to solve the problem, call it B. So

$$B(p^a q^x) \text{ halts with result 1 if } A(X) \text{ halts}$$
$$\text{and halts with result 0 if } A(X) \text{ does not halt.}$$

Now from B we can create a program C by appending to the beginning of B a sequence of instructions that converts the contents i of the input register to $(pq)^i$. Then

$$C(a) = B(p^a q^a) = 1 \text{ if } A(a) \text{ halts}$$
$$= 0 \text{ if } A(a) \text{ does not halt.}$$

C therefore solves the self-applicability problem which we know is unsolvable. Hence C does not exist and so neither does B. //

The halting problem is the classic undecidability result of computer science theory. The proof given here uses a general reduction technique in which one problem is shown to be unsolvable by demonstrating that if a solution were possible then it could be used to solve another problem which is known to be unsolvable.

Using this principle (and appropriate constructions, details of which cannot be given here) it can be shown that many other problems are unsolvable. These include the following:

(*a*) whether or not an arbitrary program halts when given input 0,
(*b*) whether or not an arbitrary program halts for all inputs,
(*c*) whether $L(G)=\varnothing$ for an arbitrary CSG G,
(*d*) whether $L(G_1)\cap L(G_2)=\varnothing$ where G_1 and G_2 are arbitrary CFGs,
(*e*) whether $L(G)=T^*$ where $G=(N, T, P, S)$ is an arbitrary CFG,
(*f*) whether $L(G_1)=L(G_2)$ where G_1 and G_2 are arbitrary CFGs,
(*g*) whether an arbitrary CFG is ambiguous.

9.1.4 An 'extended' machine

How realistic is the URM as a model computer?

The URM store is obviously more versatile than the store of any actual computer which would be subject to limits both on the number of registers and on the values that could be held in each register. However, this idealization within the URM *extends* rather than *limits* the capabilities of the machine. Are there any aspects of actual computing systems which cannot be simulated on a URM?

We claim that there are not. Although we cannot formally justify this claim we outline an 'extended' machine which might be more readily accepted as an abstract model and argue that a URM can itself simulate all these features.

Common features that we might consider are:

(*a*) a greater repertoire of computing operations,
(*b*) a wider range of external data types,
(*c*) arrays,
(*d*) stacks,
(*e*) more powerful control statements,
(*f*) an integral control mechanism and a stored program and
(*g*) recursion.

In Sections 9.1.1 and 9.1.2 we have shown how, for (*a*), further arithmetic operations can be added to the instruction set and for (*b*), by providing simple peripherals (to deal with characters not known to the URM), the range of input and output representations can be extended over any alphabet. Now for point (*c*).

Suppose we required an array A with ten components, $A[1]$ to $A[10]$, and that the contents at any time were represented by $a_1, \ldots, a_{10}$. By our usual coding technique we could preserve these values in a single quantity

$$a=\prod_{i=1}^{10} p_i^{a_i}$$

which could then be stored in a single register and we have already described procedures for extracting each of the values a_i from a. This method is not restricted to normal array bounds so it also deals with (d) stacks. With stacks we can keep track of intermediate results within a calculation without the need to access significantly more registers. With an appropriately sophisticated control mechanism (see below) we can also stack return addresses of subroutines.

Any URM program is finite and can be labelled by integers 1 to n for some $n \in \mathbb{N}$. Within the context of a URM flowchart we can therefore construct in (e) a 'computed goto' by using a table look-up technique to achieve 'goto (R_k)' as indicated in Figure 9.10.

Fig. 9.10

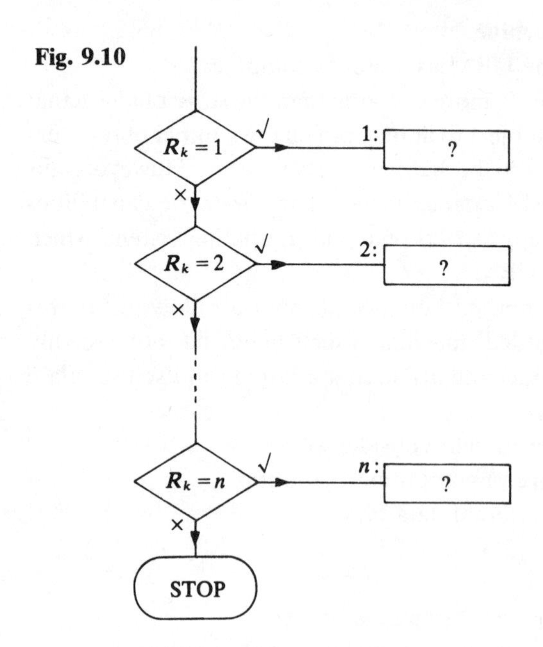

Using the simple (direct) coding scheme for programs (Φ not Ψ) we can encode the entire program into a single value stored in a register. The prime factorization routine can then be activated to extract the code of 'the next statement', decode this into its components and execute the appropriate instruction. This models feature (f) and with stacks also gives us the capability to recurse (g).

We could therefore proceed using a URM 'extended' by adding these features but in fact such extensions are only cosmetic and the resulting system can be modelled by another URM having a flow-chart program with the simple increment, decrement and 'is zero' instructions.

Exercises 9.1

1. Construct a URM flowchart segment to perform the test '$R_i = m$'.

2. Design a URM control machine which, given a program coded into R_0, will execute that program. (Base your machine on the coding schemes in Section 9.1.2.)

9.2 Finite-state machines

So much for the capabilities and limitations of universal machines which can compute *anything* that is computable and consequently, reversing the argument, any computational problem which is not soluble on such a general system is not soluble on *any* system; but what about real computers? The main objection which can be levelled at the URM described in Section 9.1.1 concerns size. The URM consisted of an unbounded number of registers each of which had an unbounded capacity for holding any value in the set V. Although, as implied by the constructions of Section 9.1.4, we can trade a 'short wide' store (having few registers, each with a large capacity) for a 'long thin' store (in which the capacity of each register is reduced, but the number of registers is increased), the essential bound enforced by physical limitations is the number of different configurations or states that the store can attain. It is this concept that forms the basis of our mathematical model of finite machines.

9.2.1 Deterministic machines

In line with most people's experience with electronic machinery (albeit low-level hardware in the form of **nand** gates etc., or microprocessor systems or general purpose digital computers) our first model embodies the notion of determinism; that is, when a certain situation is reached on more than one occasion, exactly the same behavioural pattern ensues in each case. We begin with the formal description.

Definition. A (*deterministic*) *finite-state machine* (FSM), M, is an algebraic structure

$$M = (Q, \Sigma, t, q_0, F, p)$$

where

(*a*) Q is a non-empty finite set of *states*,
(*b*) Σ is a finite *input alphabet*,
(*c*) t is a mapping $Q \times \Sigma \to Q$, called a *transition*,
(*d*) $q_0 \in Q$ is the *initial state*,
(*e*) $F \subseteq Q$, F is the set of *final states* (or *accepting states*), and

(*f*) p is a function $Q \times \Sigma \to \Sigma$ called a *print function* (or output function).

(In certain cases it is useful to modify the print function so that it has functionality $Q \times \Sigma \to \Sigma'$ where Σ' is some other alphabet. In such situations the machine is called a *finite-state transducer.*) //

The idea is that we start at state q_0 and if $q_i \in t(q_0, s)$ then inputting the character s enables the automaton to move into state q_i. Also, if $(q_0, s) \in \mathscr{D}_p$ then $p(q_0, s)$ will be output as this change of state takes place. Thus we continue, reading more input and changing from state to state until we read a terminator (any symbol not in Σ) or the input is exhausted in which case processing stops. The input is said to have been *accepted* if the state reached is in F. Hopefully the mechanics will become clear when we look at an example; however, before doing so it is useful to describe a diagrammatic representation of M.

First, we represent elements of Q by nodes of a directed graph, drawn as small circles with the state name written inside. Elements of F have a further circle drawn around them. If $((q_i, s_j), q_k) \in t$ then denote this by a directed edge from q_i to q_k and label it with s_j, further, if $((q_i, s_j), s_l) \in p$ label the edge by $s_j : s_l$. Several labels may be appended to the same edge. Finally, designate q_0 by an arrow pointing to q_0.

Example 9.2.1

Consider the machine in Figure 9.11. Suppose that we read the string '*abbaa*'. The edges denoting the transition function cause the machine to pass through the states q_0, q_1, q_2, q_1, q_0 and q_1 in that order. Explicitly, this is because starting at q_0, under the action of t,

$$(q_0, a) \mapsto q_1,$$
$$(q_1, b) \mapsto q_2,$$
$$(q_2, b) \mapsto q_1,$$
$$(q_1, a) \mapsto q_0,$$
$$(q_0, a) \mapsto q_1.$$

But q_1 is not an accepting state so this particular input is rejected. Inspection of the diagram will disclose that the only strings acceptable to this machine are those over the alphabet $\{a, b\}$ in which there are an even number of as and an even number of bs. This machine produces no explicit output, the information required being extracted from the resultant state after the machine has halted. //

The next two examples relate to arithmetic hardware and do generate intermediate output.

Fig. 9.11

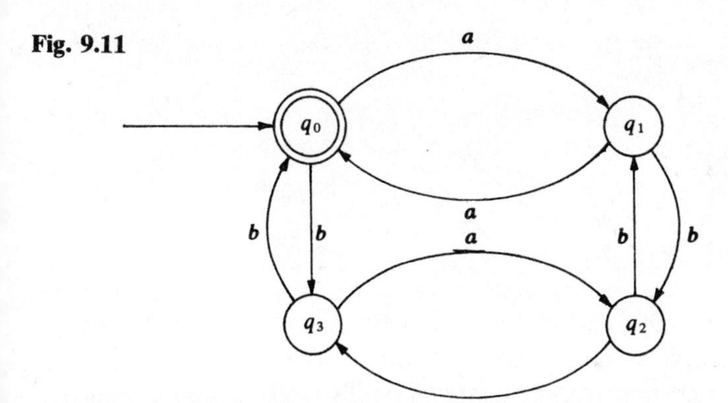

Example 9.2.2

The machine in Figure 9.12 accepts pairs of strings over $\{0, 1\}$ and computes and outputs their sum. The input starts with the lowest valued bits and the pairs are read from right to left. Given two n-bit numbers this machine will calculate their n-bit sum but will not recognise overflow conditions or deal with negative forms. //

Fig. 9.12

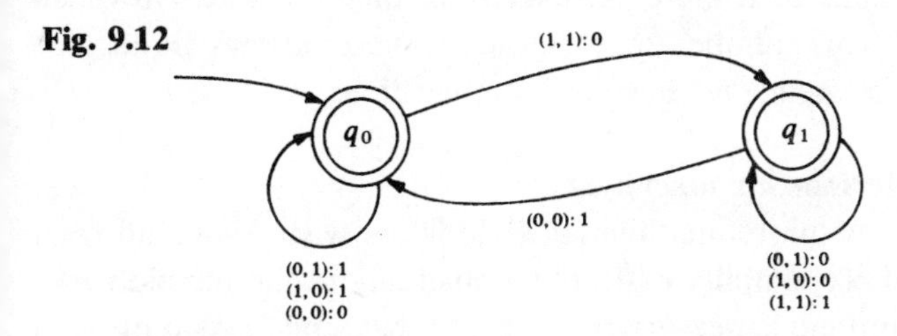

In checking the characteristics of this transducer (its print function is of type $Q \times \{0, 1\}^2 \rightarrow \{0, 1\}$) you may find it useful to refresh your memory concerning 2s complement binary arithmetic as described in Section 4.3.

Knowledge of error conditions within the binary addition process can be included within a more sophisticated model.

Example 9.2.3

The machine depicted in Figure 9.13 also performs binary addition in the same manner as the previous machine except that it includes two states, q_2 and q_3, which reflect the presence of errors associated with carry *into* and *out of* the sign bit respectively. The accepting states q_0 and q_1 indicate outputs in which neither or both of these carries have occurred and hence are arithmetically correct. //

These examples show that some useful manipulations and computations can be performed but how powerful are finite-state machines? In answer

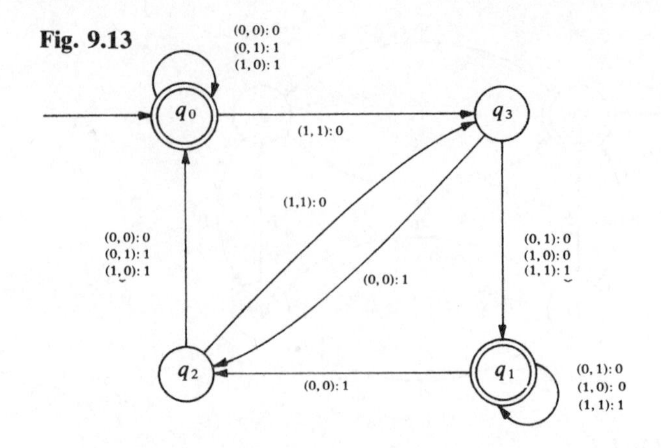

Fig. 9.13

to this question we make two points. First, although not typical, the reader may have noticed that the alphabets used in the examples are each based on only two symbols. We can use any finite alphabet; however, as noted in Section 8.1, an alphabet of size 2 is always sufficient. Second, the total storage available to a specific computer at a given time is finite. If it comprises n bits then there are exactly 2^n different states in which store can exist. This number may be large but is always finite (see Exercises 9.2.) and hence we have a finite machine.

9.2.2 Non-deterministic machines

Now for a more mathematical look at what these abstract machines can do. To simplify matters we shall ignore the function p in our earlier definition; hence we shall strictly be concerned only with *finite-state acceptors* (FSAs). Also we will find it convenient to widen the range of the transition function so that

$$t: Q \times \Sigma \to \mathcal{P}(Q).$$

Hence from a given state and a given input the machines may have a choice of where next to go. This adds nothing to their scope but provides a simple way of constructing FSAs which will be used extensively throughout the section. A machine of this type is called a *non-deterministic* FSA.

To make matters more explicit, for a given FSA, M, we define the set of strings accepted by M, denoted by $A(M)$, as follows.

If $M = (Q, \Sigma, t, q_0, F)$ and $s \in \Sigma^*$, write $s = s_1 s_2 \ldots s_n$ and define $T(s)$ inductively on the length of s by

$$T(\Lambda) = \{q_0\}$$

and

$$T(\sigma s_k) = \bigcup_{q \in T(\sigma)} t(q, s_k) \quad \text{where } \sigma = s_1 \ldots s_{k-1} \text{ for all } k: 1 \leqslant k \leqslant n.$$

Thus $T(s)$ is the set of all terminal states of M that can be reached by reading s. If $T(s) \cap F \neq \varnothing$ then s is accepted, so the set of strings accepted by M is

$$A(M) = \{s: T(s) \cap F \neq \varnothing\}.$$

Non-deterministic FSAs are useful in that they simplify the task of building complex machines from more basic ones. Essentially, what we want to be able to do is to put machines together in such a way that we need not, at least in the first instance, be concerned about whether or not the resultant transition is a function or a relation over $(Q \times \Sigma) \times Q$. (Recall that a *function* $Q \times \Sigma \to \mathscr{P}(Q)$ corresponds to a *relation* $Q \times \Sigma \to Q$.)

The introduction of non-determinism does not increase the computing potential of an FSA and can be removed by virtue of the following result.

Theorem. For any non-deterministic FSA, M_1, there is a deterministic FSA, M_2, such that $A(M_1) = A(M_2)$.

Outline proof. Let $M_1 = (Q_1, \Sigma_1, t_1, q_0, F_1)$ and $M_2 = (Q_2, \Sigma_2, t_2, q_0', F_2)$. First set $Q_2 = \mathscr{P}(Q_1)$ (in general we shall not need all of these states but we never need more) and $\Sigma_2 = \Sigma_1$. What we shall do in essence is, starting from $q_0 \in Q_1$, take the set of all states in $t(q_0, s_1)$ (for $s_1 \in \Sigma_1$) and call this set the image of (q_0', s_1) under t_2; i.e. we bunch different possibilities in M_1 into a single state in M_2. Suppose that we next read the character s_2. In M_1 this would cause a change from a state in $t_1(q_0, s_1)$ to a state in the set $t_1(t_1(q_0, s_1), s_2)$. The set of all states obtained in this way now gives the single state in M_2 resulting from the application of t_2 to the image of (q_0', s_1) and s_2. Thus we proceed with the construction of transitions between states in Q_2.

Since Q_1 is finite so is Q_2, hence we must ultimately loop back to previously constructed states and the process eventually terminates. A state in Q_2 is an accepting state, and hence in F_2, if one of the 'constituent' states from Q_1 was in F_1.

Proof. With the construction described above $q_0' = \{q_0\}$,

$$t_2: (\{q_{i1}, \ldots, q_{ij}\}, s_k) \mapsto \left\{ \bigcup_{l=1}^{j} t_1(q_{il}, s_k) \right\}$$

and

$$\{q_{i1}, \ldots, q_{ij}\} \in F_2 \quad \text{iff } q_{ik} \in F_1 \text{ for some } k: 1 \leqslant k \leqslant j.$$

Referring back to the definition of the set $A(M)$ for a given machine

M, we can extend t_1 and t_2 to give T_1 and T_2 as follows:

$$T_1(\Lambda) = \{q_0\},$$

$$T_1(\sigma s_k) = \bigcup_{q \in T_1(\sigma)} t_1(q, s_k)$$

and

$$T_2(\Lambda) = \{q_0'\},$$

$$T_2(\sigma s_k) = \bigcup_{q \in T_2(\sigma)} t_2(q, s_k)$$

$$= t_2(q, s_k)$$

where $T_2(\sigma) = \{q\}$, and $\sigma = s_1 \ldots s_{k-1}$.

Now since $A(M) = \{s: T(s) \cap F \neq \varnothing\}$ and there is a natural correspondence between F_1 and F_2 we need only demonstrate the same correspondence between $T_1(s)$ and $T_2(s)$ for any s; i.e.

$$\{\ldots, q_i, \ldots\} = T_1(s) \quad \text{iff } \{\{\ldots, q_i, \ldots\}\} = T_2(s).$$

We do this by induction on the length of s.

If $|s| = 0$ then $s = \Lambda$ whence

$$T_1(\Lambda) = \{q_0\}$$

and

$$T_2(\Lambda) = \{q_0'\} = \{\{q_0\}\}.$$

Now suppose that the correspondence holds for all strings $\sigma: |\sigma| \leqslant k-1$ and consider the string σs_k. By the inductive hypothesis

$$\{\ldots, q_i, \ldots\} = T_1(\sigma) \quad \text{iff } \{\{\ldots, q_i, \ldots\}\} = T_2(\sigma)$$

but then if $q_j \in t_1(q_i, s_k)$ it follows that $q_j \in T_1(\sigma s_k)$ and, by definition of t_2, that

$$\{\ldots, q_j, \ldots\} \in t_2(\{\ldots, q_i, \ldots\}, s_k)$$

$$= T_2(\sigma s_k).$$

Again, by the definition of T_2 and t_2, all elements of $T_2(\sigma s_k)$ must be derived in this way and hence

$$\{\ldots, q_j, \ldots\} = T_1(\sigma s_k) \quad \text{iff } \{\{\ldots, q_j, \ldots\}\} = T_2(\sigma s_k).$$

Therefore, for any $s \in \Sigma^*$, $T_1(s)$ and $T_2(s)$ 'correspond' and so

$$A(M_1) = \{s: T_1(s) \cap F_1 \neq \varnothing\}$$

$$= \{s: T_2(s) \cap F_2 \neq \varnothing\} = A(M_2). \quad //$$

The above argument was necessarily complicated and involved the somewhat dubious (but well-defined) notion of 'correspondence' because of the level of nesting of sets involved. In particular instances we can circumvent this by devising suitable names for the resulting states in the deterministic machine M_2.

The construction of M_2 from M_1 is straightforward as will be shown in a subsequent example but to keep the mathematics tidy we first recall the common notational convention associated with functions. By virtue of the constructions used such a convention was inappropriate here but brief mention of this lapse in precision will serve to explain 'where some of the sets came from'.

Given

$$f: A \to B$$

such that

$$f: x \mapsto y$$

we *should* write $f(x) = \{y\}$ but we commonly reduce this to $f(x) = y$.

Similarly, we usually denote a deterministic transition

$$t: Q \times \Sigma \to Q$$

where

$$t: (q_i, s) \mapsto q_j$$

by

$$t(q_i, s) = q_j.$$

However, when we have $t: Q \times \Sigma \to \mathscr{P}(Q)$ (as in M_1 *and* M_2) we must include the set brackets even when $|t(q_i, s)| = 1$, and hence write

$$t(q_i, s) = \{q_j\}.$$

So much for the theory, now for an example.

Example 9.2.4

Take M_1 as in Figure 9.14(a). Now

$$t_1(q_0, x) = \{q_1, q_2\}$$

and $q_1 \in F_1$ hence we obtain $\{q_1, q_2\}$ in F_2. Similarly,

$$t_1(q_1, x) = \{q_0\} \quad \text{and} \quad t_1(q_2, x) = \{q_2, q_3\}$$

hence

$$t_2(\{q_1, q_2\}, x) = \{q_0, q_2, q_3\} \text{ in } M_2,$$

and so on, to give Figure 9.14(b). A relabelling then gives us Figure 9.14(c). //

Henceforth we need not bother about whether or not our machines are deterministic; they can always be made so.

9.2.3 Composite machines

We are now in a position to describe how, given a collection of FSAs, they can be 'plugged together' to accept certain well-defined sets

Fig. 9.14

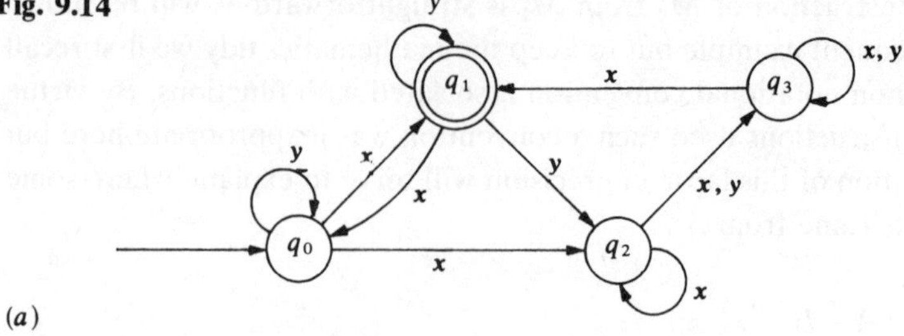

(*a*)

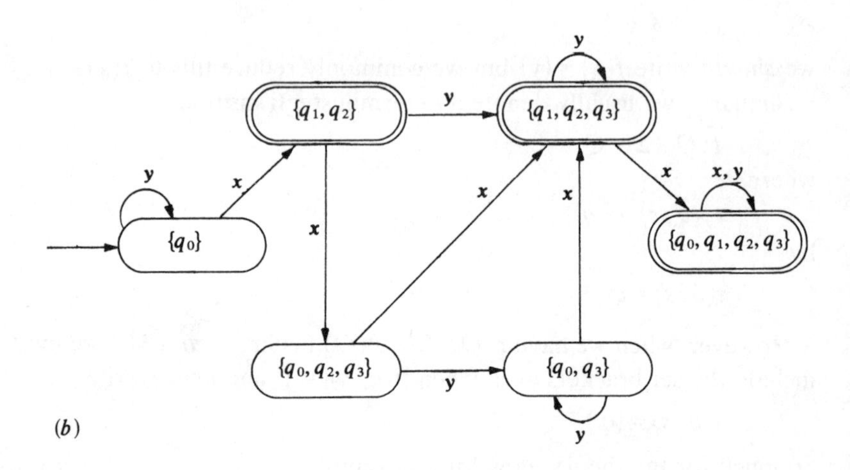

(*b*)

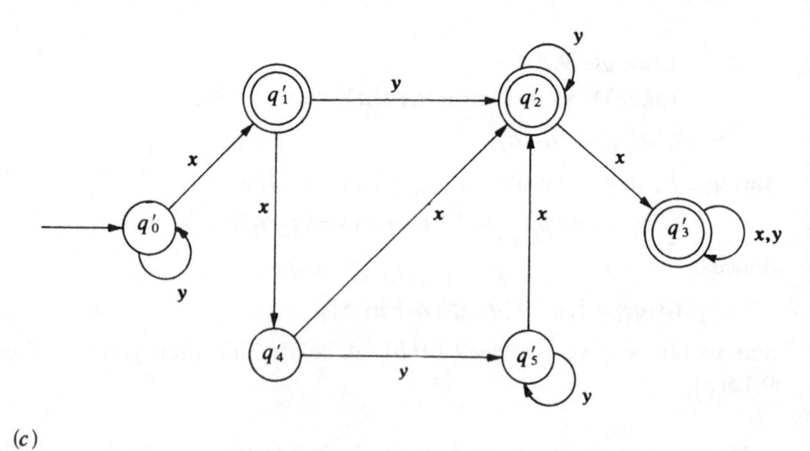

(*c*)

of strings. The main capabilities are given as a proposition but instead of formal proofs we give descriptions of the constructions involved.

Proposition. Given machines $M_1 = (Q_1, \Sigma, t_1, q, F_1)$ and $M_2 = (Q_2, \Sigma, t_2, p, F_2)$ we can mechanically construct machines

$M_3, \ldots, M_8$ such that

$$A(M_3) = \Sigma^* \backslash A(M_1),$$
$$A(M_4) = A(M_1) \cap A(M_2),$$
$$A(M_5) = A(M_2) \backslash A(M_1),$$
$$A(M_6) = A(M_1) \cup A(M_2),$$
$$A(M_7) = A(M_1)A(M_2) \quad \text{(concatenation of strings)},$$

and

$$A(M_8) = A^*(M_1).$$

Constructions. M_3: This is obtained by interchanging all accepting and rejecting states of M_1. Therefore

$$M_3 = (Q_1, \Sigma, t_1, q, Q_1 \backslash F_1).$$

M_4: Take $Q_4 = Q_1 \times Q_2$ – possibly with some suitable renaming – then let $r = (q, p)$, define t_4 by

$$t_4 : ((q_i, p_j), s) \mapsto (t_1(q_i, s), t_2(p_j, s))$$

and set

$$F_4 = \{(q_i, p_j) : q_i \in F_1 \text{ and } p_j \in F_2\}.$$

Then

$$M_4 = (Q_4, \Sigma, t_4, r, F_4) \quad \text{(see Example 9.2.5)}.$$

M_5: The construction of M_5 follows from M_2 and M_3 since

$$A(M_2) \backslash A(M_1) = A(M_2) \cap (\Sigma^* \backslash A(M_1)) = A(M_2) \cap A(M_3).$$

M_6: Similar to M_4; $Q_6 = Q_4$, $t_6 = t_4$, but

$$F_6 = \{(q_i, p_j) : q_i \in F_1 \text{ or } p_j \in F_2\}.$$

Then

$$M_6 = (Q_6, \Sigma, t_6, r, F_6).$$

M_7: In concatenation, the idea is to join the output of M_1 to the input of M_2. However, care must be taken so as not to involve an extra step while moving from M_1 to M_2. Thus take $Q_7 = Q_1 \sqcup Q_2$ (the disjoint union: change names so that Q_1 and Q_2 have no common states and use $Q_1 \cup Q_2$), set

$$F_7 = \begin{cases} F_2 & \text{if } p \notin F_2 \\ F_1 \cup F_2 & \text{if } p \in F_2 \end{cases}$$

and add transitions from elements of F_1 to the second stages of M_2, viz.

$$t_7 = t_1 \cup t_2 \cup \{((q_i, s), p_j) \quad \text{where } q_i \in F_1 \text{ and } p_j \in t_2(p, s) \text{ for some } s \in \Sigma\}.$$

Then

$$M_7 = (Q_7, \Sigma, t_7, q, F_7).$$

Again the examples below help.

M_8: Similarly, care needs to be taken at the start of M_8 to ensure that $A^0(M_1) = \varnothing$ can be accepted. To do this we append a new starting state u ($u \notin Q_1$) so that $Q_8 = Q_1 \cup \{u\}$, $F_8 = F_1 \cup \{u\}$ and t_8 is extended, as was t_7, to give

$$t_8 = t_1 \cup \{((u, s), q_j) \quad \text{where } q_j \in t_1(q, s) \text{ for some } s \in \Sigma\}$$
$$\cup \{((q_i, s), q) \quad \text{where } t_8(q_i, s) \in F_1 \text{ for some } s \in \Sigma\}.$$

Then

$$M_8 = \{Q_8, \Sigma, t_8, u, F_8\}. \quad //$$

Example 9.2.5

In Figure 9.15 we give a sequence of state diagrams for machines M_1 and M_2 and resulting composite machines M_3 to M_8 as constructed above. Notice that some of these machines are non-deterministic but the non-determinism is removable by the previous results. //

The constructions involving union, concatenation and closure (the 'star' operation) provide the basis for an algebraic system for describing sets acceptable to FSAs. This algebra is discussed in Section 9.3 but for the remainder of the current section we turn our attention to models of *real* computers.

9.2.4 Modelling 'real' computers

As already noted, given a computer whose total memory consists of n bits then potentially we need an FSM with 2^n states in order to model its behaviour. With the initial state configured so as to have the program and data values 'in core' there is no external stimulus which will trigger a sequence of transitions. However, viewing the machine in a slightly different way provides an instructive realization of the dynamic processes involved, based on the finite state machines already defined.

Suppose that the store is made up of m l-bit words; so $n = m * l$. Now isolate the program pointer (also called the program counter, or next instruction pointer) from the rest of the machine to give a machine with $n' = l * (m - 1)$ bits. (For simplicity this location is assumed to be the same length as other store components.) The program pointer can now be referenced after each state change and its value used to drive the next transition. Thus the transition function becomes:

$$t : Q \times P \rightarrow Q \times P$$

Fig. 9.15(a)

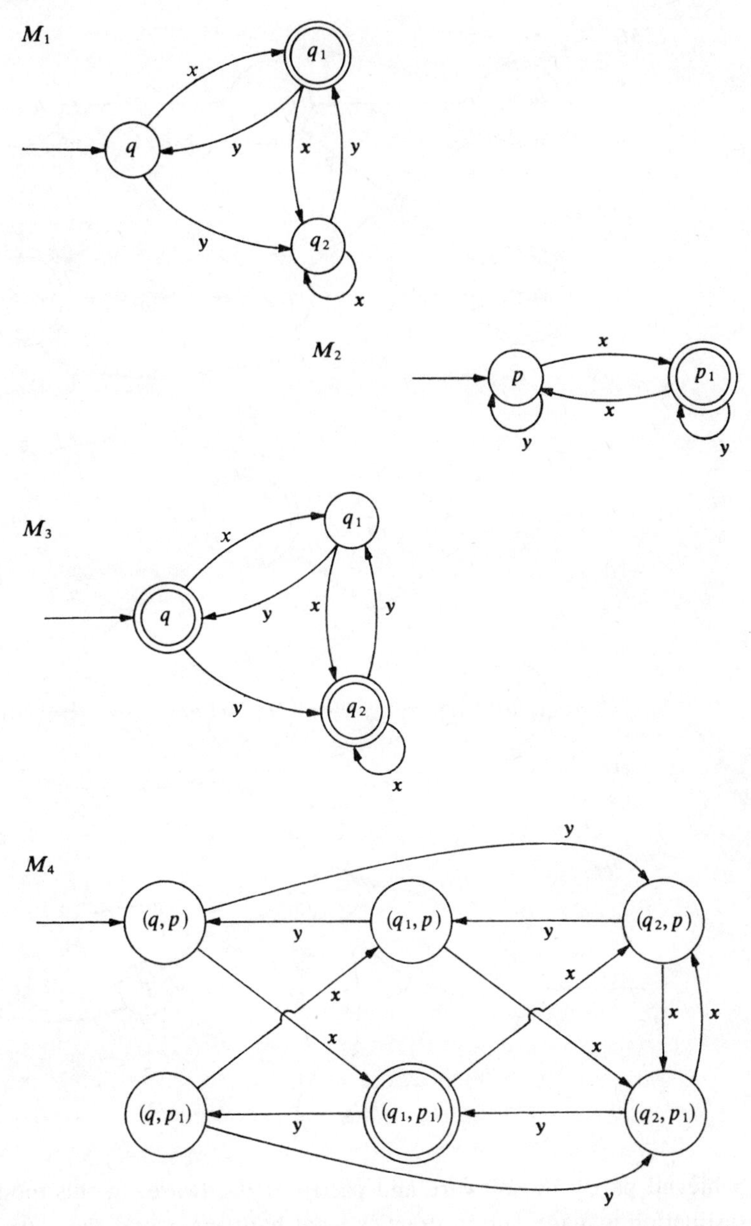

where

$$t:(q_a, p_b) \mapsto (q_c, p_d)$$

causes the pointer value p_b to extract an instruction from state q_a thus causing the state q_c to be created and the value p_d to be placed in the program pointer. (The action here is conventionally regarded as being

Fig. 9.15(b)

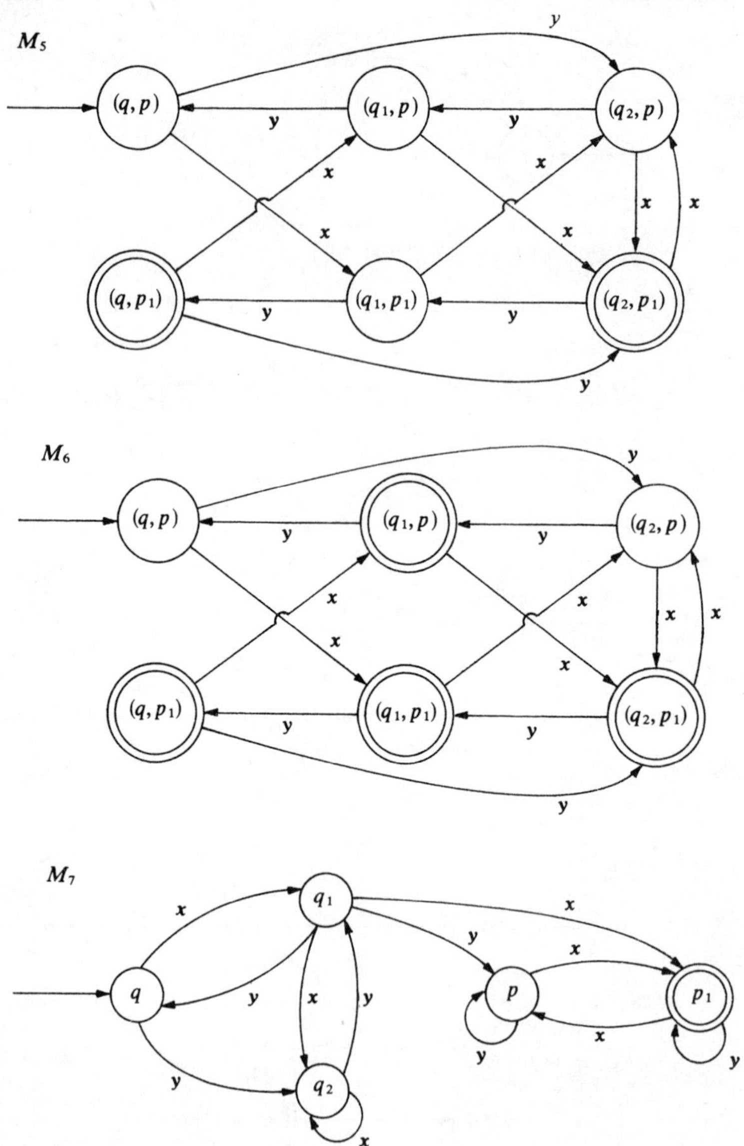

achieved partly by software and partly by hardware. In this model no distinction is made and in practice what happens is that the software is merely data that controls the action of the hardware.)

Conventionally, programs are started by setting the program pointer to a fixed value. From this initial value subsequent values are generated automatically by the process described above until a stop instruction is executed.

Fig. 9.15(c)

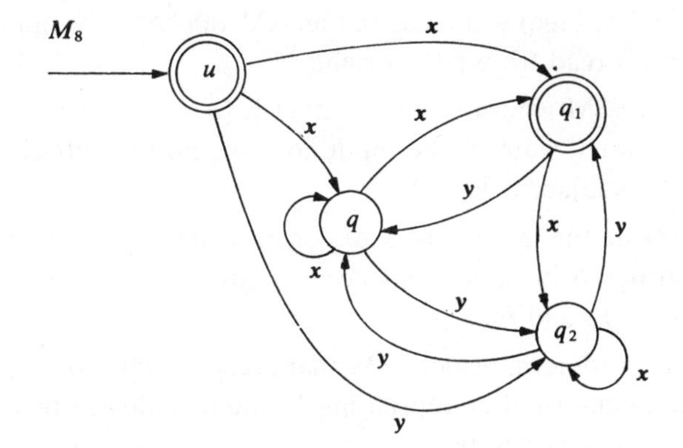

Of course, not all the 2^l values which can be held in the program pointer need be valid. Often 2^l will be greater than m and hence $2^l - m$ values will refer to illegal addresses and will give rise to an error state causing processing to halt.

The number of legal values which the program pointer may attain determines the number of transition arrows that must be drawn from each state; it does not, however, restrict the number of states which is $2^{n'}$.

Without more specific information about the system being modelled we can say little which is of general application but one final point should be appreciated.

If the FSM we have constructed halts in a 'short' time, either by executing a stop instruction or by generating an illegal pointer value, then the result is known; but if the machine goes on (and on) will it ever halt and give a result? Since our machine *is* finite (and hence not a URM) this problem is decidable, but how long will it take to determine the result? By virtue of the way in which the machine is configured the state-pointer pair (q, p) uniquely determines the sequence of states to follow and hence repetition of any given pair implies an infinite loop. Hence, to guarantee that a computation sequence will never halt we need to ensure that it must traverse a given transition arc *twice.* This requires sufficient time to execute $2^{n'} * m = 2^{l*(m-1)} * m$ instructions. Even with modest values of m and l this number is colossal and multiplying by a realistic time for the execution of a typical (or ideally, the slowest) instruction yields a total time in tens, hundreds or even thousands of years. The lesson to be learned from this exercise is that it is completely impracticable to justify *correctness* of a reasonably sized program by using test data; too many runs are needed and it takes too long. Such test runs can only find errors.

Exercises 9.2

1. Devise an FSM that will recognize an odd number if the number, in binary, is read from left to right.
2. Devise an FSM whose input alphabet is $\{a, b\}$ and which halts in an accepting state iff the input contains no two adjacent *a*s and no two adjacent *b*s.
3. Demonstrate the impossibility of constructing an FSM that will input strings over $\{a, b\}$ and accept only those having equal numbers of *a*s and *b*s.
4. Construct a (deterministic) FSM that accepts strings over $\{0, 1\}$ which are comprised of alternating 1s and 0s followed by alternating pairs of 1s and 0s.
5. Devise an FSM capable of recognizing real numbers written in the form

   $\pm dd^* . d^* E \pm dd$

   where $d \in \{0, 1, \ldots, 9\}$.

9.3 Regular algebra

The ways in which finite-state acceptors may be combined can be used to justify sets of axioms from which algebraic systems can be developed. The classic system, known as a *regular algebra*, is based on the three operations of union, closure and concatenation defined over sets of strings. Equations within the algebra are capable of specifying certain sets of strings, these sets being solutions of the equations (Section 9.3.1). Results emanating from the solutions then have direct application to aspects of language theory (Section 9.3.2).

9.3.1 Expressions and equations

Before giving the formal definition of the algebra we must introduce the two constants within the system; these are the sets of strings $\varnothing$ and $\{\Lambda\}$ and are denoted by 0 and 1 respectively. We justify their use by giving the FSAs in Figures 9.16(a) and (b).

Fig. 9.16

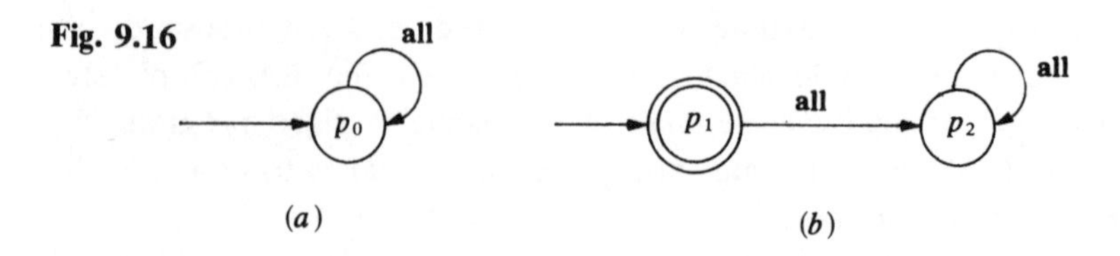

Definition. If X is a set of strings acceptable to an FSA then X is a *regular expression.* If X and Y denote the sets of strings acceptable to two specific machines then, by virtue of FSA constructions, $X \cup Y$ (henceforth written $X+Y$), X^* and XY are also regular expressions.

The allowable constructions of composite finite-state machines suggest the following axioms for the manipulation of arbitrary regular expressions A, B and C:

(1) $A+B=B+A$,
(2) $A+(B+C)=(A+B)+C$,
(3) $A+A=A$,
(4) $A+0=A$,
(5) $A(BC)=(AB)C$,
(6) $A1=A=1A$,
(7) $A0=0=0A$,
(8) $A(B+C)=AB+AC$,
(9) $(A+B)C=AC+BC$,
(10) $0^*=1$,
(11) $A^*=A+A^*$,
(12) $(A^*)^*=A^*$.

Such an algebraic system is called a *regular algebra.* //

A proper justification of the veracity of these axioms requires consideration of perfectly *general* cases and in consequence is difficult to describe properly. Nevertheless, to indicate the underlying ideas we give two examples based on two of the axioms applied to a *specific* regular expression.

Example 9.3.1

Let $A=\{ab^n : 0 \leq n\}$. This is regular since it is accepted by the machine M in Figure 9.17(a).

Fig. 9.17

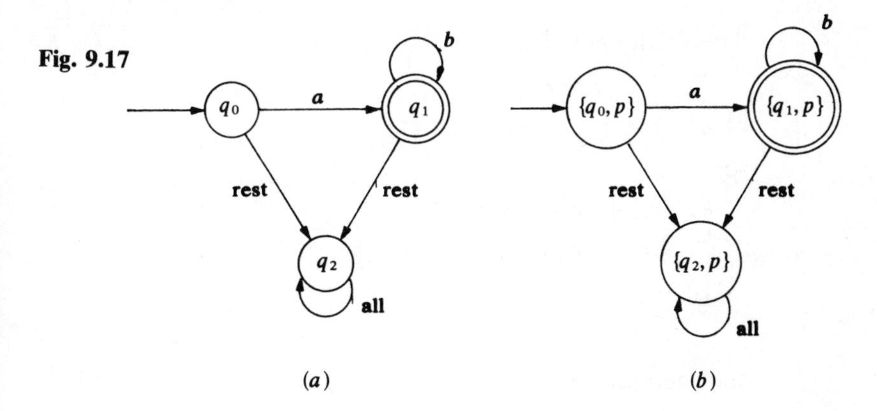

(a) (b)

Now using the machine given in Figure 9.16(a) which recognizes 0, and applying the relevant construction yields the machine in Figure 9.17(b) which is obviously equivalent to M. Hence $A = A + 0$. //

Example 9.3.2

Using A and 1 defined by the machines given above and the standard concatenation construction we have, in Figure 9.18(a), a non-deterministic FSA which will accept the strings in the regular expression $1A$. This machine is made deterministic (Figure 9.18(b)) and rationalized, resulting (Figure 9.18(c)) in a machine equivalent to M, thus justifying (in this case) part of axiom 6, namely $1A = A$. //

Having defined a regular algebra, how can it be used?

Suppose that A, B and C are given regular expressions (sets of strings) and that X and Y are unknown regular expressions such that the following equivalences hold:

$$X = AX + BY,$$

$$Y = XC + B.$$

Are there any solutions to these 'equations' and if so how do we find them? As we shall see shortly, regular equations do not have unique solutions, hence we first introduce the concept of approximation between regular expressions.

Definition. For regular expressions X and Y we define a relation '$\leqslant$' by $X \leqslant Y$ (X *approximates* Y) if $X + Y = Y$. //

Theorem. $\leqslant$ defined over regular expressions is an order relation.

Proof.

(i) Transitivity holds since

$$X \leqslant Y, \quad Y \leqslant Z$$

$$\Rightarrow X + Y = Y, \quad Y + Z = Z \quad (*)$$

so

$$\begin{aligned} X + Z &= X + (Y + Z) && \text{by } (*) \\ &= (X + Y) + Z && \text{associativity} \\ &= Y + Z && \text{by } (*) \\ &= Z && \text{by } (*) \end{aligned}$$

and therefore $X \leqslant Z$.

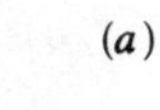

Fig. 9.18

(*a*)

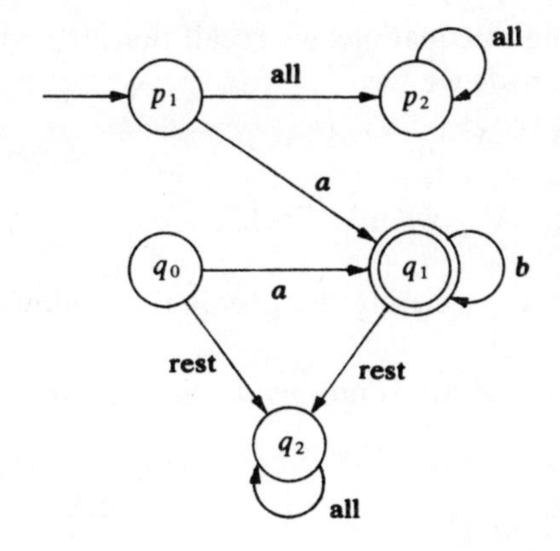

(*b*)

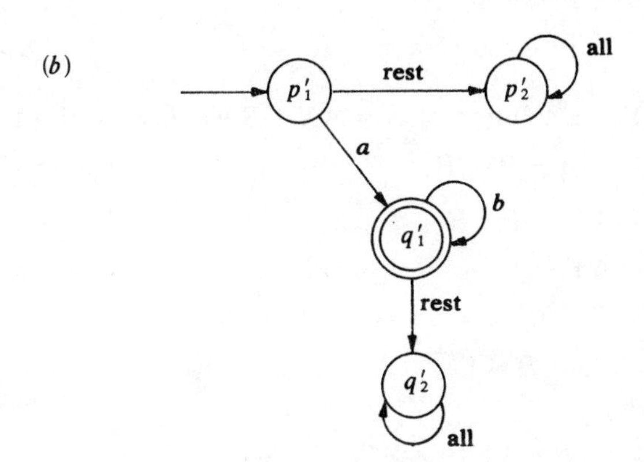

(*c*)

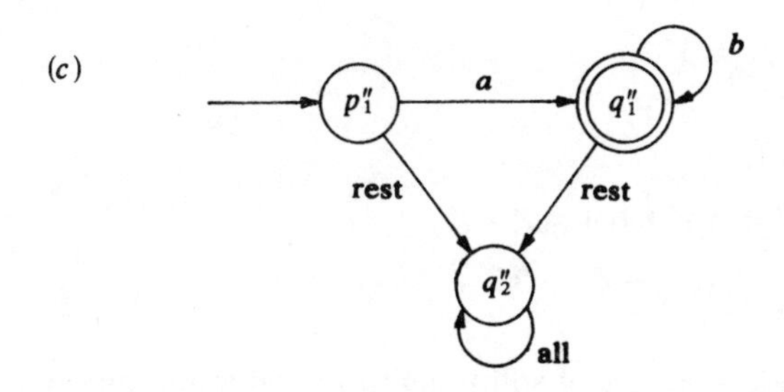

(ii) Antisymmetry holds since

$$X \leqslant Y, \quad Y \leqslant X$$

$$\Rightarrow Y = X + Y = Y + X = X.$$

(iii) Reflexivity follows from $X + X = X$. //

It also follows that $X \leqslant Y \Rightarrow ZX \leqslant ZY$, etc. for regular expressions X, Y and Z but details of the proofs are left as an exercise.

Before looking at equations we recall that, from the definition of the closure operation, *, we have

$$A^* = A^0 + A^1 + \ldots + A^n + \ldots$$

$$= \sum_{n=0}^{\infty} A^n \quad \text{where } A^0 = 1.$$

This is useful in appreciating the proofs that follow.

Lemma. If Y and Z are solutions of the regular equation $X = AX + B$ then,

(i) $B \leqslant Y$,

(ii) $AY \leqslant Y$, and

(iii) $Y + Z$ is a solution.

Proof.

(i) Since Y is a solution of $X = AX + B$ we have $Y = AY + B$, so

$$\begin{aligned} Y + B &= (AY + B) + B \\ &= AY + (B + B) \\ &= AY + B \\ &= Y \end{aligned}$$

and therefore $B \leqslant Y$.

(ii) Similarly,

$$\begin{aligned} Y + AY &= (B + AY) + AY \\ &= B + AY \\ &= Y \end{aligned}$$

hence $AY \leqslant Y$.

(iii) Trivially,

$$\begin{aligned} Y + Z &= AY + B + AZ + B \\ &= A(Y + Z) + B. \quad // \end{aligned}$$

So much for properties of solutions to the equation but are there any actual solutions?

Lemma. A^*B is a solution of $X = AX + B$.

Proof. By substitution we have

$$\begin{aligned} A(A^*B) + B &= (AA^* + A^0)B \\ &= \left(A \sum_{n=0}^{\infty} A^n + A^0\right) B \end{aligned}$$

$$= \left(\sum_{n=1}^{\infty} A^n + A^0 \right) B$$

$$= \left(\sum_{n=0}^{\infty} A^n \right) B$$

$$= A^*B. \quad //$$

Theorem. A^*B is the least solution (wrt $\leqslant$) of the regular equation $X = AX + B$.

Proof. Using the lemmas, A^*B is a solution and if Y were any other solution then $B \leqslant Y$, $AB \leqslant AY \leqslant Y$ so $AB \leqslant Y$, $A^nB \leqslant Y$ and, by adding inequalities for all $n \in \mathbb{N} \cup \{0\}$, $A^*B \leqslant Y$. Hence the result follows. //

Note that A^*B is a set of strings each of which is a valid solution of the equation; if $A = \{a\}$ and $B = \{b\}$ then all strings of the form a^nb are solutions.

Notice also that since A^*B is the least solution of $X = AX + B$ it is the least expression such that $X \mapsto AX + B$ coincides with the identity mapping; hence it is also called the *least fixed point* of the equation (or the *minimal fixed point*).

Having considered at length one particular equation we can now immediately obtain similar results for other related equations but in these cases proofs will be omitted.

Theorem. Given regular expressions A and B, the equation $X = AX + B$ and the pair of equations

$$X = YB,$$
$$Y = YA + 1$$

are equivalent in the sense that they have the same least solution and that solution is $X = A^*B$. //

Theorem. Given regular expressions A and B, the equation $X = XA + B$ and the pair of equations

$$X = BY,$$
$$Y = AY + 1$$

are equivalent. (In this case both systems have the least solution $X = BA^*$.) //

These two theorems allow us to transform right-linear forms ($X = AX + B$) into left-linear forms ($X = XA + B$) which correspond to right-

and left-recursion within regular grammars. The details of the correspondence are given in Section 9.3.2 but first we generalize the above results to systems of simultaneous linear equations.

Suppose that $X_1, \ldots, X_n$ are unknown regular expressions and that $A_{ij}(1 \leqslant i,j \leqslant n)$ and $B_1, \ldots, B_n$ are constant regular expressions such that

$$X_1 = X_1A_{11} + X_2A_{21} + \ldots + X_nA_{n1} + B_1,$$
$$\vdots$$
$$X_i = X_1A_{1i} + X_2A_{2i} + \ldots + X_nA_{ni} + B_i,$$
$$\vdots$$
$$X_n = X_1A_{1n} + X_2A_{2n} + \ldots + X_nA_{nn} + B_n.$$

We can now use the operations defined over regular expressions to induce operations on matrices of regular expressions by

$$(\mathbf{C}+\mathbf{D})_{ij} = \mathbf{C}_{ij} + \mathbf{D}_{ij},$$
$$(\mathbf{CD})_{ij} = \sum_k \mathbf{C}_{ik}\mathbf{D}_{kj},$$
$$\mathbf{C}^* = \sum_{n=0}^{\infty} \mathbf{C}^n,$$
$$\mathbf{C} \leqslant \mathbf{D} \quad \text{iff} \quad \mathbf{C}_{ij} \leqslant \mathbf{D}_{ij} \quad \text{for all } i, j,$$

where **C** and **D** are compatible matrices of regular expressions. Adopting the notation **X** to denote a 'regular matrix' we can represent the aforementioned equations by:

$$\mathbf{X} = \mathbf{XA} + \mathbf{B}.$$

Applying similar arguments to those used in the 1-equation situation we can show that this system of equations has $\mathbf{BA}^*$ as its minimal fixed point, as does $\mathbf{X} = \mathbf{BY}$ where $\mathbf{Y} = \mathbf{AY} + \mathbf{I}$ and **I** is the identity matrix. The corresponding results hold for a right-linear *system* for **X** but of special note is the application of the equivalence of *these* two systems to aid the removal of left-recursion in context-free grammars.

Explicitly, given $\mathbf{X} = \mathbf{XA} + \mathbf{B}$, a solution can be specified by

$$\mathbf{X}_i = (\mathbf{BY})_i = \sum_k \mathbf{B}_k \mathbf{Y}_{ki}$$

where

$$\mathbf{Y}_{ij} = (\mathbf{AY} + \mathbf{I})_{ij}$$
$$= \sum_k \mathbf{A}_{ik}\mathbf{Y}_{kj} + \mathbf{I}_{ij}.$$

An illustration of how this result transfers to CFGs can be seen in Example 8.4.4. The manipulations used in that example (although well-founded) are by analogy. We conclude this chapter by giving a formal link with the regular algebra of finite-state acceptors and regular grammars.

9.3.2 Regular grammar representations

Recall from Chapter 8 that a phrase structure grammar $G = (N, T, P, S)$ is called Chomsky type 3 or a (Λ-free) *regular grammar* if all the elements of P are of the form

$$A \to x$$

or

$$A \to xB \quad \text{where } x \in T, \text{ and } A, B \in N$$

or (if $\Lambda \in L(G)$)

$$S \to \Lambda \quad \text{in which case } S \text{ does not occur in any right hand side.}$$

The set of sentences generated by a regular grammar is called a *regular set* or a *regular language.* We are now in a position to justify the two uses of the same terminology and hence the use of regular algebra in grammar modifications.

The key result is in two parts and is given as a theorem with a constructive proof. Part (i) is straightforward since each transition is defined over the whole of its domain; part (ii) is more involved because the grammar (N, T, P, S), given $X \in N$ and $a \in T$, need not necessarily have productions of the form $X \to a$ or $X \to aY$ for some $Y \in N$.

Theorem. The sets accepted by finite-state automata are exactly those derivable from regular grammars.

Constructions.

(i) First consider the finite-state machine $M = (Q, \Sigma, t, q, F)$. From this we construct a regular grammar $G = (N, T, P, S)$. If $\Lambda \notin A(M)$ we can proceed as follows: let $N = Q$, $T = \Sigma$ and $S = q$, and construct P such that

$$P = \{X \to aY \text{ where } Y \in t(X, a)\}$$
$$\cup \{X \to a \text{ where } t(X, a) \cap F \neq \varnothing\}.$$

If $\Lambda \in A(M)$ we extend the construction by creating a new non-terminal $\bar{q}$, setting $S = \bar{q}$ and adding to P the productions

$S \to \Lambda$

and

$S \to aY \quad (\text{if } Y \in t(q, a))$

and

$S \to a \quad (\text{if } t(q, a) \cap F \neq \varnothing)$.

It is now easy to show for each $x \in \Sigma^*$ that $x \in A(M)$ iff $x \in L(G)$.

Trivially, if $q \in F$ then Λ is accepted by M and $S \to \Lambda$ is a production of G and vice versa. Now if also $a \in A(M)$ and $|a| = n$ so

$a = a_1 a_2 \dots a_n$ with each $a_i \in \Sigma$

then there is a path in the state graph of M as shown in Figure 9.19. Here each $A_i \in N$ (not necessarily all distinct) and $B \in F$,

Fig. 9.19

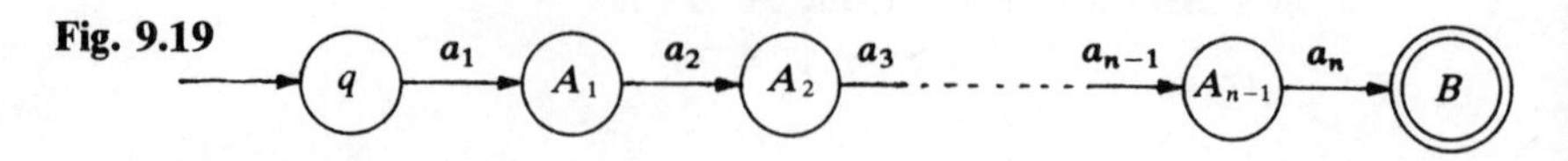

and by construction

$$\{S \to a_1 A_1,$$
$$A_1 \to a_2 A_2,$$
$$\vdots$$
$$A_{n-2} \to a_{n-1} A_{n-1},$$
$$A_{n-1} \to a_n \quad \} \subseteq P$$

(see Figure 9.20).

Fig. 9.20

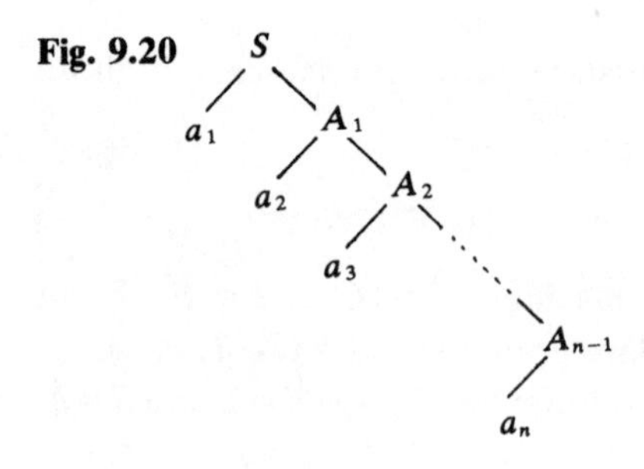

Thus $a \in L(G)$. Similarly, by reversing the argument $L(G) \subseteq A(M)$ and hence equality holds.

(ii) Now we must show that given a (Λ-free) regular grammar, the sentences of that grammar can be accepted by some finite-state automaton and that no other strings are accepted by this machine.

We take $G_1 = (N_1, T_1, P_1, S_1)$ and construct $M_1 = (Q_1, \Sigma_1, t_1, q_1, F_1)$ as follows.

Let $\Sigma_1 = T_1$ and $Q_1 = N_1 \cup \{\tau\} \cup \{\pi\}$ (where τ and π are special symbols not in N_1; τ represents a valid terminal string and π represents an error), $q_1 = S_1$ and $F_1 = \{\tau\}$. Also if $\Lambda \in L(G_1)$, i.e. $S_1 \to \Lambda$ is in P_1, then $F_1 = \{\tau, q_1\}$.

Finally,

$$\begin{aligned}
t_1 = &\{((X, a), Y)\text{: } X \to aY \text{ is in } P_1\} \\
&\cup\{((X, a), \tau)\text{: } X \to a \text{ is in } P_1\} \\
&\cup\{((\tau, a), \pi) \quad \text{for all } a \in \Sigma_1\} \\
&\cup\{((X, b), \pi)\text{: if } X \in N_1 \text{ and neither} \\
&\qquad\qquad X \to bY \text{ for any } Y \in N_1 \text{ nor} \\
&\qquad\qquad X \to b \text{ is in } P_1\} \\
&\cup\{((\pi, a), \pi) \quad \text{for all } a \in \Sigma_1\}.
\end{aligned}$$

Justification that this machine accepts $L(G_1)$ is left as an exercise. //

Example 9.3.3

Given the regular grammar $G = (\{A, B, C, D\}, \{x, y, z\}, P, A)$ where

$$\begin{aligned}
P = \{&A \to \Lambda|xB|yC, \\
&B \to zB|y|yC, \\
&C \to xD, \\
&D \to yD|x\}
\end{aligned}$$

using the prescribed construction yields the machine given in Figure 9.21. //

Fig. 9.21

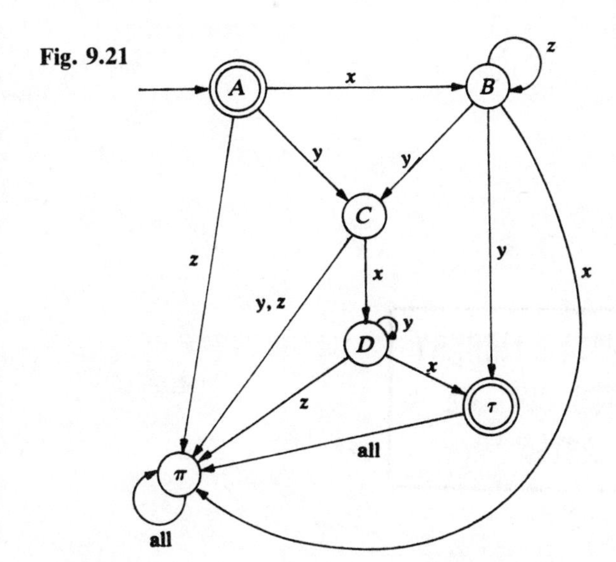

Exercises 9.3

1. Given regular expressions A, B, C and D, such that $A \leq C$ and $B \leq C$, show that
   (i) $A+D \leq C+D$,
   (ii) $AD \leq CD$,
   (iii) $DA \leq DC$,
   (iv) $A^* \leq C^*$,
   (v) $A+B \leq C$.
2. Investigate to what extent the results of Question 1 hold when $\mathbf{A}$, $\mathbf{B}$, $\mathbf{C}$ and $\mathbf{D}$ are $n \times n$ regular matrices.
3. Design a finite automaton for the language L defined by 11^*01^* (i.e. $L = \{w \in \{0, 1\}^* : w$ begins with a 1 and has exactly one 0$\}$). Write down the right-linear grammar determined by your automaton and the sequence of moves made in accepting the string 111011.

10 COMPUTER GEOMETRY

This chapter presents some geometrical techniques that are widely used in computer graphics and computer aided design software in a more rigorous and unified manner than is customary in many texts aimed at computer scientists. The approach is not to present a series of 'tricks of the trade' in some apparently random order, but to use the concepts developed in earlier chapters (particularly Chapters 1 to 3 and 5 and 6) to provide a framework for a discussion of selected topics including homogeneous coordinates and curves and surfaces. A full treatment of the four-dimensional representation of three-dimensional geometric data, so widely used in practice, is also given.

Before embarking on a discussion of these topics it will be helpful if the reader has some understanding of the difference between the terms 'topology' and 'geometry'; geometry is concerned with distances and angles, whereas topology is concerned with more general properties. For example, two spheres of different radii are topologically equivalent but not geometrically equivalent. Two objects are equal topologically if one can be obtained from the other by bending and stretching without tearing (to be precise, by using a continuous mapping with continuous inverse) whereas in geometry, equivalent objects have to be identical in all respects apart from their position and orientation in space. Graph theory is a branch of topology, since vertices do not have the attribute of position in space and the topology of a graph is the edge relation.

It is usually convenient when representing geometrical objects on a computer to store two distinct but related sets of data for the object, the *topological* data and the *geometrical* data. As an example a wide class of objects may be usefully represented by wire frame models. Figure 10.1 shows a wire frame representation of a finite cylinder. The topology of this type of model may be regarded as a graph and represented as a linked list structure. The geometrical data may simply be a list of vectors defining the vertex positions in $\mathbb{R}^3$. The purpose in keeping the geometry and topology separate is that we may wish to transform our model in

Fig. 10.1

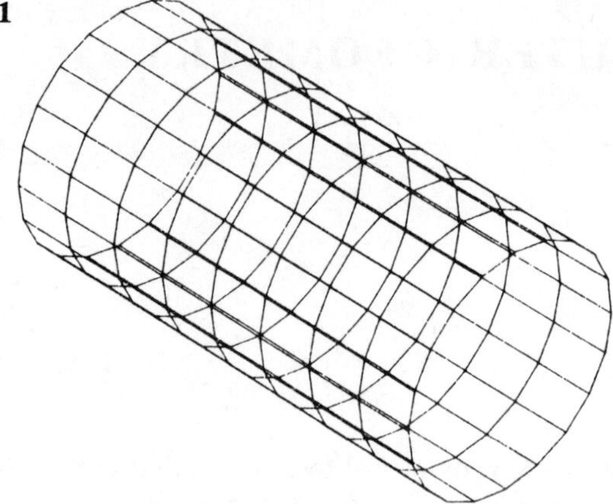

some way, for example make it physically smaller, or move it to some new position and orientation in space. Such transformations do not alter the topology of the model and we need simply transform the geometric data set in the appropriate way. The following two sections deal with coordinate systems for representation of geometric data and with some of the more useful sets of transformations that may be applied to geometric data sets.

10.1 Coordinate systems for subsets of $\mathbb{R}^n$

Generally, a coordinate system or parameterization for a set $S \subseteq \mathbb{R}^n$ is an identification of each point in S by a unique ordered set of real numbers $(\xi_1, \ldots, \xi_q) \in \mathbb{R}^q$ for some $q \in \mathbb{N}$. In the terminology of mappings, a coordinate system for S may be defined as a continuous bijection $\Phi: S \to \mathcal{S}$ where $\mathcal{S} \subseteq \mathbb{R}^q$. Any transformation or other calculation to be performed on S is then done in terms of the coordinates $(\xi_1, \ldots, \xi_q)$. The particular mapping chosen for a given problem will often be determined by the geometry of the space S. It may be shown that for a fixed S, q is constant for all coordinate systems and is called the *dimension* of S. Unfortunately, one of our examples, the homogeneous coordinate systems, requires $\mathcal{S}$ to be a more general space and in addition there are many useful subsets for which no such mappings exist (for example, the circle and the sphere). To make good these deficiencies would involve us in a long excursion into topology; instead we hope that the treatment given below provides a good understanding despite the lack of a completely general working definition.

Our interest lies chiefly in spaces of dimension three or less; for example, curves are one dimensional and surfaces have dimension two. The following examples present some of the more common coordinate systems for $\mathbb{R}^2$ and $\mathbb{R}^3$ but where it is easy to keep the discussion completely general we do so.

Example 10.1.1

Cartesian coordinates for $\mathbb{R}^n$ ($n \in \mathbb{N}$). If $B = \{\mathbf{e}_1, \ldots, \mathbf{e}_n\}$ is a basis for $\mathbb{R}^n$ then each $\mathbf{x} \in \mathbb{R}^n$ may be written uniquely in the form

$$\mathbf{x} = \sum_{i=1}^{n} a_i \mathbf{e}_i \quad \text{for } a_i \in \mathbb{R}, \quad 1 \leqslant i \leqslant n.$$

The mapping

$$\Phi : \mathbb{R}^n \to \mathbb{R}^n$$

given by

$$\Phi(\mathbf{x}) = (a_1, \ldots, a_n)$$

defines a coordinate system for $\mathbb{R}^n$. If B is orthonormal the corresponding coordinates are called *Cartesian coordinates* for $\mathbb{R}^n$. In $\mathbb{R}^2$ and $\mathbb{R}^3$ it is usual to interpret an orthonormal basis geometrically as right-handed in the sense of Section 5.4. //

Our next example for $\mathbb{R}^n$ is conceptually and mathematically more difficult. Some preliminary discussion and notation is necessary before the coordinate systems can be described.

We begin by defining a relation $\sim$ on $\mathbb{R}^{n+1}$ in the following way:

$$\mathbf{x} \sim \mathbf{y} \quad \text{if } \mathbf{x} = \alpha \mathbf{y} \quad \text{for some } \alpha \in \mathbb{R}\backslash\{0\}.$$

The important properties of this relation are summarized in the following proposition, the proof of which is left as an exercise.

Proposition. $\sim$ is an equivalence relation on $\mathbb{R}^{n+1}$. //

The equivalence classes (or partition of $\mathbb{R}^{n+1}$) determined by $\sim$ are

$$\mathbb{P}^n = \{L\backslash\{\mathbf{0}\} : L \text{ is a 1-dimensional vector subspace of } \mathbb{R}^{n+1} \text{ and } \mathbf{0} \text{ is the zero vector of } \mathbb{R}^{n+1}\}$$

and

$$\{\mathbf{0}\}.$$

Clearly, $\mathbb{R}^{n+1} = \mathbb{P}^n \cup \{\mathbf{0}\}$.

An important subset of $\mathbb{P}^n$ may be identified.

Definition. If $\mathbf{x} \in \mathbb{R}^{n+1}$ has the form $(x_1, x_2, \ldots, x_n, 0)$ where $x_i \neq 0$ for some $1 \leqslant i \leqslant n$ then $[\mathbf{x}] \in \mathbb{P}^n$ is called a *point at infinity.* We denote the set of all points at infinity in $\mathbb{P}^n$ by L_∞^n and define $\mathbb{H}^n$ to be $\mathbb{P}^n \backslash L_\infty^n$. //

The following result is crucial and establishes the coordinate mapping for $\mathbb{R}^n$.

Proposition. There exists a bijection $Q^n : \mathbb{H}^n \to \mathbb{R}^n$.

Proof. We define a relation $Q^n : \mathbb{H}^n \to \mathbb{R}^n$ by

$$Q^n([(x_1, \ldots, x_{n+1})]) = \frac{1}{x_{n+1}}(x_1, \ldots, x_n).$$

The first point to note is that the right hand side is always defined as L^n_∞ has been excluded from the domain of Q^n, hence $\mathscr{D}(Q^n) = \mathbb{H}^n$. Three things need to be established:

(i) Q^n is a mapping on $\mathbb{H}^n$,

(ii) Q^n is injective and

(iii) Q^n is surjective.

Q^n is a mapping if

$$(x_1, \ldots, x_{n+1}) \sim (y_1, \ldots, y_{n+1}) \Rightarrow Q^n([(x_1, \ldots, x_{n+1})]) = Q^n([(y_1, \ldots, y_{n+1})]).$$

Well

$$(x_1, \ldots, x_{n+1}) \sim (y_1, \ldots, y_{n+1}) \Rightarrow \text{there exists } \alpha \in \mathbb{R}\backslash\{0\} \text{ with } x_i = \alpha y_i \quad \text{for all } 1 \leqslant i \leqslant n+1.$$

Now

$$\begin{aligned} Q^n([(x_1, \ldots, x_{n+1})]) &= Q^n([(\alpha y_1, \ldots, \alpha y_{n+1})]) \\ &= \frac{1}{\alpha y_{n+1}}(\alpha y_1, \ldots, \alpha y_n) \\ &= \frac{1}{y_{n+1}}(y_1, \ldots, y_n) \\ &= Q^n([(y_1, \ldots, y_{n+1})]) \quad \text{by definition of } Q^n. \end{aligned}$$

Therefore Q^n is a mapping on $\mathbb{H}^n$.

(ii) Q^n is injective if $Q^n([\mathbf{x}]) = Q^n([\mathbf{y}]) \Rightarrow \mathbf{x} \sim \mathbf{y}$.

Writing $\mathbf{x} = (x_1, \ldots, x_n, x_{n+1})$ and $\mathbf{y} = (y_1, \ldots, y_n, y_{n+1})$,

$$\begin{aligned} Q^n([\mathbf{x}]) = Q^n([\mathbf{y}]) &\Rightarrow \frac{1}{x_{n+1}}(x_1, \ldots, x_n) = \frac{1}{y_{n+1}}(y_1, \ldots, y_n) \\ &\Rightarrow x_i = (x_{n+1}/y_{n+1})y_i \\ &\Rightarrow \mathbf{x} \sim \mathbf{y} \quad \text{with} \quad \alpha = x_{n+1}/y_{n+1}. \end{aligned}$$

Hence Q^n is injective.

(iii) Q^n is surjective if for all $\mathbf{x} = (x_1, \ldots, x_n) \in \mathbb{R}^n$ there is an $\mathbf{x}^* \in \mathbb{R}^{n+1}$ such that $Q^n([\mathbf{x}^*]) = \mathbf{x}$.

If we define $\mathbf{x}^* = (x_1, \ldots, x_n, 1) \in \mathbb{R}^{n+1}$ then clearly

$$Q^n([\mathbf{x}^*]) = \mathbf{x}$$

and Q^n is surjective. //

Example 10.1.2

Homogeneous coordinates for $\mathbb{R}^n$ $(n \in \mathbb{N})$. The *homogeneous coordinates* for $\mathbb{R}^n$ are defined by mapping $\mathbb{R}^n$ onto $\mathbb{H}^n$ using the inverse of Q^n, so that

$$\Phi : \mathbb{R}^n \to \mathbb{H}^n \quad \text{where } \Phi = (Q^n)^{-1}.$$

In other words the homogeneous coordinates of the point $(x_1, \ldots, x_n) \in \mathbb{R}^n$ are the $(n+1)$ tuples of the equivalence class

$$[(x_1, \ldots, x_n, 1)] = \{(px_1, \ldots, px_n, p) : p \neq 0\}.$$

A particular element of $[(x_1, \ldots, x_n, 1)]$ is called a *homogeneous representation* for $(x_1, \ldots, x_n)$.

It is often convenient to represent geometrical data on a computer in homogeneous form (see Section 10.3). To obtain the *physical* coordinates from a homogeneous representation we perform the mapping

$$(x_1, \ldots, x_{n+1}) \mapsto \frac{1}{x_{n+1}}(x_1, \ldots, x_n).$$

The proposition above assures us that *all* homogeneous representations of a given physical point produce the *same* physical coordinates.

It is illuminating to consider the geometrical interpretation of homogeneous coordinates for $\mathbb{R}$ with the aid of pictures. The higher dimensional cases have a similar geometrical interpretation but these are not easily represented by pictures. Elements of $\mathbb{P}^1$ are infinite lines in $\mathbb{R}^2$ through the origin but with the origin removed; see Figure 10.2. Clearly, the only point at infinity is the x_1 axis with the origin removed. Figure 10.3 clarifies the concept of a point at infinity in $\mathbb{P}^1$. The points

Fig. 10.2. Geometrical interpretation of $\mathbb{P}^1$.

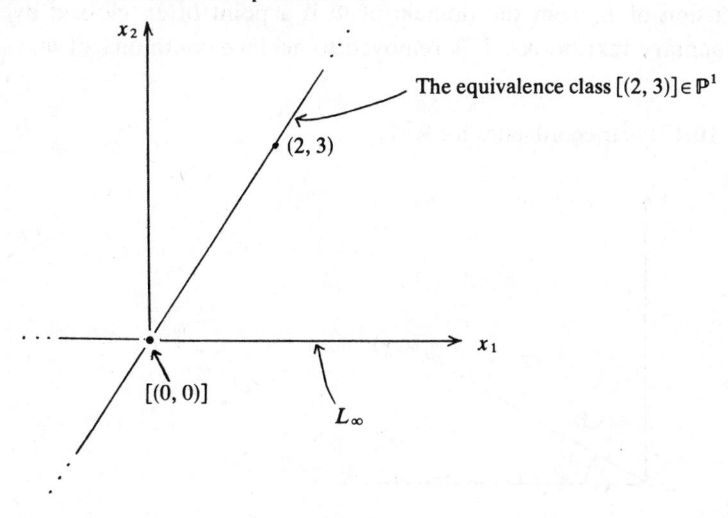

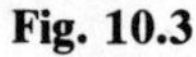
Fig. 10.3

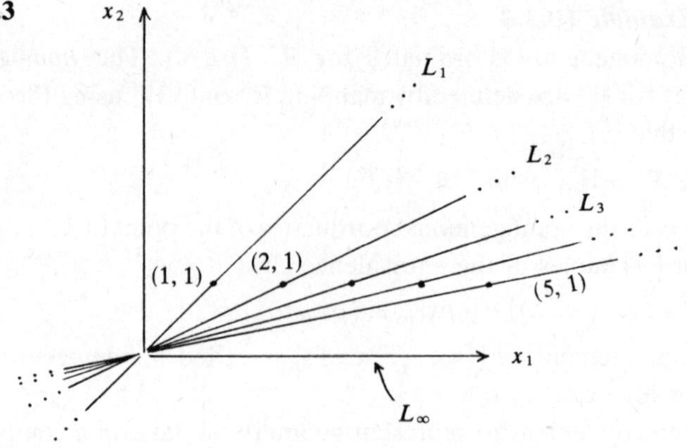

on the line $L_n = [(n, 1)]$ are the homogeneous representations of $n \in \mathbb{R}$. From Figure 10.3 we observe that as $n \to \infty$ the line L_n tends to the point at infinity L_∞. //

Example 10.1.3

Polar coordinates for $\mathbb{R}^2$. Let $L \subseteq \mathbb{R}^2$ be the semi-infinite line $L = \{(x, 0): x \geqslant 0\}$ and $(x, y) \in \mathbb{R}^2 \backslash L$ then, if $\Phi: \mathbb{R}^2 \backslash L \to]0, \infty[\times]0, 2\pi[$ is defined by

$$\Phi(x, y) = (r, \theta) \quad \text{where } r = (x^2 + y^2)^{1/2} \quad \text{and } \theta = \tan^{-1}(y/x),$$

then Φ defines a set of *polar coordinates* on $\mathbb{R}^2$. The inverse of Φ is given by

$$\Phi^{-1}(r, \theta) = (r \cos \theta, r \sin \theta).$$

See Figure 10.4 for the geometrical interpretation of r and θ. The exclusion of L from the domain of Φ is a point often glossed over in elementary text books. L is removed to achieve continuity of Φ.

Fig. 10.4. Polar coordinates for $\mathbb{R}^2 \backslash L$.

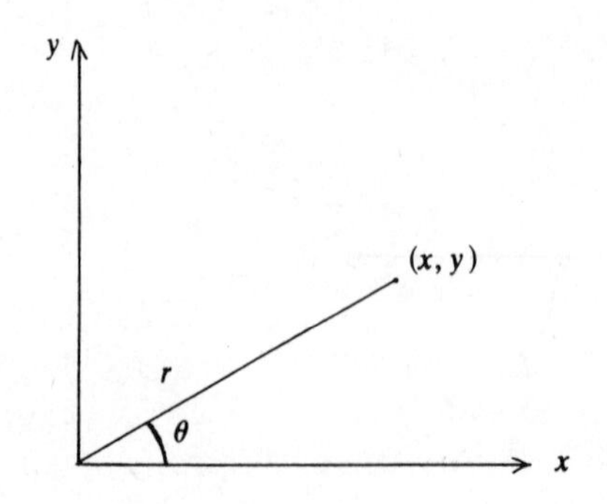

Example 10.1.4

Cylindrical polar coordinates for $\mathbb{R}^3$. If $P \subset \mathbb{R}^3$ is the semi-infinite plane $P = \{x, 0, z): x \geq 0\}$ and $(x, y, z) \in \mathbb{R}^3 \backslash P$ then the mapping $\Phi: \mathbb{R}^3 \backslash P \to]0, \infty[\times]0, 2\pi[\times \mathbb{R}$ defined by

$$\Phi(x, y, z) = (\rho, \phi, z) \quad \text{where } \rho = (x^2 + y^2)^{1/2}$$
$$\text{and } \phi = \tan^{-1}(y/x)$$

defines a set of *cylindrical polar* coordinates on $\mathbb{R}^3 \backslash P$. The inverse mapping is given by

$$\Phi^{-1}(\rho, \phi, z) = (\rho \cos \phi, \rho \sin \phi, z).$$

Figure 10.5 depicts the quantities ρ, ϕ and z relative to the Cartesian axis system. P is excluded from the domain of Φ to achieve continuity. //

Fig. 10.5. Polar coordinates for $\mathbb{R}^3 \backslash P$.

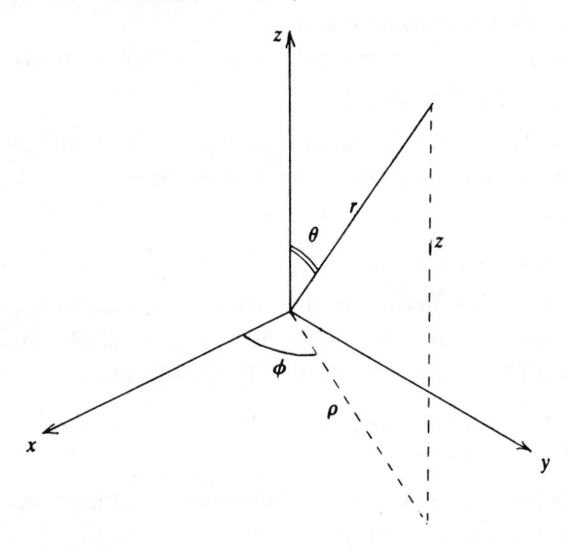

Example 10.1.5

Spherical polar coordinates on $\mathbb{R}^3 \backslash P$, where P is as defined in Example 10.1.4. The *spherical polar coordinates* are defined by the mapping

$$\Phi: \mathbb{R}^3 \backslash P \to]0, \infty[\times [0, \pi] \times]0, 2\pi[\quad \text{where } \Phi(x, y, z) = (r, \theta, \phi)$$

and $r = (x^2 + y^2 + z^2)^{1/2}$, $\theta = \cos^{-1}(z/r)$ and $\phi = \tan^{-1}(y/x)$. See Figure 10.5.

10.2 Transformations

One of the most frequently performed operations on geometric data is transformation. For example rotation of three-dimensional space

$\mathbb{R}^3$ enables a three-dimensional geometric object to be viewed from any vantage point. This may be useful to check the geometric integrity of the object by inspection, or to examine a particular feature in detail, or for one of many other possible reasons. Some approaches to computer based component design are based on libraries of basic geometric 'building blocks' from which more complex geometries may be synthesized by transformation, followed by some set-theoretic operation with further objects similarly constructed. The algorithms for this type of geometric synthesis are highly specialized and not discussed in this book, but we mention in passing that a generalized form of Euler's theorem for graphs may be used to check the topology of the object under construction. Transformations are more widely used and some of the more common ones and their linear representations are now described.

10.2.1 Some useful transformations of $\mathbb{R}^2$

A *translation* of $\mathbb{R}^2$ is a mapping $T:\mathbb{R}^2 \to \mathbb{R}^2$ of the form

$$T\mathbf{r} = \mathbf{r} + \mathbf{r}_0 \quad \text{for all } \mathbf{r} \in \mathbb{R}^2$$

where $\mathbf{r}_0 \in \mathbb{R}^2$ is a fixed vector. Translations displace the whole plane by a fixed vector and may be written in component form as

$$\begin{aligned} T(x, y) &= (x, y) + (x_0, y_0) \\ &= (x + x_0, y + y_0). \end{aligned}$$

We denote the set of all translations of $\mathbb{R}^2$ by $T(2)$ and define a product law in $T(2)$ by composition of mappings. If T_1 is a translation by $\mathbf{r}_1$ and T_2 is a translation by $\mathbf{r}_2$ then the product $T_2 \circ T_1$ is given by

$$\begin{aligned} T_2 \circ T_1\mathbf{r} &= T_2(T_1\mathbf{r}) \\ &= \mathbf{r} + \mathbf{r}_1 + \mathbf{r}_2. \end{aligned}$$

Clearly, $T_2 \circ T_1$ is a translation by $\mathbf{r}_1 + \mathbf{r}_2$ and hence $\circ$ is a binary operation on $T(2)$. In fact much more is true.

Proposition. $(T(2), \circ)$ is a commutative group of non-linear transformations of $\mathbb{R}^2$.

Proof. We have already shown closure under $\circ$. The group axioms follow from the group structure of $(\mathbb{R}^2, +)$; for example, if T_i: $1 \leqslant i \leqslant 3$ translates by $\mathbf{r}_i$, then

$$\begin{aligned} (T_3 \circ (T_2 \circ T_1))\mathbf{r} &= T_3((T_2 \circ T_1)\mathbf{r}) \\ &= T_3(\mathbf{r} + \mathbf{r}_1 + \mathbf{r}_2) \\ &= \mathbf{r} + \mathbf{r}_1 + \mathbf{r}_2 + \mathbf{r}_3 \\ &= (\mathbf{r} + \mathbf{r}_1) + (\mathbf{r}_2 + \mathbf{r}_3) \end{aligned}$$

$$= (T_3 \circ T_2)(T_1\mathbf{r})$$
$$= ((T_3 \circ T_2) \circ T_1)\mathbf{r}$$

so that $(T(2), \circ)$ is associative. The identity is the translation by $(0, 0)$, and the inverse of

$$T\mathbf{r} = \mathbf{r} + \mathbf{r}_0$$

is the translation

$$T'\mathbf{r} = \mathbf{r} - \mathbf{r}_0$$

and so

$$T \circ T'\mathbf{r} = T' \circ T\mathbf{r} = \mathbf{r} \quad \text{for all } \mathbf{r} \in \mathbb{R}^2.$$

Commutativity also holds since

$$(T_2 \circ T_1)\mathbf{r} = \mathbf{r} + \mathbf{r}_1 + \mathbf{r}_2$$
$$= \mathbf{r} + \mathbf{r}_2 + \mathbf{r}_1$$
$$= (T_1 \circ T_2)\mathbf{r}.$$

In general if $T \in T(2)$ then

$$T(0, 0) \neq (0, 0)$$

so that the origin is moved and hence translations are non-linear. //

In practice this means that any number of translations may be applied, *without regard to order*, to a geometric object with the same resulting position in the plane. In addition, the non-linear nature of the elements of $T(2)$ means that we may not implement $T(2)$ by elements of $\mathcal{M}(2, \mathbb{R})$; or in other words the equation

$$\begin{pmatrix} a_{11} & a_{12} \\ a_{21} & a_{22} \end{pmatrix}\begin{pmatrix} x \\ y \end{pmatrix} = \begin{pmatrix} x + x_0 \\ y + y_0 \end{pmatrix} \quad \text{for all } (x, y) \in \mathbb{R}^2$$

has no solution. The consequences of this will be discussed later.

We describe the rotation transformations of $\mathbb{R}^2$ with the aid of Figure 10.6. Intuitively, a rotation through an angle θ, denoted $W(\theta)$, maps the point (x, y) to the point (x', y') where

$$\|\mathbf{r}\| = \|\mathbf{r}'\|$$
$$= r \quad \text{say.}$$

From Figure 10.6 we have

$$x = r \cos \theta_1, \quad y = r \sin \theta_1,$$

and

$$x' = r \cos (\theta + \theta_1), \quad y' = r \sin (\theta + \theta_1).$$

Applying the compound angle formulae for cos and sin we obtain

$$x' = r(\cos \theta \cos \theta_1 - \sin \theta \sin \theta_1)$$
$$= x \cos \theta - y \sin \theta$$

Fig. 10.6

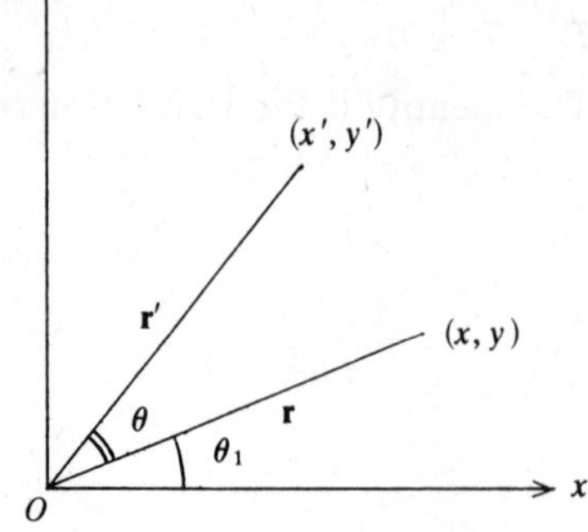

and, similarly,

$$y' = x \sin\theta + y \cos\theta$$

so that

$$W(\theta)(x, y) = (x\cos\theta - y\sin\theta,\ x\sin\theta + y\cos\theta)$$

which may be written in matrix form as

$$W(\theta)(x, y) = \begin{pmatrix} \cos\theta & -\sin\theta \\ \sin\theta & \cos\theta \end{pmatrix}\begin{pmatrix} x \\ y \end{pmatrix}$$

and $W(\theta)$ corresponds geometrically to a rotation about O.

The next proposition follows immediately from this representation in $\mathcal{M}(2, \mathbb{R})$ and summarizes the important properties of rotations from our point of view.

Proposition.

(*a*) The transformation $W(\theta)$ is
 (i) linear,
 (ii) orthogonal with $\det W(\theta) = 1$

(*b*) The set $\{W(\theta): 0 \leq \theta < 2\pi\}$ is a commutative group under composition of mappings (matrix multiplication). //

The implications of part (*a*) from a practical point of view are that to compute the inverse of a rotation one merely has to transpose its matrix, no elaborate inversion procedure need be invoked. Part (*b*) means that any number of rotations may be applied to a geometric data set in *any* order with equal results. In fact the converse of (*a*) is true; in particular if $W \in \mathcal{M}(2, \mathbb{R})$ and satisfies (i) and (ii) then W is a rotation transformation. In other words we can define a *rotation* of $\mathbb{R}^2$ to be an element of the group $SO(2)$.

Although $T(2)$ and $SO(2)$ are commutative groups, the transformations of $T(2)$ do not commute with those of $SO(2)$ for if $W \in SO(2)$

and $T \in T(2)$ translates by $\mathbf{r}_0$ we have

$$TW\mathbf{r} = W\mathbf{r} + \mathbf{r}_0$$

and

$$\begin{aligned} WT\mathbf{r} &= W(\mathbf{r} + \mathbf{r}_0) \\ &= W\mathbf{r} + W\mathbf{r}_0 \\ &\neq TW\mathbf{r} \quad \text{unless } \mathbf{r}_0 = (0, 0). \end{aligned}$$

In practice this means that consideration must be given to order when both translation and rotation transformations are applied to a given geometrical object. The non-commutativity is illustrated graphically in Figure 10.7 where the results of applying $T \circ W(\pi/2)$ and $W(\pi/2) \circ T$ to a semicircular shape are shown. Referring to Figure 10.7, if we imagine that the area

$$\{(x, y): 0 \leqslant x \leqslant a, 0 \leqslant y \leqslant b\}$$

Fig. 10.7

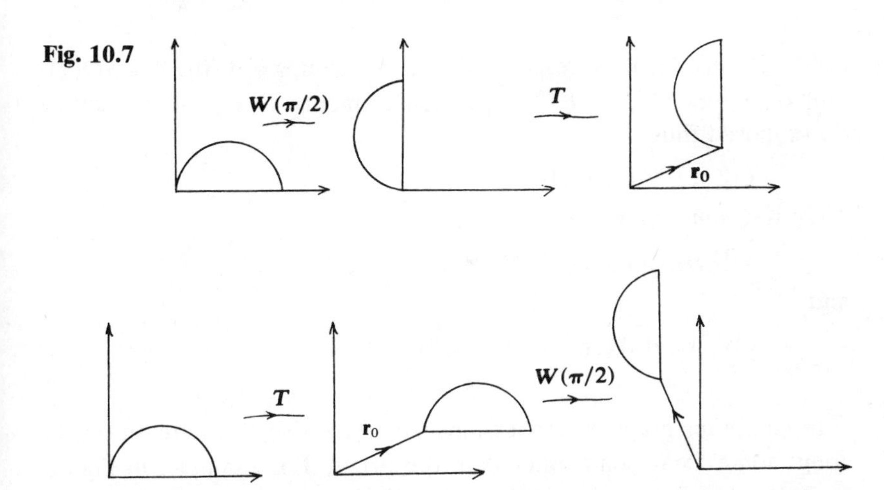

corresponds to the screen of a graphics display terminal, then the result of applying $W(\pi/2) \circ T$ to the semicircle would result in a blank screen!

The groups $T(2)$ and $SO(2)$ may be combined to form a third set, $E(2)$, of transformations of the plane, called the *Euclidean* transformations. Elements $U \in E(2)$ have the form

$$U\mathbf{r} = W\mathbf{r} + \mathbf{r}_0 \quad \text{for all } \mathbf{r} \in \mathbb{R}^2$$

where $W \in SO(2)$ and $\mathbf{r}_0 \in \mathbb{R}^2$ is a fixed vector. These transformations are often written as pairs $(W, \mathbf{r}_0)$. If $(W_1, \mathbf{r}_1) \in E(2)$ and $(W_2, \mathbf{r}_2) \in E(2)$ then

$$\begin{aligned} (W_2, \mathbf{r}_2) \circ (W_1, \mathbf{r}_1)\mathbf{r} &= (W_2, \mathbf{r}_2)(W_1\mathbf{r} + \mathbf{r}_1) \\ &= W_2(W_1\mathbf{r} + \mathbf{r}_1) + \mathbf{r}_2 \\ &= (W_2 W_1, W_2\mathbf{r}_1 + \mathbf{r}_2)\mathbf{r} \end{aligned}$$

so that

$$(W_2, \mathbf{r}_2)\circ(W_1, \mathbf{r}_1) = (W_2 W_1,\ W_2\mathbf{r}_1 + \mathbf{r}_2) \in E(2)$$

and $E(2)$ is closed under composition. In fact the following is true.

Proposition. $E(2)$ is a group under composition.

Sketch proof. We have shown closure. Associativity follows by repeated application of the product law above. If $I \in GL(2, \mathbb{R})$ is the identity element then

$$(I, \mathbf{0})\circ(W, \mathbf{r}_0) = (W, \mathbf{r}_0)\circ(I, \mathbf{0}) = (W, \mathbf{r}_0) \quad \text{for all } (W, \mathbf{r}_0) \in E(2),$$

hence $(I, \mathbf{0})$ is the identity of $E(2)$. The inverse $(W, \mathbf{r}_0)^{-1}$ of $(W, \mathbf{r}_0)$ is $(W^T, -W^T\mathbf{r}_0)$ for $(W^T, -W^T\mathbf{r}_0) \in E(2)$ and

$$\begin{aligned}(W, \mathbf{r}_0)\circ(W^T, -W^T\mathbf{r}_0) &= (I, \mathbf{0}) \\ &= (W^T, -W^T\mathbf{r}_0)\circ(W, \mathbf{r}_0). \quad /\!/\end{aligned}$$

$SO(2)$ and $T(2)$ are respectively the subgroups $\{(W, \mathbf{0})\colon W \in SO(2)\}$ and $\{(I, \mathbf{r}_0)\colon \mathbf{r}_0 \in \mathbb{R}^2\}$ of $E(2)$ and each element $(W, \mathbf{r}_0) \in E(2)$ may be decomposed thus

$$(W, \mathbf{r}_0) = (I, \mathbf{r}_0)\circ(W, \mathbf{0}).$$

$E(2)$ is a *non*-commutative group, for

$$(W_2, \mathbf{r}_2)\circ(W_1, \mathbf{r}_1) = (W_2 W_1,\ W_2\mathbf{r}_1 + \mathbf{r}_2)$$

and

$$\begin{aligned}(W_1, \mathbf{r}_1)\circ(W_2, \mathbf{r}_2) &= (W_1 W_2,\ W_1\mathbf{r}_2 + \mathbf{r}_1) \\ &\neq (W_2, \mathbf{r}_2)\circ(W_1, \mathbf{r}_1).\end{aligned}$$

In computer graphics applications we often wish to rotate an object about some fixed point other than the origin. For example, in Figure 10.8 the semicircle is rotated through $\pi/2$ about the point P. A common error is to try and perform this transformation by applying $W(\pi/2)$; however, this will produce the result illustrated in Figure 10.9. We illustrate the sequence of transformations for rotation about a fixed point

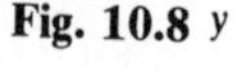

Fig. 10.8

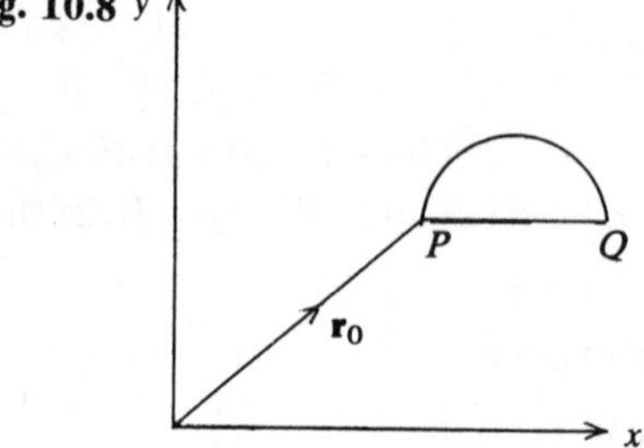

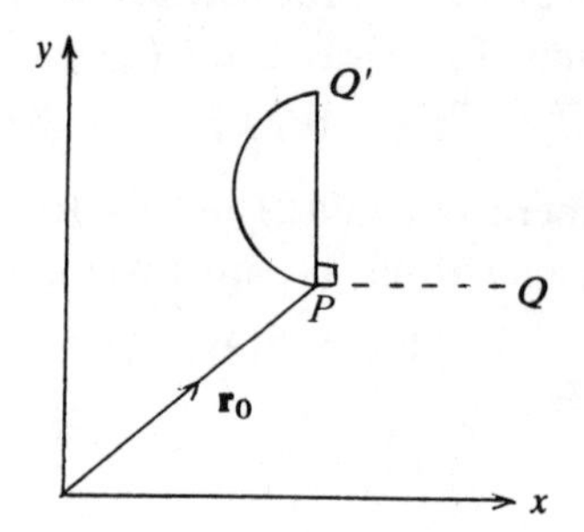

Fig. 10.9

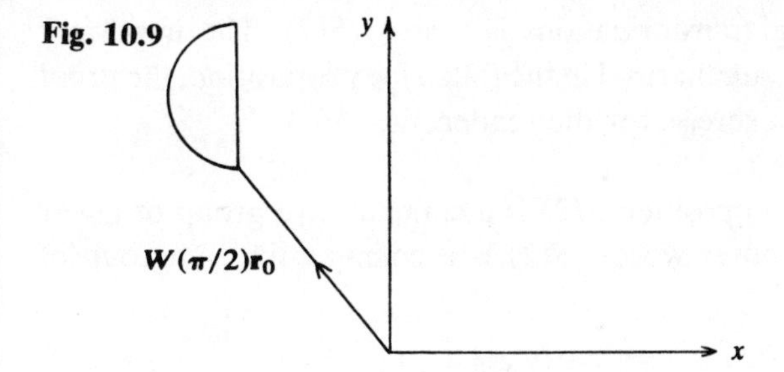

$\mathbf{r}_0 \in \mathbb{R}^2$ in Figure 10.10 using the notation of $E(2)$. The required transformation is thus

$$(I, \mathbf{r}_0)\circ(W(\pi/2), \mathbf{0})\circ(I, -\mathbf{r}_0)$$

Fig. 10.10

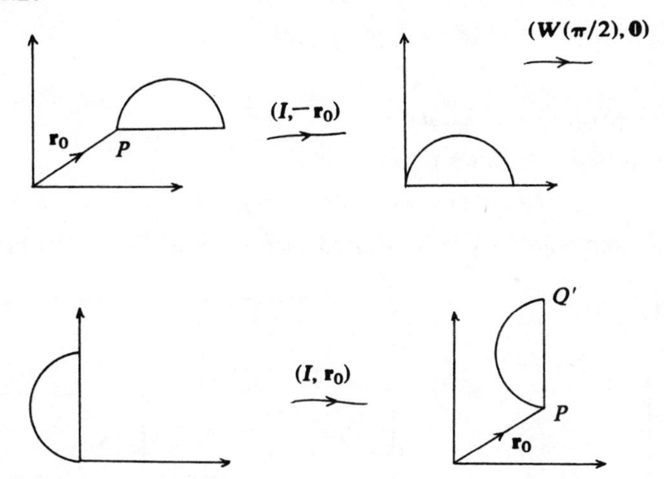

which may be simplified using the product rule of $E(2)$ to

$$(W(\pi/2), -W(\pi/2)\mathbf{r}_0+\mathbf{r}_0).$$

It is more efficient to apply a composed transformation to a geometric data set than to apply an equivalent series of individual transformations. $E(2)$ enables an object to be positioned at any point in the plane in any required orientation. A further set of useful operations on the plane comprises the scale transformations. A *scale* transformation of $\mathbb{R}^2$ is a mapping $S:\mathbb{R}^2\to\mathbb{R}^2$ of the form

$$S(x, y) = (\lambda x, \mu y) \quad \text{where } \lambda,\mu > 0.$$

In matrix form

$$S(x, y) = \begin{pmatrix} \lambda & 0 \\ 0 & \mu \end{pmatrix}\begin{pmatrix} x \\ y \end{pmatrix}.$$

The set of all scaling transformations is written $S(2)$. The important properties of $S(2)$ are summarized in the following proposition, the proof of which is left as an exercise for the reader.

Proposition. Under composition $S(2)$ is a commutative group of linear transformations. (In other words, $S(2)$ is a commutative subgroup of $GL(2, \mathbb{R})$.) //

Scale transformations do not commute with translations for if T is a translation by $\mathbf{r}_0$ and

$$S(x, y) = (\lambda x, \mu y)$$

then

$$\begin{aligned} S \circ T(x, y) &= S(x + x_0, y + y_0) \\ &= (\lambda(x + x_0), \mu(y + y_0)) \end{aligned}$$

and

$$\begin{aligned} T \circ S(x, y) &= T(\lambda x, \mu y) \\ &= (\lambda x + x_0, \mu y + y_0) \\ &\neq S \circ T(x, y) \quad \text{unless } \lambda = \mu = 1. \end{aligned}$$

This non-commutativity is illustrated pictorially in Figure 10.11 using

Fig. 10.11

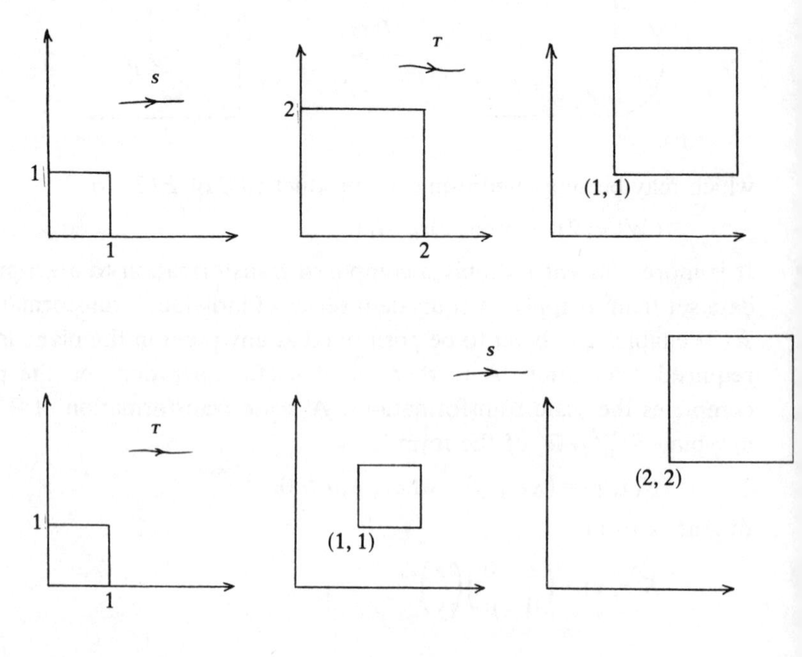

the scaling

$$S(x, y) = (2x, 2y)$$

and the translation

$$T(x, y) = (x+1, y+1)$$

on a unit square object at the origin. Similarly, we may show that scale transformations do not commute with rotations. The proof of this and the construction of a suitable pictorial demonstration is left as an exercise.

In practice these facts mean that if scaling is to be combined with either translation or rotation then respect must be given to the order of application of the transformations if (apparently) unpredictable results are to be avoided! Figure 10.12 summarizes the main results of this section.

Fig. 10.12

Type	Linear	Group	Commute	Other
Translation	No	Yes	Yes	
Rotation	Yes	Yes	Yes	Orthogonal matrix
Euclidean	No	Yes	No	
Scale	Yes	Yes	Yes	Diagonal matrix

Most high-level computer languages provide arrays as a data structure, hence a matrix representation of all the transformations discussed would greatly simplify their implementation. For example, all product transformations could then be computed by matrix multiplication (see Chapter 6) instead of working from the definitions. The non-linear nature of translations prevents a representation in $\mathcal{M}(2, \mathbb{R})$; however, it is possible to obtain a suitable representation of all the transformations discussed above in $\mathcal{M}(3, \mathbb{R})$ where matrices operate in the space $\mathbb{H}^2$ of homogeneous coordinates of $\mathbb{R}^2$. Because the technique for achieving the representation in homogeneous coordinates is entirely general and in particular may be applied to transformations of $\mathbb{R}^3$ (where a representation in $\mathcal{M}(4, \mathbb{R})$ is obtained), we postpone the discussion until some transformations of $\mathbb{R}^3$ have been described.

10.2.2 Some useful transformations of $\mathbb{R}^3$

A *translation* of $\mathbb{R}^3$ is a mapping $T: \mathbb{R}^3 \to \mathbb{R}^3$ of the form

$$T\mathbf{r} = \mathbf{r} + \mathbf{r}_0 \quad \text{for all } \mathbf{r} \in \mathbb{R}^3$$

where $\mathbf{r}_0 \in \mathbb{R}^3$ is a fixed vector. If $\mathbf{r}_0 = (x_0, y_0, z_0)$ then a translation by $\mathbf{r}_0$ may be written in component form as,

$$T(x, y, z) = (x + x_0, y + y_0, z + z_0).$$

If $T(3)$ denotes the set of all translations of $\mathbb{R}^3$ then under composition of mappings $T(3)$ forms a commutative group of non-linear transformations of $\mathbb{R}^3$ isomorphic to the group $(\mathbb{R}^3, +)$. The implications of this are that a sequence of translations of $\mathbb{R}^3$ may be applied in any order with equal result and that we may not implement $T(3)$ by elements of $\mathcal{M}(3, \mathbb{R})$.

The rotation group of $\mathbb{R}^3$ has a more complex structure than $SO(2)$. Consider a rotation $W_{\hat{\mathbf{n}}}(\theta)$ through an angle θ about an axis defined by a unit vector $\hat{\mathbf{n}} \in \mathbb{R}^3$ as pictured in Figure 10.13. We then have

$$W_{\hat{\mathbf{n}}}(\theta)\mathbf{r} = \mathbf{r}'$$

Fig. 10.13

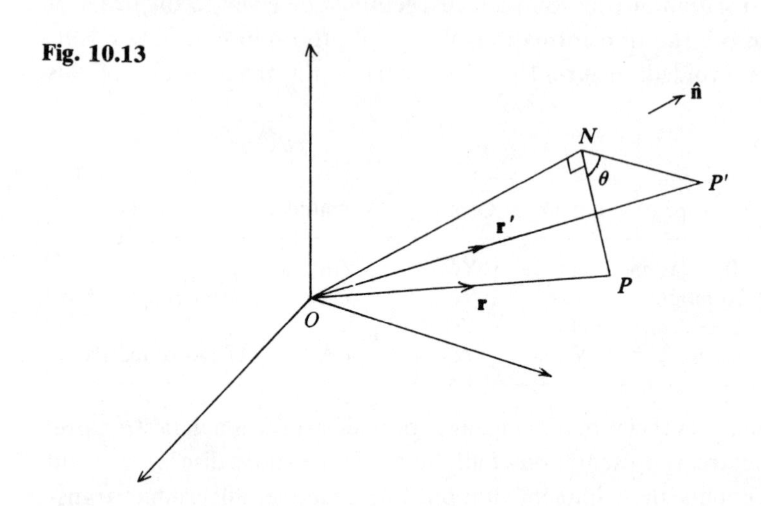

where $\mathbf{r}$ is the position vector of the point P and $\mathbf{r}'$ is the position vector of the transformed point P'. To determine $W_{\hat{\mathbf{n}}}(\theta)$ explicitly we need to express $\mathbf{r}'$ as a function of $\mathbf{r}$, $\hat{\mathbf{n}}$ and θ. Clearly, if $\mathbf{r}_N$ is the position vector of the point N then

$$\mathbf{r}_N = (\hat{\mathbf{n}} \cdot \mathbf{r})\hat{\mathbf{n}}$$

and if $\mathbf{r}_{NP}$ is the vector joining N to P then

$$\begin{aligned}\mathbf{r}_{NP} &= \mathbf{r} - \mathbf{r}_N \\ &= \mathbf{r} - (\hat{\mathbf{n}} \cdot \mathbf{r})\hat{\mathbf{n}} \\ &= (\hat{\mathbf{n}} \times \mathbf{r}) \times \hat{\mathbf{n}}.\end{aligned}$$

Figure 10.14 shows a view looking from N towards the origin O; Q is the point obtained by rotating the point P through $\pi/2$ about ON. Now $W_{\hat{\mathbf{n}}}(\theta)$ is a rotation, hence

$$\|\mathbf{r}_{NQ}\| = \|\mathbf{r}_{NP}\| = \|\mathbf{r}_{NP'}\|$$

and

$$\mathbf{r}_{NP'} = \mathbf{r}_{NQ} \sin \theta + \mathbf{r}_{NP} \cos \theta$$

Fig. 10.14

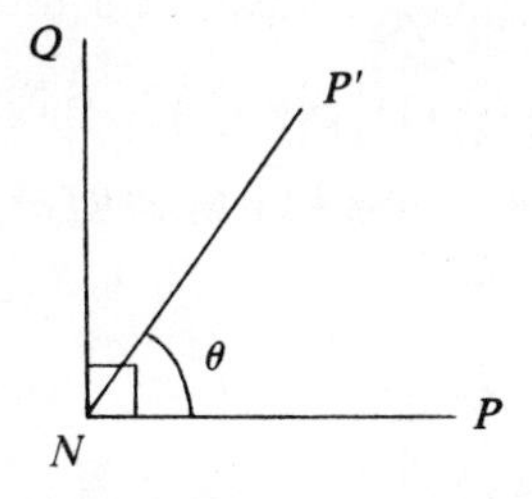

where $\mathbf{r}_{NQ}$ is the position vector of Q' relative to N. It follows that

$$\begin{aligned}\mathbf{r}_{NQ} &= \hat{\mathbf{n}} \times \mathbf{r}_{NP} \\ &= \hat{\mathbf{n}} \times ((\hat{\mathbf{n}} \times \mathbf{r}) \times \hat{\mathbf{n}}) \\ &= \hat{\mathbf{n}} \times \mathbf{r} \quad \text{(see Exercises 10.2)}\end{aligned}$$

therefore

$$\begin{aligned}\mathbf{r}_{NP'} &= (\hat{\mathbf{n}} \times \mathbf{r}) \sin\theta + ((\hat{\mathbf{n}} \times \mathbf{r}) \times \hat{\mathbf{n}}) \cos\theta \\ &= (\hat{\mathbf{n}} \times \mathbf{r}) \sin\theta - (\hat{\mathbf{n}} \times (\hat{\mathbf{n}} \times \mathbf{r})) \cos\theta\end{aligned}$$

but

$$\begin{aligned}W_{\hat{\mathbf{n}}}(\theta)\mathbf{r} &= \mathbf{r}' \\ &= \mathbf{r}_N + \mathbf{r}_{NP'} \\ &= \mathbf{r} + (\hat{\mathbf{n}} \times (\hat{\mathbf{n}} \times \mathbf{r}))(1 - \cos\theta) + (\hat{\mathbf{n}} \times \mathbf{r}) \sin\theta.\end{aligned}$$

Proposition. $W_{\hat{\mathbf{n}}}(\theta) \in SO(3)$.

Proof. Some of the details of this proof are left as exercises. It is a little easier to proceed if we first write $W_{\hat{\mathbf{n}}}(\theta)$ in a slightly different form. If $\hat{\mathbf{n}} \in \mathbb{R}^3$ determines the axis of rotation we define a transformation $A_{\hat{\mathbf{n}}}: \mathbb{R}^3 \to \mathbb{R}^3$ by

$$A_{\hat{\mathbf{n}}}\mathbf{r} = \hat{\mathbf{n}} \times \mathbf{r};$$

$A_{\hat{\mathbf{n}}}$ is linear and antisymmetric (see Exercises 10.2). Now

$$\begin{aligned}A_{\hat{\mathbf{n}}}^2\mathbf{r} &= A_{\hat{\mathbf{n}}}(A_{\hat{\mathbf{n}}}\mathbf{r}) \\ &= A_{\hat{\mathbf{n}}}(\hat{\mathbf{n}} \times \mathbf{r}) \\ &= \hat{\mathbf{n}} \times (\hat{\mathbf{n}} \times \mathbf{r});\end{aligned}$$

$A_{\hat{\mathbf{n}}}^2$ is linear and symmetric. We can now write $W_{\hat{\mathbf{n}}}(\theta)$ as the sum,

$$W_{\hat{\mathbf{n}}}(\theta) = I + (1 - \cos\theta)A_{\hat{\mathbf{n}}}^2 + \sin\theta A_{\hat{\mathbf{n}}}$$

in $\mathcal{M}(3, \mathbb{R})$. But $\mathcal{M}(3, \mathbb{R})$ is a vector space, hence $W_{\hat{\mathbf{n}}}(\theta)$ is linear. Orthogonality follows from the properties of $A_{\hat{\mathbf{n}}}$, for

$$\begin{aligned}W_{\hat{\mathbf{n}}}(\theta)^T &= I + (1 - \cos\theta)(A_{\hat{\mathbf{n}}}^2)^T + \sin\theta(A_{\hat{\mathbf{n}}})^T \\ &= I + (1 - \cos\theta)A_{\hat{\mathbf{n}}}^2 - \sin\theta A_{\hat{\mathbf{n}}}\end{aligned}$$

and

$$W_{\hat{\mathbf{n}}}(\theta)\,W_{\hat{\mathbf{n}}}(\theta)^T=[I+(1-\cos\theta)A_{\hat{\mathbf{n}}}^2+\sin\theta A_{\hat{\mathbf{n}}}]$$

$$*[I+(1-\cos\theta)A_{\hat{\mathbf{n}}}^2-\sin\theta A_{\hat{\mathbf{n}}}]$$

$$=I+[2(1-\cos\theta)-\sin^2\theta]A_{\hat{\mathbf{n}}}^2+(1-\cos\theta)^2A_{\hat{\mathbf{n}}}^4$$

$$=I+(1-\cos\theta)^2(A_{\hat{\mathbf{n}}}^2+A_{\hat{\mathbf{n}}}^4)$$

but (see Exercises 10.2)

$$A_{\hat{\mathbf{n}}}^4=-A_{\hat{\mathbf{n}}}^2$$

hence

$$W_{\hat{\mathbf{n}}}(\theta)\,W_{\hat{\mathbf{n}}}(\theta)^T=I$$

and $W_{\hat{\mathbf{n}}}(\theta)$ is orthogonal.

The proof that det $W_{\hat{\mathbf{n}}}(\theta)=1$ is left as an exercise. //

The converse of this proposition is also true (see Question 12 of Exercises 10.2); in particular if $W\in SO(3)$ it can be shown that there is a unit vector $\hat{\mathbf{n}}\in\mathbb{R}^3$ and a θ in the range $0\leqslant\theta<2\pi$ such that

$$W=W_{\hat{\mathbf{n}}}(\theta).$$

It therefore makes sense to define the *rotations* of $\mathbb{R}^3$ as the group $SO(3)$.

Some useful subgroups of $SO(3)$ are now easily identified. It is often required to rotate a geometric object about one of the Cartesian axes. The corresponding matrices may be obtained from the general form $W_{\hat{\mathbf{n}}}(\theta)$; for example, to rotate through an angle θ about the z-axis we choose $\hat{\mathbf{n}}=\hat{\mathbf{k}}=(0,0,1)$ for which

$$A_{\hat{\mathbf{k}}}=\begin{pmatrix}0&-1&0\\1&0&0\\0&0&0\end{pmatrix}\quad\text{and}\quad A_{\hat{\mathbf{k}}}^2=\begin{pmatrix}-1&0&0\\0&-1&0\\0&0&0\end{pmatrix}$$

so that

$$W_{\hat{\mathbf{k}}}(\theta)=\begin{pmatrix}\cos\theta&-\sin\theta&0\\\sin\theta&\cos\theta&0\\0&0&1\end{pmatrix}.$$

Similarly, the matrices

$$W_{\hat{\mathbf{i}}}(\phi)=\begin{pmatrix}1&0&0\\0&\cos\phi&-\sin\phi\\0&\sin\phi&\cos\phi\end{pmatrix}\quad\text{and}$$

$$W_{\hat{\mathbf{j}}}(\chi)=\begin{pmatrix}\cos\chi&0&\sin\chi\\0&1&0\\-\sin\chi&0&\cos\chi\end{pmatrix}$$

correspond to rotations through ϕ and χ about the x- and y-axes respectively; the derivation of these is left as an exercise. It is now easy to show that $SO(3)$ is *not* a commutative group for in general

$$[W_{\hat{\mathbf{i}}}(\phi), W_{\hat{\mathbf{j}}}(\chi)] \neq 0$$

$$[W_{\hat{\mathbf{i}}}(\phi), W_{\hat{\mathbf{k}}}(\theta)] \neq 0$$

and

$$[W_{\hat{\mathbf{j}}}(\chi), W_{\hat{\mathbf{k}}}(\theta)] \neq 0;$$

however, all the subgroups

$$\{W_{\hat{\mathbf{n}}}(\theta): 0 \leq \theta < 2\pi\} \quad \text{for } \hat{\mathbf{n}} \in \mathbb{R}^3$$

are commutative (see Exercises 10.2).

From a practical point of view these results on rotations of $\mathbb{R}^3$ imply that to invert a rotation one simply takes the transpose of the matrix and that if several rotations are to be applied to a geometric data set then the order of application is important.

Following the two-dimensional case we combine $T(3)$ and $SO(3)$ to form the set $E(3)$ of *Euclidean* transformations of $\mathbb{R}^3$. Elements of $E(3)$ have the form

$$U\mathbf{r} = W\mathbf{r} + \mathbf{r}_0 \quad \text{for all } \mathbf{r} \in \mathbb{R}^3$$

where $W \in SO(3)$ and $\mathbf{r}_0 \in \mathbb{R}^3$. The law of composition in $E(3)$ is

$$(W_2, \mathbf{r}_2) \circ (W_1, \mathbf{r}_1) = (W_2 W_1, W_2\mathbf{r}_1 + \mathbf{r}_2)$$

under which $E(3)$ becomes a non-commutative group of non-linear transformations of $\mathbb{R}^3$. The implications of this are analogous to the two-dimensional case; in particular we stress that $E(3)$ may *not* be implemented in $\mathcal{M}(3, \mathbb{R})$.

A *scale* transformation of $\mathbb{R}^3$ is a linear mapping of the form

$$S(x, y, z) = (\lambda x, \mu y, \sigma z) \quad \text{for all } (x, y, z) \in \mathbb{R}^3$$

where $\lambda, \mu, \sigma > 0$. The set $S(3)$ of all scale transformations determines a commutative group under composition.

The remaining transformations discussed in this section have both a different purpose and different properties to the ones above. Our problem is how to represent a three-dimensional geometric object on a two-dimensional graphics display terminal. Mathematically, we seek transformations from $\mathbb{R}^3$ onto a two-dimensional subspace that provide a useful pictorial representation of our object. Recall from Chapter 5 that a transformation P of a vector space is called a projection if $P^2 = P$.

Let us assume that our object is described in a Cartesian coordinate system (x, y, z) and that in this system our graphics display screen corresponds to the rectangular subset

$$\{(x, y): 0 \leqslant x \leqslant a, 0 \leqslant y \leqslant b\}$$

of the xy plane. The simplest method of obtaining an image of an object on the screen is to apply the transformation

$$P_1\mathbf{r} = \mathbf{r}'$$

where $\mathbf{r} = (x, y, z)$ and $\mathbf{r}' = (x, y, 0)$, to its geometric data set. Provided that the x and y values of the data set are all within the screen bounds a full picture of the object will result. If some x and/or y values are outside the screen limits then an appropriate scale transformation should first be applied.

P_1 is a linear projection for

$$P_1(x, y, z) = \begin{pmatrix} 1 & 0 & 0 \\ 0 & 1 & 0 \\ 0 & 0 & 0 \end{pmatrix} \begin{pmatrix} x \\ y \\ z \end{pmatrix}$$

and

$$P_1^2\mathbf{r} = P_1\mathbf{r} \quad \text{for all } \mathbf{r} \in \mathbb{R}^3,$$

but P_1^{-1} does not exist for

$$\det \begin{pmatrix} 1 & 0 & 0 \\ 0 & 1 & 0 \\ 0 & 0 & 0 \end{pmatrix} = 0.$$

Geometrically, P_1 may be obtained as follows: referring to Figure 10.15 we draw a line segment (called a projection line) from the point Q with position vector $\mathbf{r}$ to the point Q' with position vector $\mathbf{r}'$ in the xy plane such that the line segment QQ' is orthogonal to the xy plane,

Fig. 10.15. Orthogonal parallel projection.

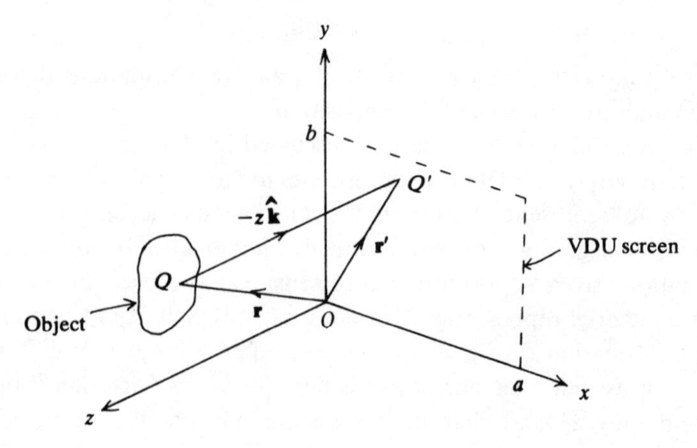

then clearly

$$\begin{aligned}\mathbf{r}' &= \mathbf{r} + (-z\hat{\mathbf{k}}) \\ &= (x, y, 0) \\ &= P_1\mathbf{r}.\end{aligned}$$

P_1 is called a *parallel orthogonal projection* of $\mathbb{R}^3$.

A parallel projection imparts no impression of depth to the resulting image, the height of an object appears the same irrespective of its distance from the xy plane. The final transformation which we describe is designed to give depth or 'perspective' to the image; instead of having parallel projection lines we construct projection lines radiating from some fixed point. We determine the projection

$$P_2\mathbf{r} = \mathbf{r}'$$

where the fixed point is at $z_p\hat{\mathbf{k}}$ (as shown in Figure 10.16) and hence the point Q is mapped to the point Q'. Figure 10.16(ii) is a view of Figure 10.16(i) from along the x-axis. Clearly,

$$\frac{y'}{z_p} = \frac{y}{z_p - z}$$

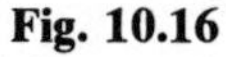

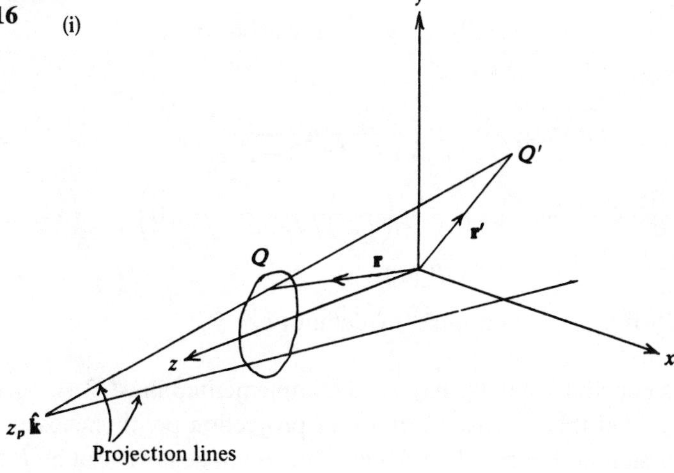

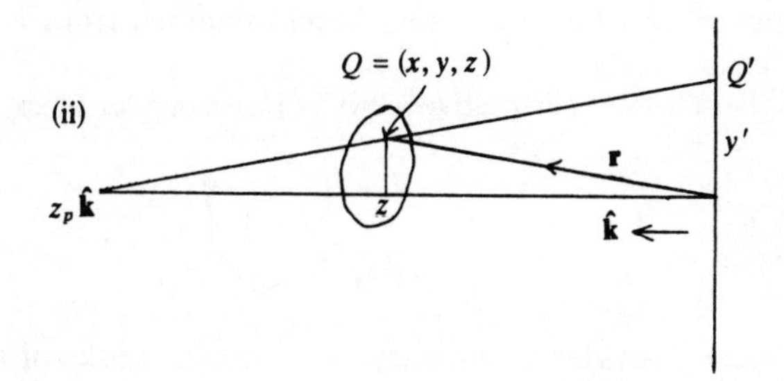

or

$$y' = \frac{y}{1 - z/z_p}.$$

Similarly, we obtain

$$x' = \frac{x}{1 - z/z_p}$$

and the explicit form for P_2 is

$$P_2(x, y, z) = \frac{1}{1 - z/z_p}(x, y, 0).$$

Proposition. P_2 is a non-linear projection of $\mathbb{R}^3$.

Proof.

$$\begin{aligned} P_2\lambda(x, y, z) &= P_2(\lambda x, \lambda y, \lambda z) \\ &= \frac{1}{1 - \lambda z/z_p}(\lambda x, \lambda y, 0) \\ \lambda P_2(x, y, z) &= \frac{1}{1 - z/z_p}(\lambda x, \lambda y, 0) \\ &\neq P_2\lambda(x, y, z) \quad \text{for arbitrary } \lambda, \end{aligned}$$

therefore P_2 is non-linear.

$$\begin{aligned} P_2(P_2(x, y, z)) &= P_2\left(\frac{x}{1 - z/z_p}, \frac{y}{1 - z/z_p}, 0\right) \\ &= \frac{1}{1-0}\left(\frac{x}{1 - z/z_p}, \frac{y}{1 - z/z_p}, 0\right) \\ &= P_2(x, y, z) \end{aligned}$$

and P_2 is a projection of $\mathbb{R}^3$ by definition. //

We conclude that P_2 may not be implemented in $\mathcal{M}(3, \mathbb{R})$.

Provided that a sensible choice of projection point is chosen, a good impression of depth will be achieved in an image produced by P_2. Figure 10.17 shows views of a rectangular block obtained from P_1 and P_2.

Fig. 10.17. Parallel and perspective views of a rectangular block.

Figure 10.18 provides a summary of the main results of this section.

Fig. 10.18

Transformation	Linear	Group	Commute	Other
Translation	No	Yes	Yes	
Rotation	Yes	Yes	No	Orthogonal matrix
Euclidean	No	Yes	No	
Scale	Yes	Yes	Yes	Diagonal matrix
P_1	Yes	No		
P_2	No	No		

10.2.3 Homogeneous coordinates and the linear representation

In this section we describe the technique for representing the transformations of $\mathbb{R}^2$ and $\mathbb{R}^3$, discussed above, in $\mathcal{M}(3, \mathbb{R})$ and $\mathcal{M}(4, \mathbb{R})$ respectively. The ideas are completely general and are described in $\mathbb{R}^n$; we seek an implementation of some special classes of transformation of $\mathbb{R}^n$ (in particular those in Figures 10.12 and 10.18), not all of which are linear, in the matrix algebra $\mathcal{M}(n+1, \mathbb{R})$.

In Section 10.1 we showed that Q^n is a bijection $\mathbb{H}^n \to \mathbb{R}^n$ and may be used to determine a coordinate system on $\mathbb{R}^n$. This means that if T is a transformation of $\mathbb{R}^n$ we may express it as a transformation of $\mathbb{H}^n$. In other words, there is a mapping

$$\tilde{T}: \mathbb{H}^n \to \mathbb{H}^n$$

such that the diagram of Figure 10.19 is commutative. We have

Fig. 10.19

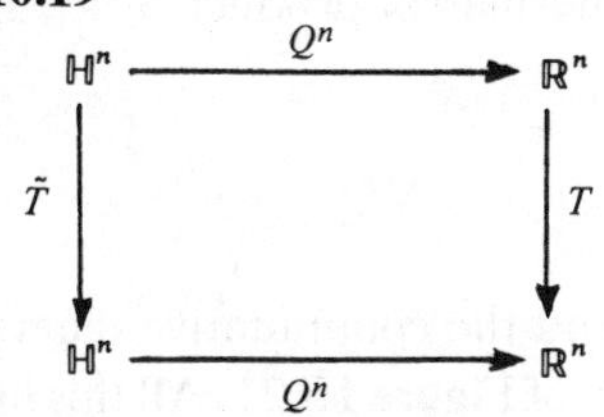

$Q^n \circ \tilde{T} = T \circ Q^n$ but Q^n is invertible so that we may write $\tilde{T} = (Q^n)^{-1} \circ T \circ Q^n$. If S is a second transformation of $\mathbb{R}^n$ then

$$\begin{aligned}\widetilde{S \circ T} &= (Q^n)^{-1} \circ S \circ T \circ Q^n \\ &= (Q^n)^{-1} \circ S \circ Q^n \circ (Q^n)^{-1} \circ T \circ Q^n \\ &= \tilde{S} \circ \tilde{T}.\end{aligned}$$

In other words we may piece together diagrams without destroying commutativity as in Figure 10.20. For some transformations T of $\mathbb{R}^n$ there is a linear transformation

$$T_L: \mathbb{R}^{n+1} \to \mathbb{R}^{n+1}$$

Fig. 10.20

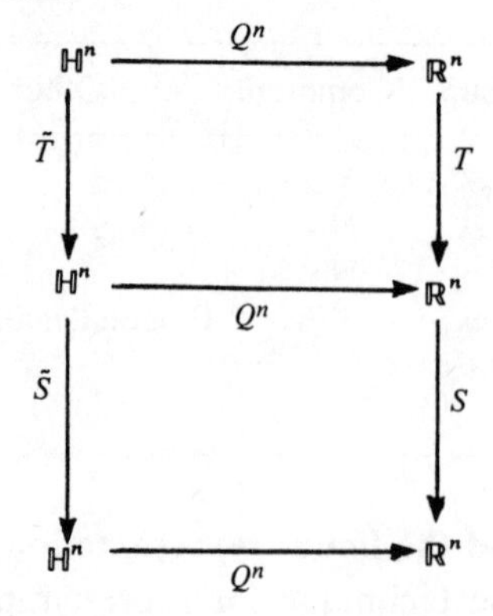

such that

$$\tilde{T}[\mathbf{x}] = [T_L\mathbf{x}] \quad \text{for all } \mathbf{x} \in \mathbb{R}^{n+1}.$$

T_L exists for all the transformations discussed in this chapter and T_L is the required representation of T in $\mathbb{R}^{n+1}$, for if $\mathbf{r} \in \mathbb{R}^n$ we have

$$\begin{aligned} T\mathbf{r} &= Q^n \circ \tilde{T} \circ (Q^n)^{-1}\mathbf{r} \\ &= Q^n \circ \tilde{T}[\mathbf{r}, 1] \\ &= Q^n[T_L(\mathbf{r}, 1)]. \end{aligned}$$

Thus the transformation T is achieved by applying T_L to a homogeneous representation and then applying Q^n to return to physical coordinates.

To make sure that everything works properly we need to show that the composition $\tilde{S} \circ \tilde{T}$ corresponds to the matrix product $S_L T_L$; in fact this is easy for

$$\begin{aligned} \tilde{S} \circ \tilde{T}[\mathbf{x}] &= \tilde{S}[T_L\mathbf{x}] \\ &= [S_L T_L\mathbf{x}]. \end{aligned}$$

Pictorially, this means that we may extend the commutative diagram of Figure 10.20 to the commutative diagram of Figure 10.21. All this implies

Fig. 10.21

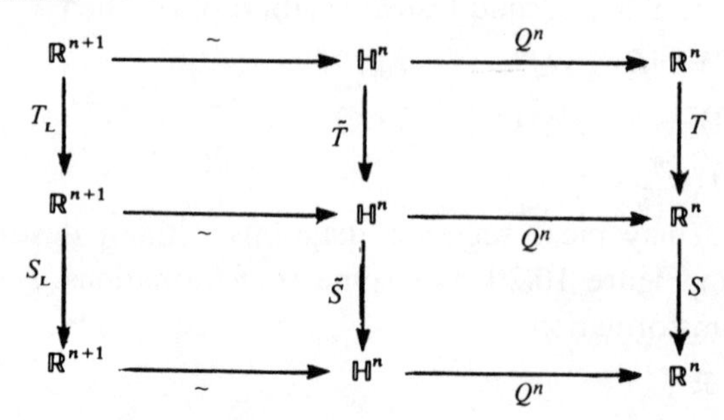

that once we have determined matrices T_L, S_L, ... corresponding to the transformations T, S, ... which we wish to perform then it is safe to drop the equivalence class notation and work with a representative vector. Products of transformations in $\mathbb{R}^n$ are obtained by matrix multiplication in $\mathcal{M}(n+1, \mathbb{R})$ applied to the representative vector. All representatives of a given class will produce the same physical coordinates under the correspondence

$$(x_1, \ldots, x_n, x_{n+1}) \mapsto \frac{1}{x_{n+1}}(x_1, \ldots, x_n).$$

It is usual to choose the vector $(\mathbf{r}, 1) \in \mathbb{R}^{n+1}$ to represent $\mathbf{r} \in \mathbb{R}^n$. Such a choice is clearly always possible.

In the following examples we obtain the T_L matrices for the transformations described in Sections 10.2.1 and 10.2.2. All matrices are computed with respect to standard bases.

Example 10.2.1

If $T: \mathbb{R}^n \to \mathbb{R}^n$ is a *linear* transformation with matrix A_T then the matrix of T_L is

$$\begin{pmatrix} & & 0 \\ & A_T & \vdots \\ & & 0 \\ 0 & \ldots 0 & 1 \end{pmatrix}$$

for if $\mathbf{x} = (\mathbf{r}, 1) \in \mathbb{R}^{n+1}$ then

$$\begin{aligned} \tilde{T}[\mathbf{x}] &= (Q^n)^{-1} \circ T \circ Q^n[\mathbf{x}] \\ &= (Q^n)^{-1} \circ T\mathbf{r} \\ &= (Q^n)^{-1} A_T \mathbf{r} \\ &= [(A_T\mathbf{r}, 1)] \\ &= \left[\begin{pmatrix} & & 0 \\ & A_T & \vdots \\ & & 0 \\ 0 & \ldots 0 & 1 \end{pmatrix} \begin{pmatrix} \\ \mathbf{r} \\ \\ 1 \end{pmatrix} \right]. \end{aligned}$$

The notation

$$\begin{pmatrix} A_T & 0 \\ 0 & 1 \end{pmatrix} \text{ is used for } \begin{pmatrix} & & 0 \\ & A_T & \vdots \\ & & 0 \\ 0 & \ldots 0 & 1 \end{pmatrix}$$

when it is more convenient. Expanding by the bottom row we see that

$$\det\begin{pmatrix} A_T & 0 \\ 0 & 1 \end{pmatrix} = 0 \quad \text{iff } \det A_T = 0$$

and clearly

$$\begin{pmatrix} A_T & 0 \\ 0 & 1 \end{pmatrix}^{-1} = \begin{pmatrix} A_T^{-1} & 0 \\ 0 & 1 \end{pmatrix}. \quad //$$

Example 10.2.1 shows how to implement our linear transformations; for example, applying this to $SO(2)$ we obtain the representation

$$\begin{pmatrix} \cos\theta & -\sin\theta & 0 \\ \sin\theta & \cos\theta & 0 \\ 0 & 0 & 1 \end{pmatrix}$$

in $\mathcal{M}(3, \mathbb{R})$ for the matrix

$$\begin{pmatrix} \cos\theta & -\sin\theta \\ \sin\theta & \cos\theta \end{pmatrix}.$$

Similarly if $W \in SO(3)$ then W_L has the form

$$\begin{pmatrix} W & 0 \\ 0 & 1 \end{pmatrix}$$

in $\mathcal{M}(4, \mathbb{R})$ with

$$\begin{pmatrix} W & 0 \\ 0 & 1 \end{pmatrix}^{-1} = \begin{pmatrix} W^T & 0 \\ 0 & 1 \end{pmatrix}.$$

Example 10.2.2

If $T: \mathbb{R}^n \to \mathbb{R}^n$ is the translation

$$T\mathbf{r} = \mathbf{r} + \mathbf{a} \quad \text{for all } \mathbf{r} \in \mathbb{R}^n$$

where $\mathbf{a} = (a_1, \ldots, a_n) \in \mathbb{R}^n$ is a fixed vector, then the matrix of T_L is

$$\begin{pmatrix} & & a_1 \\ & I & \vdots \\ & & a_n \\ 0 \ldots 0 & & 1 \end{pmatrix}$$

which may be written in the shorthand form

$$\begin{pmatrix} I & \mathbf{a} \\ 0 & 1 \end{pmatrix}.$$

This is easy to show for if $\mathbf{x} = (\mathbf{r}, 1)$ then

$$\begin{aligned} \tilde{T}[\mathbf{x}] &= (Q^n)^{-1} \circ T \circ Q^n[\mathbf{x}] \\ &= (Q^n)^{-1} \circ T\mathbf{r} \\ &= (Q^n)^{-1}(\mathbf{r} + \mathbf{a}) \end{aligned}$$

$$=[(\mathbf{r}+\mathbf{a},1)]$$

$$=\left[\begin{pmatrix} I & \mathbf{a}\\ 0 & 1\end{pmatrix}\begin{pmatrix}\mathbf{r}\\ 1\end{pmatrix}\right]. \quad //$$

Example 10.2.3

Similar calculations show that $(W,\mathbf{a})\in E(2)$ may be implemented by

$$\begin{pmatrix} W & \mathbf{a}\\ 0 & 1\end{pmatrix}\in GL(3,\mathbb{R}) \quad \text{for } W\in SO(2) \text{ and } \mathbf{a}\in\mathbb{R}^2$$

and $(W,\mathbf{a})\in E(3)$ by

$$\begin{pmatrix} W & \mathbf{a}\\ 0 & 1\end{pmatrix}\in GL(4,\mathbb{R}) \quad \text{where } W\in SO(3) \text{ and } \mathbf{a}\in\mathbb{R}^3.$$

The inverses have the form

$$\begin{pmatrix} W & \mathbf{a}\\ 0 & 1\end{pmatrix}^{-1}=\begin{pmatrix} W^T & -W^T\mathbf{a}\\ 0 & 1\end{pmatrix}$$

in both $E(2)$ and $E(3)$. //

Example 10.2.4

The perspective projection $P_2:\mathbb{R}^3\to\mathbb{R}^3$ defined by

$$P_2(x,y,z)=\frac{1}{1-z/z_p}(x,y,0)$$

may be implemented in $\mathcal{M}(4,\mathbb{R})$ as

$$\begin{pmatrix} 1 & 0 & 0 & 0\\ 0 & 1 & 0 & 0\\ 0 & 0 & 0 & 0\\ 0 & 0 & -1/z_p & 1\end{pmatrix}$$

for if $\mathbf{x}=(\mathbf{r},1)\in\mathbb{R}^4$ with $\mathbf{r}=(x,y,z)$ then

$$\tilde{P}_2[\mathbf{x}]=(Q^n)^{-1}\circ P_2\circ Q^n[(\mathbf{r},1)]$$

$$=(Q^n)^{-1}\circ P_2\mathbf{r}$$

$$=(Q^n)^{-1}\left(\frac{x}{1-z/z_p},\frac{y}{1-z/z_p},0\right)$$

$$=\left[\left(\frac{x}{1-z/z_p},\frac{y}{1-z/z_p},0,1\right)\right]$$

$$=[(x,y,0,1-z/z_p)]$$

$$=\left[\begin{pmatrix} 1 & 0 & 0 & 0\\ 0 & 1 & 0 & 0\\ 0 & 0 & 0 & 0\\ 0 & 0 & -1/z_p & 1\end{pmatrix}\begin{pmatrix} x\\ y\\ z\\ 1\end{pmatrix}\right]. \quad //$$

Exercises 10.2

1. Show that the groups $(T(2), \circ)$ and $(\mathbb{R}^2, +)$ are isomorphic.
2. Show that if

   $$J = \begin{pmatrix} 0 & -1 \\ 1 & 0 \end{pmatrix}$$

   then the rotation matrix $W(\theta) \in SO(2)$ may be written in exponential form as

   $$W(\theta) = \exp(\theta J).$$

3. Determine the Euclidean transformation mapping the triangle PQR of Figure 10.22(i) to triangle $P'Q'R'$ of Figure 10.22(ii).

Fig. 10.22

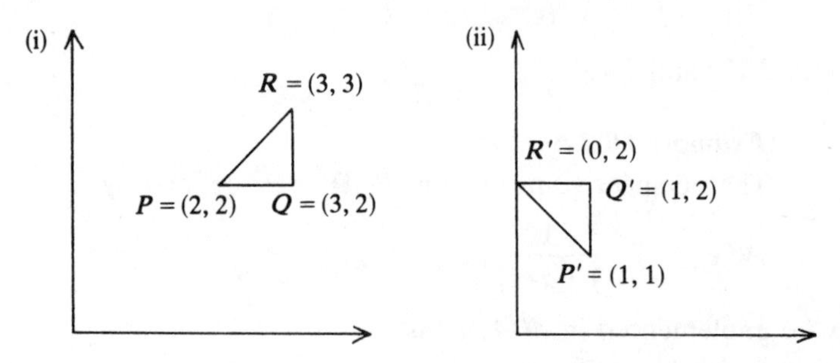

4. Show that $S(2)$ is a commutative subgroup of linear transformations of $\mathbb{R}^2$.
5. Show that, in general, the elements of $S(2)$ do not commute with the elements of $SO(2)$ and construct a suitable pictorial demonstration of this fact.
6. Identify a non-trivial subgroup of $S(2)$ that does commute with $SO(2)$.
7. It is required to scale $\mathbb{R}^2$ by the transformation

   $$S(x, y) = (3x, 5y)$$

   leaving the point $(2, 2)$ fixed. Determine the appropriate transformation.
8. Show that if $\hat{\mathbf{n}} \in \mathbb{R}^3$ is a unit vector then

   $$\hat{\mathbf{n}} \times ((\hat{\mathbf{n}} \times \mathbf{r}) \times \hat{\mathbf{n}}) = \hat{\mathbf{n}} \times \mathbf{r} \quad \text{for all } \mathbf{r} \in \mathbb{R}^3.$$

9. If $\mathbf{n} \in \mathbb{R}^3$ and $A_{\mathbf{n}}: \mathbb{R}^3 \to \mathbb{R}^3$ is defined by

   $$A_{\mathbf{n}}\mathbf{r} = \mathbf{n} \times \mathbf{r} \quad \text{for all } \mathbf{r} \in \mathbb{R}^3$$

show that $A_{\mathbf{n}}$ is a linear transformation of $\mathbb{R}^3$ and determine the matrix of $A_{\mathbf{n}}$ with respect to the standard basis $\{\hat{\mathbf{i}}, \hat{\mathbf{j}}, \hat{\mathbf{k}}\}$ of $\mathbb{R}^3$.

10. Use Question 9 and the expression for $W_{\hat{\mathbf{n}}}(\theta)$ in terms of $A_{\hat{\mathbf{n}}}$ to obtain the explicit matrix form of the rotation $W_{\hat{\mathbf{n}}}(\theta)$ of $\mathbb{R}^3$. Hence show that

    $\det W_{\hat{\mathbf{n}}}(\theta) = 1.$

11. In the notation of Question 9 show that for all $\mathbf{r} \in \mathbb{R}^3$ we have

    $A_{\hat{\mathbf{n}}}^3 \mathbf{r} = -A_{\hat{\mathbf{n}}} \mathbf{r}$

    $A_{\hat{\mathbf{n}}}^4 \mathbf{r} = -A_{\hat{\mathbf{n}}}^2 \mathbf{r}$

    $A_{\hat{\mathbf{n}}}^5 \mathbf{r} = A_{\hat{\mathbf{n}}} \mathbf{r}$

    $A_{\hat{\mathbf{n}}}^6 \mathbf{r} = A_{\hat{\mathbf{n}}}^2 \mathbf{r}$

    $A_{\hat{\mathbf{n}}}^{6+k} \mathbf{r} = A_{\hat{\mathbf{n}}}^{2+k} \mathbf{r} \quad \text{for all } k \in \mathbb{N}$

    and use these results to prove that

    $W_{\hat{\mathbf{n}}}(\theta) = \exp(\theta A_{\hat{\mathbf{n}}}).$

12. (i) Use the exponential form of $W_{\hat{\mathbf{n}}}(\theta)$ to obtain an alternative proof that

    $W_{\hat{\mathbf{n}}}(\theta)(W_{\hat{\mathbf{n}}}(\theta))^T = I.$

    (ii) If $A \in \mathcal{M}(n, \mathbb{R})$ it may be shown that

    $\det(\exp A) = \mathrm{e}^{\sum_{i=0}^{n} A_{ii}};$

    use this to obtain a second proof of the fact that

    $\det W_{\hat{\mathbf{n}}}(\theta) = 1.$

13. Show that $\hat{\mathbf{n}}$ is an eigenvector of $W_{\hat{\mathbf{n}}}(\theta)$ for all $0 \leq \theta < 2\pi$. What is the eigenvalue? Interpret this result geometrically.

14. Derive the formulae quoted in the text for the rotation matrices $W_{\hat{\mathbf{i}}}(\phi)$ and $W_{\hat{\mathbf{j}}}(\chi)$.

15. Show that the subgroups

    $S_{\hat{\mathbf{n}}} = \{W_{\hat{\mathbf{n}}}(\theta) : 0 \leq \theta < 2\pi\}$

    are all isomorphic to $SO(2)$.

16. Let $\mathbf{a} \in \mathbb{R}^3$ be a fixed vector. A transformation $T : \mathbb{R}^3 \to \mathbb{R}^3$ is defined by

    $$T\mathbf{r} = \frac{\mathbf{r}}{\mathbf{a} \cdot \mathbf{r}} \quad \text{when } \mathbf{a} \cdot \mathbf{r} \neq 0$$
    $$= \mathbf{0} \quad \text{otherwise.}$$

    (i) Show that T is a non-linear projection of $\mathbb{R}^3$.

    (ii) Show that

    $\tilde{T}[(\mathbf{r}, 1)] = [(\mathbf{r}, \mathbf{a} \cdot \mathbf{r})] \quad \text{for all } \mathbf{r} \in \mathbb{R}^3 \text{ with } \mathbf{r} \cdot \mathbf{a} \neq 0.$

(iii) Determine a matrix $A \in \mathcal{M}(4, \mathbb{R})$ such that

$$\tilde{T}[\mathbf{x}] = [A\mathbf{x}] \quad \text{for all } \mathbf{x} \in \mathbb{R}^4$$

where

$$[\mathbf{x}] \in \mathbb{H}^3.$$

10.3 Curves and surfaces

10.3.1 Mathematical representation

Curves and surfaces form the basis of much computer aided design and computer graphics software. Alternative mathematical descriptions of the same geometrical shape are possible; some of these are discussed and assessed from the point of view of computing applications.

If $I \subset \mathbb{R}$ is an interval we denote by κ^n the set of all C^1 mappings

$$c: I \to \mathbb{R}^n$$

with $c' \neq 0$ on I. We define a relation $\sim$ on κ^n in the following way: if

$$c_1: I_1 \to \mathbb{R}^n$$

and

$$c_2: I_2 \to \mathbb{R}^n$$

then c_1 and c_2 are related if there is a C^1 mapping

$$\phi: I_1 \to I_2$$

such that

$$c_1 = c_2 \circ \phi$$

and $\phi' \neq 0$ on I_1.

Proposition. $\sim$ is an equivalence relation on κ^n. //

The proof of this is left as an exercise for the reader.

Let $\mathscr{C}^n$ denote the equivalence classes $\kappa^n/\sim$.

Definition. A *curve* in $\mathbb{R}^n$ is an element of $\mathscr{C}^n$. //

If $c \in \kappa^n$ and $c: I \to \mathbb{R}^n$ then I is called the *parameter space* of c and $t \in I$ the *parameter* or coordinate of c. The *graph* of c is defined to be the set of points

$$\{c(t): t \in I\}.$$

If $[c]$ denotes the equivalence class of c in $\mathscr{C}^n$ then for all $c_1 \in [c]$ we have

$$\text{graph of } c_1 = \text{graph of } c$$

so that it makes sense to speak of the graph of a curve $[c]$. The converse of this is not true; we may have graph of $[c_1]$ = graph of $[c_2]$ but $[c_1] \neq [c_2]$. The equivalence relation $\sim$ groups together elements of κ^n that are parameterized in a 'similar' way.

For computational purposes we select an element $c \in [c]$ and abuse our terminology by calling c a curve also; this is alright provided we understand the distinction and exercise caution when necessary. The graph of a curve in $\mathscr{C}^2$ is the set of points we see when it is displayed at a graphics terminal.

$\mathscr{C}^2$ is the set of *plane* curves and elements of $\mathscr{C}^3$ are called *space* or *twisted* curves. $\mathscr{C}^2$ and $\mathscr{C}^3$ are the important cases for computing applications but where it is easy to keep the discussion completely general we do so.

In computer graphics literature the terms 'parametric', 'explicit' and 'implicit' are applied to different methods of specifying curves. If $(\xi_1, \ldots, \xi_n)$ is a system of coordinates on $\mathbb{R}^n$ and an element $c \in \kappa^n$ is specified by defining n functions $t \mapsto \xi_i(t)$, $1 \leq i \leq n$, then this is called an *explicit parametric description.* Two types are sometimes identified: 'symmetrical' and 'non-symmetrical'. A description is non-symmetrical if the parameter is one of the coordinate variables, i.e. $t = \xi_i$ for some $1 \leq i \leq n$. Non-symmetrical descriptions therefore have the general form

$$(\xi_i, (\xi_1(\xi_i), \xi_2(\xi_i), \ldots, \xi_i, \ldots, \xi_n(\xi_i))), \quad \xi_i \in I,$$

which for $\mathscr{C}^2$, using a Cartesian coordinate system (x, y), takes the more familiar form

$$(x, (x, y(x))), \quad x \in [x_0, x_1]$$

and the curve is defined by giving y explicitly as a function of x. Alternatively, we may describe a curve in $\mathbb{R}^n$ by specifying a suitable function $f: \mathbb{R}^n \to \mathbb{R}$. For a fixed $a \in \mathbb{R}$, f defines a curve with graph

$$f^{-1}(a) = \{(\xi_1, \ldots, \xi_n): f(\xi_1, \ldots, \xi_n) = a\};$$

$f(\xi_1, \ldots, \xi_n) = a$ is called the *equation* of the curve (strictly speaking it should be called the equation of the *graph* of the curve). In principle, an equation of this type may be recast in explicit form, although this is not an easy task in general. Curves defined by such equations are said to be described *implicitly.* We remark that not all mappings $f: \mathbb{R}^n \to \mathbb{R}$ will produce a curve in the way described, for example the constant mappings produce the whole of $\mathbb{R}^n$. Further, a function $f: \mathbb{R}^n \to \mathbb{R}$ may produce a curve for some values of the constant a, but not for others as we shall see later. A plane curve with implicit equation of the form

$$ax^2 + by^2 + cxy + dx + ey + f = 0$$

is called a *quadratic* curve.

If $c_1:[a, b] \to \mathbb{R}^n$ and $c_2:[c, d] \to \mathbb{R}^n$ with

$$a < b = c < d$$

and

$$c_1(b) = c_2(c)$$

then $c_1 \vee c_2$ is defined by

$$\begin{aligned} c_1 \vee c_2 &= c_1 \quad \text{on } [a, b] \\ &= c_2 \quad \text{on } [c, d] \end{aligned}$$

and is called the *join* of c_1 and c_2. For symmetrical descriptions if $c \neq b$, c_2 can be 'reparameterized' to meet this condition; this means that we can construct a $c_2^* \in [c_2]$ such that the interval for c_2^* begins at b. Hence provided $c_1(b) = c_2(c)$ the join $c_1 \vee c_2$ makes sense but may not, strictly speaking, be a curve. The problem is that $c_1 \vee c_2$ may not be in C^1.

Example 10.3.1

Straight line curves may be defined in $\mathbb{R}^2$ in the usual way as

$$y(x) = mx + c \quad x_0 \leqslant x \leqslant x_1$$

where m is the slope of the line and c the intercept with the y-axis; see Figure 10.23. In our curve notation the line segment is

$$(x, (x, mx + c)), \quad x_0 \leqslant x \leqslant x_1;$$

Fig. 10.23

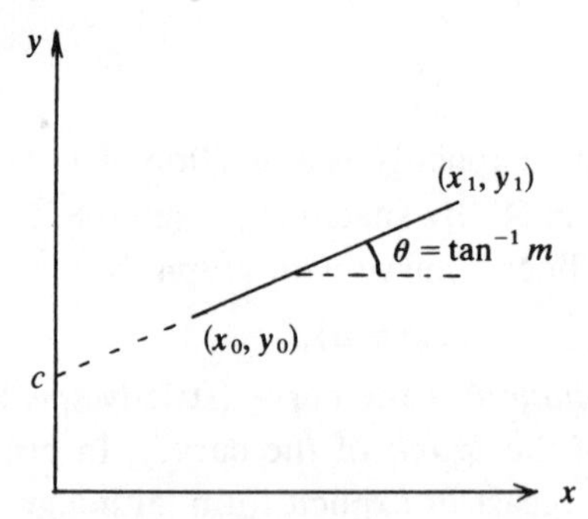

m and c may be given directly but more usually the end points $\mathbf{r}_0 = (x_0, y_0)$ and $\mathbf{r}_1 = (x_1, y_1)$ will be given from which we obtain

$$m = (y_1 - y_0)/(x_1 - x_0)$$

and

$$c = y_1 - mx_1.$$

A problem, present in all non-symmetrical descriptions, is now apparent; this form will not describe a vertical line (when $x_0 = x_1$). If instead we use the implicit form

$$(y-y_0)(x_1-x_0)-(y_1-y_0)(x-x_0)=0, \quad x_0 \leqslant x \leqslant x_1$$

then when $x_1 = x_0$ we obtain the equation of the vertical line

$$x = x_0.$$

The general implicit equation for a line is

$$ax+by+c=0$$

and vertical lines are those for which $b=0$. The usual symmetrical description may be written in vector form as

$$(t, \mathbf{r}_0 + t\hat{\mathbf{u}}), \quad t \in [0, \|\mathbf{r}_1 - \mathbf{r}_0\|]$$

where $\hat{\mathbf{u}}$ is the unit vector $\hat{\mathbf{u}} = (\mathbf{r}_1 - \mathbf{r}_0)/\|\mathbf{r}_1 - \mathbf{r}_0\|$. //

The above example leads to a general observation on the nature of curves defined in a non-symmetrical way. Before discussing this we make some further general definitions.

Definitions. A plane curve $(u, (x(u), y(u)))$, $u \in I = [u_A, u_B]$ is said to be

(i) *single valued* if for all $u_1, u_2 \in I$ we have

$$x(u_1) = x(u_2) \Rightarrow y(u_1) = y(u_2)$$

(ii) *multivalued* if it is not single valued and

(iii) *closed* if $(x(u_A), y(u_A)) = (x(u_B), y(u_B))$. //

Figure 10.24 illustrates these ideas graphically. Clearly closed curves are multivalued. A non-symmetrical description

$$(x, (x, y(x))), \quad x_0 \leqslant x \leqslant x_1$$

Fig. 10.24

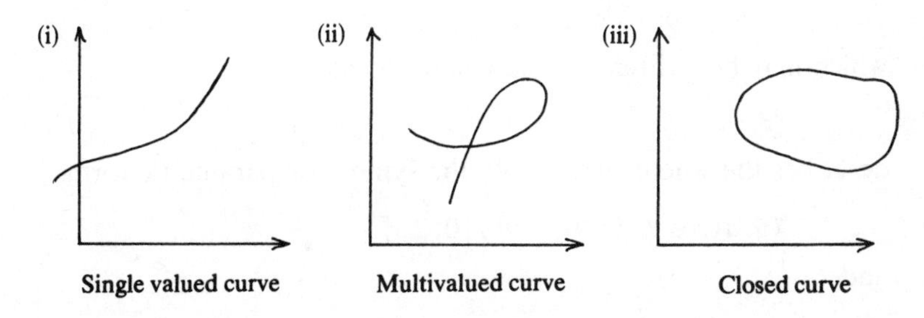

can define only a *single valued* curve; this is because y is a function of x hence

$$x_1 = x_2 \Rightarrow y(x_1) = y(x_2).$$

A vertical straight line is a multivalued curve.

Many applications in computer graphics and computer aided design require multivalued and closed curves to be defined. The only way to achieve this with non-symmetrical descriptions is to join curves as described earlier. This is inconvenient with non-symmetric forms and rarely done in practice. Symmetrical descriptions do not have this limitation and are therefore more suitable for such applications. Non-symmetrical forms are used when a single valued curve is required.

Example 10.3.2

The circle is an example of a closed curve. Consider a circle of radius a centred at the origin in $\mathbb{R}^2$ as shown in Figure 10.25. To describe

Fig. 10.25

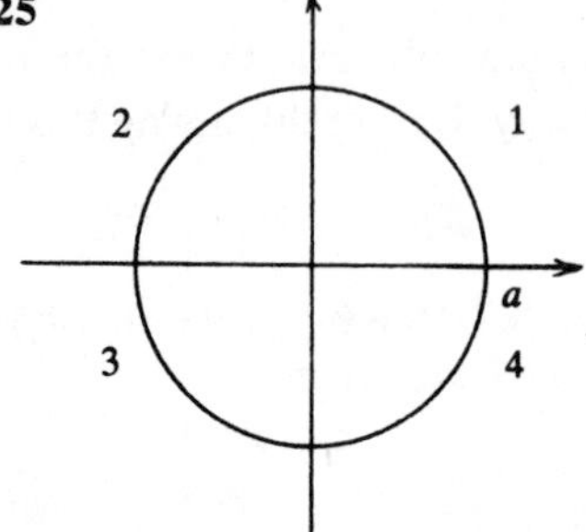

the whole circle in non-symmetrical form we need the two curves

$$y_1(x)=(a^2-x^2)^{1/2}, \quad -a\leqslant x\leqslant a \quad \text{in quadrants 1 and 2}$$

and

$$y_2(x)=-(a^2-x^2)^{1/2}, \quad -a\leqslant x\leqslant a \quad \text{in quadrants 3 and 4.}$$

The implicit form

$$x^2+y^2-a^2=0$$

which may be written in vector notation as

$$\|\mathbf{r}\|-a=0$$

describes the whole circle as do the symmetric parametric forms

$$(\theta, a(\cos\theta, \sin\theta)), \quad \theta\in[0, 2\pi[$$

and

$$\left(t, a\left(\frac{1-t^2}{1+t^2}, \frac{2t}{1+t^2}\right)\right), \quad t\in\mathbb{R}.$$

An infinity of other symmetrical forms is possible, the choice depends upon the application. For producing pictures, a parameterization with relatively uniform changes in $\mathbf{r}(t)$ is preferable; we may express this mathematically as

$$\|\mathbf{r}'(t)\| \text{ approximately constant in the interval } I.$$

The non-symmetric representation is unsatisfactory for most purposes as it requires a check on which branch of the curve one is currently on. //

The ellipse, parabola and hyperbola are further examples of quadratic curves. Their usual implicit and symmetric representations are given in Figure 10.26.

Fig. 10.26

Curve	Implicit	Symmetric
Ellipse	$\frac{x^2}{a^2}+\frac{y^2}{b^2}=1$	$(\phi, (a\cos\phi, b\sin\phi)) \quad \phi \in [0, 2\pi[$
Parabola	$y^2-4ax=0$	$(t, (at^2, 2at)), \quad t \in \mathbb{R}$
Hyperbola	$\frac{x^2}{a^2}-\frac{y^2}{b^2}=1$	$(u, \frac{1}{2}(a(e^u+e^{-u}), b(e^u-e^{-u}))), \quad u \in \mathbb{R}$

If G is a group of transformations of $\mathbb{R}^n$ then G transforms $\mathscr{C}^n$ in a natural way, for if $c \in \mathscr{C}^n$ is the curve

$$c = (t, \mathbf{r}(t)), \quad t \in I$$

and $g \in G$, we define a curve gc by

$$gc = (t, g\mathbf{r}(t)), \quad t \in I.$$

For example, when $G = SO(2)$ and $c \in \mathscr{C}^2$ the graph of $W(\theta)c$ is the graph of c rotated through θ as shown in Figure 10.27.

Fig. 10.27

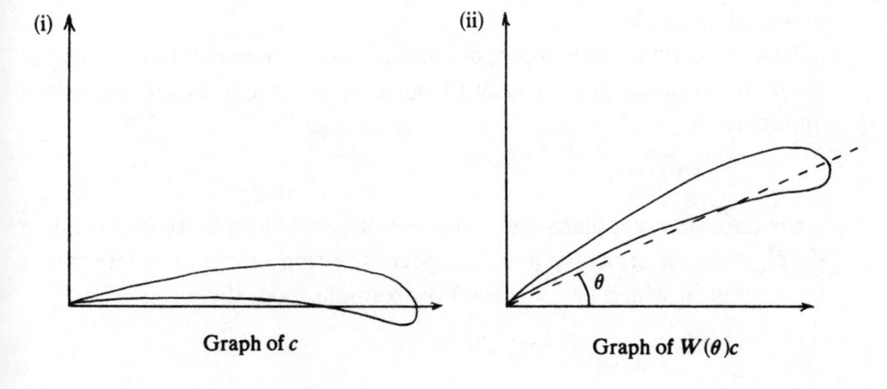

The equations in Figure 10.26 describe the curves in 'standard' position, with the x-axis as an axis of symmetry. To obtain the equation of a geometrically equivalent curve at some other position and orientation in space one simply applies the appropriate element of the group $E(2)$ to the equation. As an example we derive the equation of the ellipse with graph ε as shown in Figure 10.28. In standard position (shown dotted in the figure) the curve is

$$c = (\phi, (a \cos \phi, b \sin \phi)), \quad \phi \in [0, 2\pi[$$

Fig. 10.28

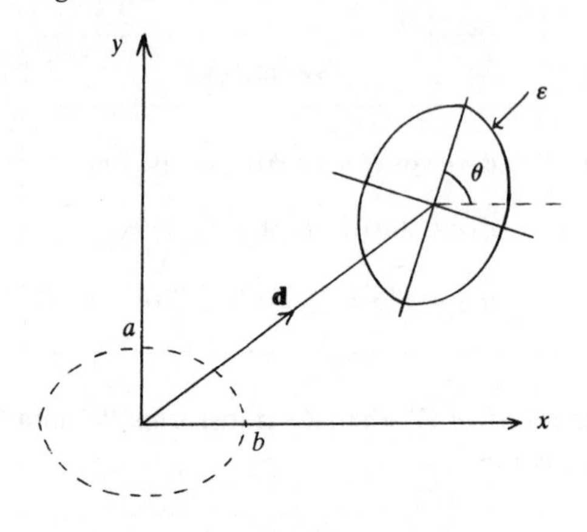

hence the curve with graph ε is

$$\begin{aligned}(W(\theta), \mathbf{d})c &= (\phi, (W(\theta), \mathbf{d})(a \cos \phi, b \sin \phi)) \\ &= (\phi, (a \cos \theta \cos \phi - b \sin \theta \sin \phi + d_1, \\ &\qquad a \sin \theta \cos \phi + b \cos \theta \sin \phi + d_2)), \\ &\qquad \phi \in [0, 2\pi[.\end{aligned}$$

The implicit equations for curves in non-standard position may be obtained similarly.

Before leaving the topic of plane curves we mention a useful geometrical check that is easy to perform when curves are expressed implicitly. If

$$f(x, y) = 0$$

is the equation of a plane curve that partitions $\mathbb{R}^2$ into three regions then if (x', y') is an arbitrary point in space, the sign of $f(x', y')$ determines the region in which (x', y') lies. For example, take the circle

$$f(x, y) = x^2 + y^2 - a^2 = 0$$

then

$$f(x', y') > 0 \Rightarrow (x', y') \text{ lies in the region outside the circle}$$
$$f(x', y') = 0 \Rightarrow (x', y') \text{ lies on the circle}$$
$$f(x', y') < 0 \Rightarrow (x', y') \text{ lies inside the circle.}$$

Checks of this type in a three-dimensional context are used in algorithms to remove hidden lines from pictures. A treatment of surfaces may be given following similar lines. The methods of representation have similar advantages and disadvantages for computing applications. Figure 10.29 is a summary applicable to both curves and surfaces.

Surfaces are two-dimensional and have parameter spaces of the form $I_1 \times I_2$ where $I_1, I_2 \subseteq \mathbb{R}$ are intervals; a parametric representation has the general form

$$((u, v), \mathbf{r}(u, v)), \quad (u, v) \in I_1 \times I_2.$$

Implicit equations for surfaces have the form $f(x, y, z) = 0$ and the *quadratic* surfaces are those for which

$$ax^2 + by^2 + cz^2 + dxy + exz + fyz + gx + hy + iz + j = 0$$

where $a, b, c, d, e, f, g, h, i, j \in \mathbb{R}$.

The following examples are all quadratic surfaces.

Fig. 10.29

Representation	Advantages	Disadvantages
Non-symmetric	Useful for fitting single valued curves to discrete data.	Describes single valued curves only.
Symmetric	Usually the most suitable representation for shape definition in computer graphics and computer aided design applications. Easy representation for plotting the graph.	
Implicit	Some geometrical checks are easy to do. Can be useful when computing intersections of surfaces to have one described by an implicit formula.	Difficult to draw the graph of the curve/surface. Necessary to solve a non-linear equation in general.

Example 10.3.3

If $D = \{\mathbf{r}_i : \mathbf{r}_i \in \mathbb{R}^3, 0 \leqslant i \leqslant 2\}$ is a set of linearly independent vectors

then

$$\mathbf{r}(u, v) = \mathbf{r}_0 + u(\mathbf{r}_1 - \mathbf{r}_0) + v(\mathbf{r}_2 - \mathbf{r}_0), \quad (u, v) \in \mathbb{R}^2$$

is a symmetric description of the plane P through the points of D as shown in Figure 10.30. If $\mathbf{q} = (\mathbf{r}_2 - \mathbf{r}_0) \times (\mathbf{r}_1 - \mathbf{r}_0)$ then $\hat{\mathbf{n}} = \mathbf{q}/\|\mathbf{q}\|$ is a unit vector orthogonal to P (called the normal); thus $\mathbf{r} \in P$ iff

$$(\mathbf{r} - \mathbf{r}_0) \cdot \hat{\mathbf{n}} = 0$$

Fig. 10.30

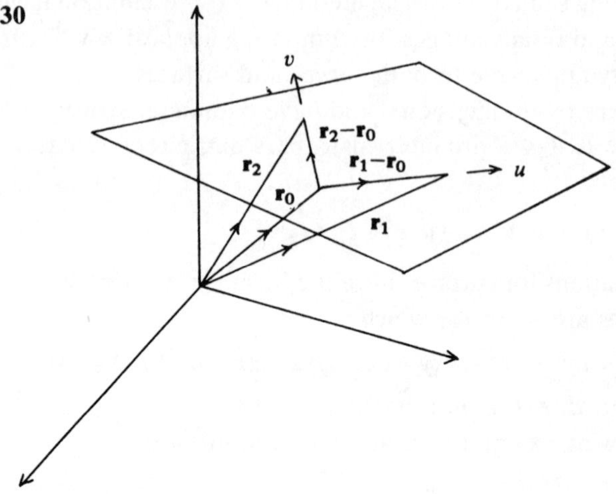

or

$$(x - x_0)n_1 + (y - y_0)n_2 + (z - z_0)n_3 = 0$$

and this is the implicit representation for P. //

Example 10.3.4

If S is a sphere of radius a centred at the origin in $\mathbb{R}^3$ then clearly $\mathbf{r} \in S$ iff

$$\|\mathbf{r}\| = a$$

or

$$x^2 + y^2 + z^2 - a^2 = 0.$$

Spherical polar coordinates provide a symmetric representation as

$$((\theta, \phi), a(\sin\theta\cos\phi, \sin\theta\sin\phi, \cos\theta)),$$
$$(\theta, \phi) \in [0, \pi] \times]0, 2\pi[. \quad //$$

Example 10.3.5

Let C be a cylinder of radius a with the z-axis as axis of symmetry, then $(x, y, z) \in C$ iff

$$x^2 + y^2 - a^2 = 0.$$

Using cylindrical polar coordinates we obtain the symmetric form

$$((\phi, z), (a \cos \phi, a \sin \phi, z)).$$

Normally one requires a 'finite' cylinder as shown in Figure 10.31; for which the equation is

$$x^2+y^2-a^2=0 \quad \text{and} \quad z_0 \leqslant z \leqslant z_1. \quad /\!/$$

Fig. 10.31

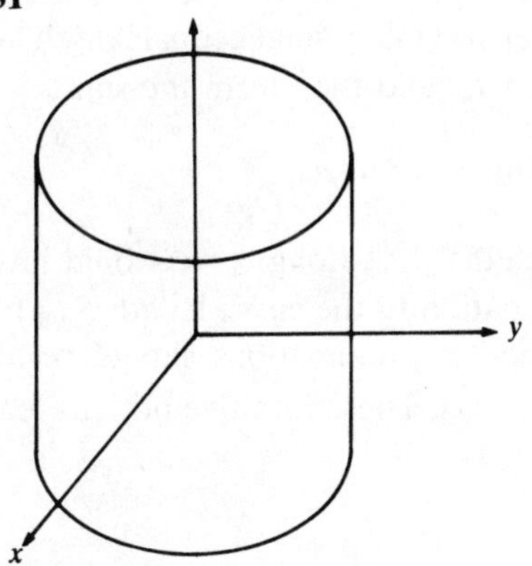

Similar remarks regarding the computation of equations for surfaces in non-standard positions apply, where of course $E(3)$ is substituted for $E(2)$.

10.3.2 Geometry of plane curves

The goal of this section is to define the concepts of length, tangent and curvature for plane curves. Figure 10.32 depicts the plane curve $c=(t, \mathbf{r}(t))$, $t \in [t_A, t_B]$; P is an arbitrary point on c with position vector

Fig. 10.32

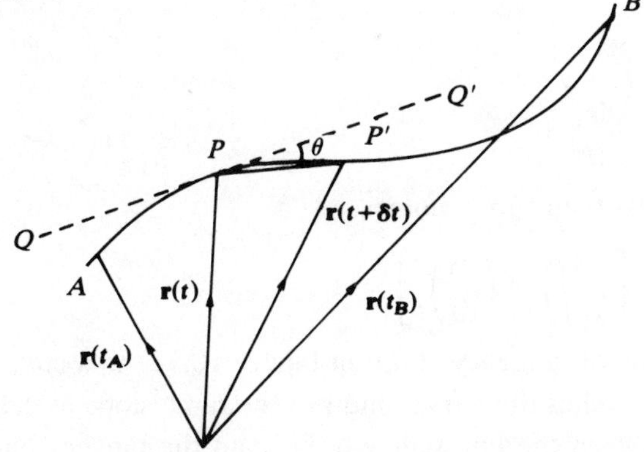

$\mathbf{r}(t)$ and P' has position vector $\mathbf{r}(t+\delta t)$ where $\delta t \in \mathbb{R}\backslash\{0\}$. If $\delta\mathbf{r}(t)$ denotes the position vector of P' relative to P then clearly

$$\delta\mathbf{r}(t) = \mathbf{r}(t+\delta t) - \mathbf{r}(t)$$

and

$$\|\delta\mathbf{r}(t)\| = \left\|\frac{\mathbf{r}(t+\delta t)-\mathbf{r}(t)}{\delta t}\right\| \delta t$$

is the length of the line segment PP'.

If we dissect the interval $[t_A, t_B]$ into many small, equal length intervals $[t_{i-1}, t_i]$, for $t_A = t_0 < t_1 < \ldots < t_n = t_B$, and then form the sum

$$S_{AB} = \sum_{i=1}^{n} \|\delta\mathbf{r}_i\|, \quad \text{where } \delta\mathbf{r}_i = \mathbf{r}(t_i) - \mathbf{r}(t_{i-1}),$$

of the resulting line segment lengths $\|\delta\mathbf{r}_i\|$ along c, we would intuitively expect to obtain a good approximation to the curve length S_{AB} between $\mathbf{r}(t_A)$ and $\mathbf{r}(t_B)$. Increasingly good approximations should result from increasingly fine dissections of $[t_A, t_B]$. These intuitive notions lead us to define the *curve length* S_{AB} as

$$S_{AB} = \operatorname*{limit}_{n\to\infty} \sum_{i=1}^{n} \|\delta\mathbf{r}_i\|.$$

Computationally, this formula is difficult to use and we can rewrite it in a more convenient way as follows

$$\begin{aligned}\operatorname*{limit}_{n\to\infty} \sum_{i=1}^{n} \|\delta\mathbf{r}_i\| &= \operatorname*{limit}_{n\to\infty} \sum_{i=1}^{n} \|\mathbf{r}(t_i) - \mathbf{r}(t_{i-1})\| \\ &= \operatorname*{limit}_{n\to\infty} \sum_{i=1}^{n} \left\|\frac{\mathbf{r}(t_{i-1}+\delta t) - \mathbf{r}(t_{i-1})}{\delta t}\right\| \delta t\end{aligned}$$

where $\delta t = (t_B - t_A)/n$

but this is by definition the (Riemann) integral

$$\int_{t_A}^{t_B} \left\|\frac{\mathrm{d}\mathbf{r}}{\mathrm{d}t}\right\| \mathrm{d}t$$

hence

$$S_{AB} = \int_{t_A}^{t_B} \left\|\frac{\mathrm{d}\mathbf{r}}{\mathrm{d}t}\right\| \mathrm{d}t.$$

If $\mathbf{r}(t) = (x(t), y(t))$ then this becomes

$$S_{AB} = \int_{t_A}^{t_B} \left[\left(\frac{\mathrm{d}x}{\mathrm{d}t}\right)^2 + \left(\frac{\mathrm{d}y}{\mathrm{d}t}\right)^2\right]^{1/2} \mathrm{d}t.$$

Intuitively, a line of tangency (QQ' in Figure 10.32) to a curve c is a straight line that touches the curve, and has the same slope as c, at P. If θ is the angle between the line segment PP' and the tangent line QQ'

then as $\delta t \to 0$ we would expect $\theta \to 0$. This means that the direction of $\delta \mathbf{r}(t)$ and $\delta \mathbf{r}(t)/\delta t$ approach the direction of QQ' as $\delta t \to 0$; in other words the vector

$$\frac{\mathrm{d}\mathbf{r}}{\mathrm{d}t} = \operatorname*{limit}_{\delta t \to 0} \frac{\delta \mathbf{r}}{\delta t}$$

is parallel to QQ'.

If

$$s = \int_{t_A}^{t} \left\| \frac{\mathrm{d}\mathbf{r}}{\mathrm{d}t} \right\| \mathrm{d}t$$

then in principle all curves may be expressed in *arc length* parameterization $(s, \mathbf{r}(s))$, $s \in [0, S_{AB}]$. Now from our definition of arc length

$$s = \int_{0}^{s} \left\| \frac{\mathrm{d}\mathbf{r}}{\mathrm{d}s'} \right\| \mathrm{d}s'$$

and differentiating both sides with respect to s we have

$$1 = \left\| \frac{\mathrm{d}\mathbf{r}}{\mathrm{d}s} \right\|;$$

in other words $\mathrm{d}\mathbf{r}/\mathrm{d}s$ is a unit vector parallel to the tangent line at $\mathbf{r}(s)$; this motivates the following definition.

Definition. If $(s, \mathbf{r}(s))$, $s \in [0, S_{AB}]$ is a curve with arc length parameterization we define the *unit tangent vector*, $\hat{\mathbf{T}}(s)$ to the curve at s by

$$\hat{\mathbf{T}}(s) = \frac{\mathrm{d}\mathbf{r}}{\mathrm{d}s}(s)$$

and the *principal normal vector* $\hat{\mathbf{N}}(s)$ to the curve by

$$\hat{\mathbf{N}}(s) = W(\pi/2)\hat{\mathbf{T}}(s). \quad //$$

The basis $(\hat{\mathbf{T}}, \hat{\mathbf{N}})$ forms a right handed system for $\mathbb{R}^2$. If the curve is not given with arc length parameterization the expression for $\hat{\mathbf{T}}$ is (see Exercises 10.3)

$$\hat{\mathbf{T}}(u) = \frac{\mathrm{d}\mathbf{r}}{\mathrm{d}u} \Big/ \left\| \frac{\mathrm{d}\mathbf{r}}{\mathrm{d}u} \right\|, \quad \text{if } \frac{\mathrm{d}s}{\mathrm{d}u} > 0$$

$$= -\frac{\mathrm{d}\mathbf{r}}{\mathrm{d}u} \Big/ \left\| \frac{\mathrm{d}\mathbf{r}}{\mathrm{d}u} \right\|, \quad \text{if } \frac{\mathrm{d}s}{\mathrm{d}u} < 0.$$

These formulae implicitly assume a convention for s. In fact the sign of $\hat{\mathbf{T}}$ is *not* an equivalence class invariant on $\mathscr{C}^2$, but is a function of the parametric representation chosen; even for arc length representation it depends upon which end of the curve we measure s from. The tangent space at a point on the curve *is* a class invariant on $\mathscr{C}^2$, where

$$\begin{aligned}&\text{tangent space of } [c] \text{ at } P\\&\quad=\{\alpha\hat{\mathbf{T}}\colon \alpha\in\mathbb{R} \text{ and } \hat{\mathbf{T}} \text{ is the tangent vector at } P \text{ of some } c\in[c]\}.\end{aligned}$$

It follows from the exercises of Section 5.4 that $\mathrm{d}\hat{\mathbf{T}}/\mathrm{d}s$ is orthogonal to $\hat{\mathbf{T}}$, hence there exists a function $\kappa\colon[0, S_{AB}]\to\mathbb{R}$ such that

$$\frac{\mathrm{d}\hat{\mathbf{T}}(s)}{\mathrm{d}s}=\kappa(s)\hat{\mathbf{N}}(s);$$

$\kappa(s)$ is called the *signed curvature* of the curve at $\mathbf{r}(s)$ and $1/\kappa(s)$ is called the *radius of curvature* at $\mathbf{r}(s)$.

Exercises 10.3

1. Obtain an explicit equation for the straight line in $\mathbb{R}^2$ passing through the points $(-1, 3)$ and $(2, -1)$. Determine a unit vector parallel to the line and write the equation of the line in parametric vector form.
2. Determine the unit tangent $\hat{\mathbf{T}}$, the principal normal $\hat{\mathbf{N}}$ and the signed curvature κ to the following plane curves
   (i) the circle $(\theta, a(\cos\theta\hat{\mathbf{i}}+\sin\theta\hat{\mathbf{j}})),\quad 0\leqslant\theta<2\pi$
   (ii) the ellipse $(\phi, (a\cos\phi\hat{\mathbf{i}}+b\sin\phi\hat{\mathbf{j}})),\quad 0\leqslant\phi<2\pi$
   (iii) the parabola $(t, (at^2\hat{\mathbf{i}}+2at\hat{\mathbf{j}})),\quad -\infty<t<\infty$
   where $a>0$ and $b>0$ are real constants.
3. A plane curve is defined by
   $(u, (a\cos u\hat{\mathbf{i}}+b(1-\mathrm{e}^{-u/2})\hat{\mathbf{j}})),\qquad 0\leqslant u<\infty$
   where $a>0$ and $b>0$ are real constants.
   (i) Show that the unit tangent $\hat{\mathbf{T}}(u)$ to the curve is given by
   $$\hat{\mathbf{T}}(u)=\frac{-2a\sin u\hat{\mathbf{i}}+b\,\mathrm{e}^{-u/2}\hat{\mathbf{j}}}{(4a^2\sin^2 u+b^2\mathrm{e}^{-u})^{1/2}}.$$
   (ii) draw a free-hand sketch of the curve in the interval $0\leqslant u\leqslant 2\pi$.
   (iii) Determine the principal normal to the curve.
4. Sketch the graph of the twisted curve defined by
   $(u, (a\cos u, b\sin u, bu)),\quad -\infty<u<\infty$
   and derive an expression for the unit tangent $\hat{\mathbf{T}}(u)$.
5. If
   $(x, (x, y(x))),\quad x_0\leqslant x\leqslant x_1$
   describes a plane curve show that the unit tangent vector $\hat{\mathbf{T}}(x)$ may be written
   $$\hat{\mathbf{T}}(x)=\frac{(1, \mathrm{d}y/\mathrm{d}x)}{(1+(\mathrm{d}y/\mathrm{d}x)^2)^{1/2}}$$

and hence show that the curvature $\kappa(x)$ at x has the form

$$\kappa(x)=\frac{d^2y/dx^2}{(1+(dy/dx)^2)^{3/2}}.$$

6. Obtain formulae analogous to those of Question 5 for a symmetric representation

   $(u, (x(u), y(u))), \quad u_0 \leqslant u \leqslant u_1.$

7. Determine a symmetric parametric equation for the plane P through the points $(0, 1, 0)$, $(3, -2, 0)$ and $(1, 3, 4)$ in $\mathbb{R}^3$. Show that the normal to P is parallel to the vector $(-4, -4, 3)$ and use this to write down an implicit equation for P.

8. A *surface of revolution* is a surface defined by rotation of a plane curve about a fixed axis in $\mathbb{R}^3$.
   (i) Show that if the plane curve

   $(u, (p(u)\hat{\mathbf{i}}+q(u)\hat{\mathbf{k}})), \quad u_A \leqslant u \leqslant u_B$

   is rotated through 2π about the z-axis then the corresponding surface of revolution has equation

   $((u, \phi), (p(u) \cos \phi\hat{\mathbf{i}}+p(u) \sin \phi\hat{\mathbf{j}}+q(u)\hat{\mathbf{k}}))$

   where $u_A \leqslant u \leqslant u_B$ and $0 \leqslant \phi < 2\pi$.
   (ii) Use (i) to obtain symmetric parametric representations for each of the following surfaces
   (*a*) the cylinder,
   (*b*) the cone,
   (*c*) the torus.

9. Part of a pipe network is to be designed by intersecting the cylinder C_1 of length $2l$ defined by

   $x^2+z^2=b^2 \quad \text{for } -l \leqslant y \leqslant l$

   with the cylinder C_2 of length h with equation

   $y^2+(z-\alpha)^2=a^3 \quad \text{for } 0 \leqslant x \leqslant h$

   where $0<\alpha<b$ and $h>b$.

   Using the parametric coordinates

   $\{(x, \theta): 0 \leqslant x \leqslant h, 0 \leqslant \theta < 2\pi\} \quad \text{on } C_2,$

   where θ parameterizes the circle of cross-section, show that the curve of intersection of C_1 and C_2 may be written $(\theta, \mathbf{r}(\theta))$, $0 \leqslant \theta < 2\pi$, where

   $\mathbf{r}(\theta)=((b^2-(\alpha+a \sin \theta)^2)^{1/2}, a \cos \theta, \alpha+a \sin \theta).$

   Show also that the unit tangent to the curve at $\theta=0$ is given by

   $$\frac{1}{ab}(-a\alpha, 0, a(b^2-\alpha^2)^{1/2}).$$

THE GREEK ALPHABET

Letters		*Names*
Α	α	alpha
Β	β	beta
Γ	γ	gamma
Δ	δ	delta
Ε	ε	epsilon
Ζ	ζ	zeta
Η	η	eta
Θ	θ	theta
Ι	ι	iota
Κ	κ	kappa
Λ	λ	lambda
Μ	μ	mu
Ν	ν	nu
Ξ	ξ	xi
Ο	ο	omicron
Π	π	pi
Ρ	ρ	rho
Σ	σ	sigma
Τ	τ	tau
Υ	υ	upsilon
Φ	φ	phi
Χ	χ	chi
Ψ	ψ	psi
Ω	ω	omega

A GLOSSARY OF SYMBOLS

Symbol	Meaning	Page
//	used in this text to indicate end of proof, example etc.	xii
⋇	contradiction, error	xii
$*$	multiplication (of numbers)	xii
$\{\ ,\ldots,\ \}$ $\{x : x\ldots\}$	set definition	1 2
$\in, \notin$	set containment and its negation	2
$\cap, \cup$	intersection, union of sets	5
$\setminus, \triangle$	difference, symmetric difference of sets	6, 7
$\varnothing, \mathscr{E}, A'$	empty set, universe of discourse, complement of A	7, 8
$\mathbb{N}$	set of natural numbers	9
$\mathbb{Z}$	set of integers	9
$\lvert A\rvert, \mathrm{card}\,(A)$	cardinality of A	9
$A+B, A-B, A*B, A/B$	induced operations on sets of numbers	12
$\subset, \subseteq, \supset, \supseteq, =$	set inclusion operators and equality	16, 17
$P \Rightarrow Q, P \Leftrightarrow Q$	implication operators	17, 18
$\mathscr{P}(X), 2^X$	power set of X	22
$(x_1, \ldots, x_n)$	n-tuple	26
$\bigtimes_{i=1}^{n} A_i,\ A_1 \times \ldots \times A_n,\ A^n$	Cartesian products	27
I_A, U_A	Identity and Universal relations on A	31
$\mathscr{D}(R), \mathscr{D}_R, \mathscr{R}(R), \mathscr{R}_R$	domain and range of the relation R	31, 33
$(a, b) \in \rho,\ a\rho b,\ b \in \rho(a)$	a ρ-related to b	32, 33
R^{-1}, ρ^{-1}	inverse relations	33
$\bigcup_{i=1}^{n} A_i = A_1 \cup \ldots \cup A_n$	indexed union	40
$\equiv, \sim$	equivalence relation	41
$[x]$	equivalence class	41
$\mathbb{Q}, \mathbb{R}$	sets of rational and real numbers	42
$\leqslant, <$	order relations	43
$(X, \leqslant)$	partially ordered set	45
inf, sup	infimum, supremum	45

INDEX

Terms marked with an asterisk (*) occur frequently and only major entries are indexed. Similarly, page numbers printed in bold indicate primary definitions.